Preparing for NEW JERSEY BIOLOGY Competency Test

Rick Hallman

Principal, Retired
Benjamin N. Cardozo High School
Bayside, New York

AMSCO SCHOOL PUBLICATIONS, INC.
315 Hudson Street, New York, N.Y. 10013

The publisher would like to thank the following people who reviewed the manuscript:

Daniel Baker
Science Teacher
Bergen County Academy
Hackensack, New Jersey

Debra Ingenito
Science Teacher/Facilitator
Harrison High School
Harrison, New Jersey

John R. Jones, Jr.
Science Teacher
Howell Middle School South
Howell, New Jersey

Leyna Williams
Science Teacher
Timber Creek High School
Erial, New Jersey

Cover Photo: Green Lynx Spider on Violets © Natural Selection
   Robert Cable/Design Pics/Corbis
Cover Design: Meghan Shupe
Text Design: Howard Petlack/A Good Thing, Inc.
Composition: Brad Walrod/Kenoza Type
Art: Hadel Studio

Please visit our Web site at: *www.amscopub.com*

When ordering this book, please specify:
either *R 075 W* or **PREPARING FOR THE NEW JERSEY BIOLOGY COMPETENCY TEST**

ISBN13: 978-1-56765-937-5

Copyright © 2009 Amsco School Publications, Inc.
No part of this book may be reproduced in any form
without written permission from the publisher.

PRINTED IN THE UNITED STATES OF AMERICA

1 2 3 4 5 6 7 8 9 10   15 14 13 12 11 10

Contents

Introduction		v
Diagnostic Test		1
Diagnostic Checklist		10

UNIT I Scientific and Laboratory Procedures (5.1, 5.2, 5.3, 5.4) — 11

Chapter 1:	The Scientific Method and Inquiry (5.1)	13
Chapter 2:	Laboratory Tools and Techniques (5.2, 5.3, 5.4)	24

UNIT II Matter, Energy, and Organization of Living Things (5.5.12 A) — 39

Chapter 3:	Organic Molecules: From Atoms to Cells (5.5.12 A1)	41
Chapter 4:	Cell Structure and Function (5.5.12 A1)	51
Chapter 5:	Homeostasis and Enzymes (5.5.12 A1)	63
Chapter 6:	Photosynthesis and Respiration (5.5.12 A2)	76
Chapter 7:	Energy and Matter in Ecosystems (5.5.12 A3)	87
Chapter 8:	Organisms and Disease (5.5.12 A4)	102

UNIT III Diversity and Biological Evolution (5.5.12 B) — 119

Chapter 9:	Classification of Organisms (5.5.12 B1)	121
Chapter 10:	The Theory of Evolution (5.5.12 B1; 5.5.12 B2)	134
Chapter 11:	Evidence for Evolution (5.5.12 B1; 5.5.12 B2)	144
Chapter 12:	Mechanisms of Evolution (5.5.12 B2)	153

UNIT IV Reproduction and Heredity (5.5.12 C) — 163

Chapter 13:	DNA and Heredity (5.5.12 C1)	165
Chapter 14:	Genes and Protein Synthesis (5.5.12 C1)	175
Chapter 15:	Asexual Reproduction and Mitosis (5.5.12 C1)	185
Chapter 16:	Sexual Reproduction and Meiosis (5.5.12 C2)	195
Chapter 17:	Patterns of Inheritance (5.5.12 C2)	206
Chapter 18:	Genetics and Biotechnology (5.5.12 C3)	221

UNIT V Environmental Systems and Interactions (5.10.12 A/B) — 233

Chapter 19:	Natural Systems and Interactions (5.10.12 A1)	235
Chapter 20:	Human Impact on the Environment (5.10.12 B1)	252
Chapter 21:	Efforts to Protect the Environment (5.10.12 B2)	265

Practice Test 1	277
Practice Test 2	283
Glossary	291
Index	301
Photo Credits	306

To the Student

About This Book

All New Jersey students enrolled in biology and seeking credit must take the Biology Competency Test (NJBCT). The purpose of this book is to help you prepare for this examination. The book's content provides you with a comprehensive review of the main content areas that are tested on the biology exam. Numerous question sets give you the opportunity to assess and reinforce your understanding of the concepts taught in your high school biology class.

The five units of the book are organized into the following content areas: scientific and laboratory procedures; matter, energy, and organization of living things; diversity and biological evolution; reproduction and heredity; and environmental systems and interactions. A Diagnostic Test and Diagnostic Checklist follow this introduction. The Diagnostic Test is modeled on the New Jersey Biology Competency Test. It has the same types and number of multiple-choice questions as on the NJBCT. The Diagnostic Checklist will help you determine your areas of strength and areas of weakness in the biology skills and content areas. This will help you focus your attention on those topics that need more review.

Many figures and tables are presented in the text to help illustrate important concepts. Each chapter ends with a Chapter Review section that contains multiple-choice and contructed-response questions. In addition, there are two complete Practice Tests at the back of the book. All questions are presented in the same format as you will encounter on the NJBCT. Throughout the book, important science terms are printed in **bold** type and are defined in text as well as in the Glossary. Many other important terms are printed in *italic* type for emphasis.

About The Competency Test

As stated in the New Jersey Department of Education's High School Biology/Life Science Course Guidance document, the Biology Competency Test is a curriculum-based achievement test designed to assess a student's skills and knowledge of the NJ High School Biology/Life Science Course. The NJBCT, which is given in May, consists of 45 multiple-choice questions and three constructed-response items. All students taking the Biology Competency Test will also take a separately scored performance assessment prompt. More information about the test is available at: *http://www.nj.gov/education/aps/cccs/science/resources*.

DIAGNOSTIC TEST

PART 1

Directions for Questions 1 through 15 For each of the questions or incomplete statements below, choose the best of the answer choices given. Write your answers on a separate sheet of paper.

1. A characteristic of a DNA molecule that is *not* a characteristic of a protein molecule is that the DNA molecule
   A. can replicate itself
   B. can be very large
   C. is found in nuclei
   D. is composed of subunits

2. Two closely related species of birds live in the same tree. Species *A* feeds on ants and termites, whereas species *B* feeds on caterpillars. The two species coexist successfully because
   A. each occupies a different niche
   B. they can interbreed
   C. they use different methods of reproduction
   D. birds compete for food

3. The diagram below shows how a coverslip should be lowered onto some microscopic organisms during the preparation of a wet mount. Why is this procedure used?

   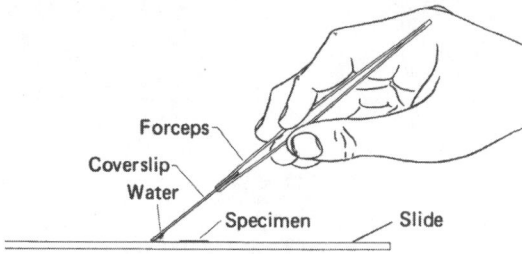

   A. The coverslip will prevent the slide from breaking.
   B. The microorganisms will be more evenly distributed.
   C. The possibility of breaking the coverslip is reduced.
   D. The possibility of trapping air bubbles is reduced.

4. In the diagram below, what does *X* most likely represent?

   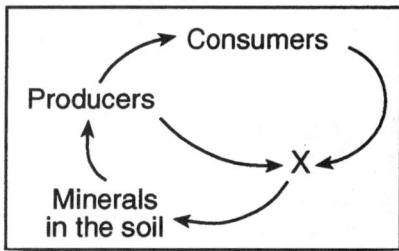

   A. autotrophs   C. herbivores
   B. decomposers   D. carnivores

5. Ten breeding pairs of rabbits are introduced onto an island having no natural predators and a good supply of food and freshwater. The rabbit population will most likely
   A. remain constant due to equal birth and death rates
   B. die out due to an increase in the mutation rate
   C. increase until it exceeds carrying capacity
   D. decrease and then increase indefinitely

6. Which statement best expresses the relationship between the three structures represented in the diagram below?

   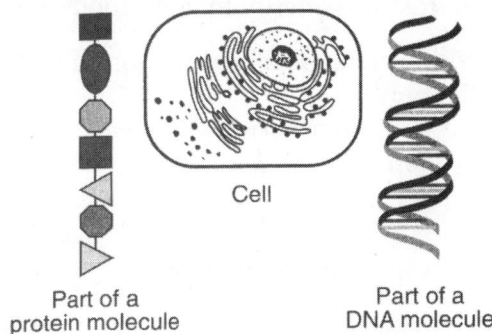

   A. DNA is produced from protein absorbed by the cell.

Diagnostic Test 1

B. Protein is composed of DNA that is produced in the cell.
C. DNA controls the production of protein in the cell.
D. Cells make DNA by digesting protein.

7. All the cells of an organism are constantly engaged in different chemical reactions. This fact is supported by the presence in each cell of thousands of different types of
   A. enzymes   C. nuclei
   B. chloroplasts   D. organelles

8. The diagram below represents an experimental setup. Two black paper discs are placed on either side of two leaves. This setup would most likely be used to show that

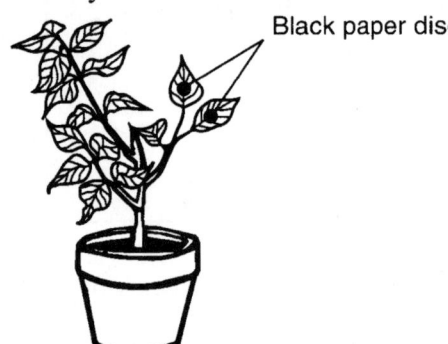
Black paper discs

   A. glucose is necessary for photosynthesis
   B. protein is a product of photosynthesis
   C. light is necessary for photosynthesis
   D. carbon dioxide is a product of photosynthesis

9. Which statement best describes cellular respiration?
   A. It occurs in animal cells but not in plant cells.
   B. It converts energy in food into a more usable form.
   C. It uses carbon dioxide and produces oxygen.
   D. It stores energy in large food molecules.

10. Which concept is best supported by this diagram?

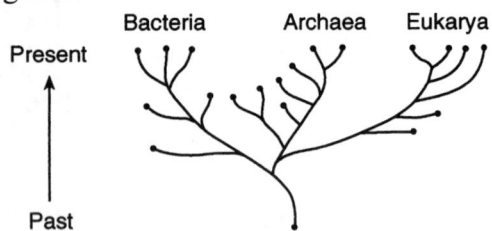

   A. Evolutionary pathways proceed in only one set direction over time.
   B. All evolutionary pathways eventually lead to present-day organisms.
   C. All evolutionary pathways will last for the same length of time.
   D. Evolutionary pathways can proceed in several directions, with only some leading to present-day organisms.

11. Which process uses carbon dioxide molecules?
    A. cellular respiration
    B. asexual reproduction
    C. active transport
    D. autotrophic nutrition

12. A new chemical was discovered and introduced into a culture that contained one species of bacteria. Within a day, most of the bacteria were dead, but a few remained. Which statement best explains why some of the bacteria survived?
    A. They had a genetic variation that gave them a resistance to the chemical.
    B. They were exposed to the chemical long enough to develop a resistance.
    C. They mutated into a different species after exposure to the new chemical.
    D. They absorbed the chemical and broke it down in their digestive systems.

Base your answer to question 13 on the diagram below, which represents nutritional relationships among organisms.

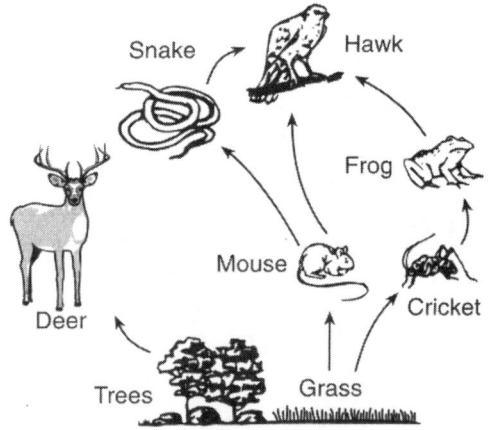

13. The mouse population would most likely *increase* if there were
    A. an increase in the frog and hawk populations
    B. a decrease in the snake and hawk populations
    C. an increase in the number of deer and snakes

D. a decrease in the amount of available sunlight

14. A mutation occurs in the liver cells of a certain field mouse. Which statement concerning the spread of this mutation throughout the mouse population is correct?
    A. It will spread because it is a beneficial trait.
    B. It will spread because it is a dominant gene.
    C. It will not spread because it is a recessive gene.
    D. It will not spread because it is not in a gamete.

Base your answer to question 15 on the following information and diagram.

An experiment was carried out to determine which mouthwash was most effective against bacteria commonly found in the mouth. Four paper discs were each dipped into a different brand of mouthwash. The discs were then placed onto the surface of a culture plate that contained food, moisture, and bacteria commonly found in the mouth. The diagram below shows the growth of bacteria on the plate after 24 hours.

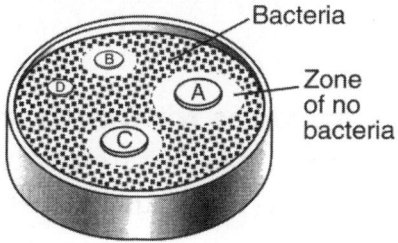

15. Which change in procedure would have improved the experiment?
    A. using a smaller plate with less food and moisture
    B. using bacteria from many habitats other than the mouth
    C. using the same size paper discs for each mouthwash
    D. using the same type of mouthwash on each disc

End of Part 1

You may check your work on this part only.

Do not go to the next page.

PART 2

Directions for Questions 16 through 30 For each of the questions or incomplete statements below, choose the best of the answer choices given. Write your answers on a separate sheet of paper.

16. The diagram below illustrates the movement of substances in a process that manufactures energy for many organisms. This process occurs within a plant cell's

    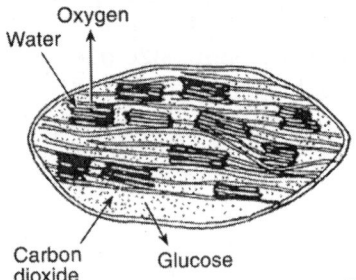

    A. chloroplasts
    B. ribosomes
    C. mitochondria
    D. vacuoles

17. When a planarian (worm) is cut in half, each half usually grows back into a complete worm over time. This situation most closely resembles
    A. asexual reproduction in which a mutation has occurred
    B. sexual reproduction in which each half represents one parent
    C. asexual reproduction of a single-celled organism
    D. sexual reproduction of a single-celled organism

18. The green aquatic plant represented in the diagram below was exposed to light for several hours. Which gas would most likely be found in the greatest amount in the bubbles?

    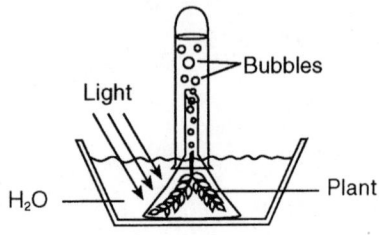

    A. oxygen
    B. ozone
    C. nitrogen
    D. carbon dioxide

19. In one variety of corn, the kernels turn red when exposed to sunlight. In the absence of sunlight, the kernels remain yellow. Based on this information, it can be concluded that the color of these corn kernels is due to the
    A. process of selective breeding
    B. rate of photosynthesis
    C. effect of environment on gene expression
    D. composition of the soil

20. Which sequence correctly shows the increasingly complex levels of organization in multicellular organisms?
    A. organelle → cell → tissue → organ → organ system → organism
    B. cell → organelle → tissue → organ → organ system → organism
    C. organelle → tissue → cell → organ → organ system → organism
    D. cell → organism → organ system → organ → tissue → organelle

21. The evolutionary pathways of seven living species are shown in the diagram below. Which two species are likely to have the most similar DNA base sequences?

    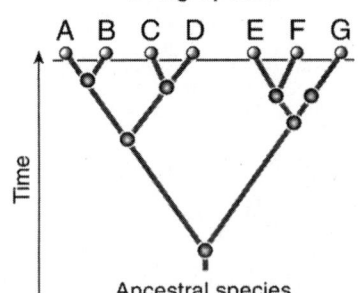

    A. D and E
    B. B and C
    C. E and G
    D. C and D

22. The production of energy-rich ATP molecules is the direct result of
    A. recycling light energy to be used in the process of photosynthesis
    B. releasing the stored energy of organic compounds through respiration
    C. breaking down starch molecules by the process of digestion
    D. copying information during the process of protein synthesis

23. An increase in acidity of a mountain lake would most likely be the result of
    A. ecological succession at the top of the mountain
    B. the introduction of new species into the lakes
    C. air pollution from smokestacks several miles away
    D. planting grasses and shrubs around the lakes

24. The diagram below represents a portion of an organic molecule. This molecule controls cellular activity by directing the synthesis of

    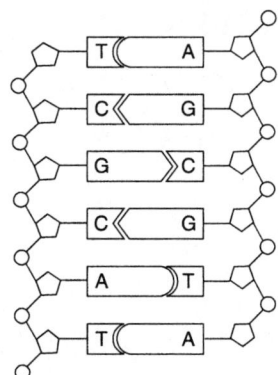

    A. carbohydrates   C. minerals
    B. lipids   D. proteins

25. Andy conducted a science experiment and was upset that the results from the experiment did not support his original hypothesis. Andy should
    A. perform a different experiment to support his hypothesis
    B. ignore the results and redo the same experiment
    C. repeat the experiment and modify his hypothesis if necessary
    D. publish the results of the experiment without repeating it

26. The diagram below, which represents a series of reactions that can occur in an organism, illustrates the relationship between

    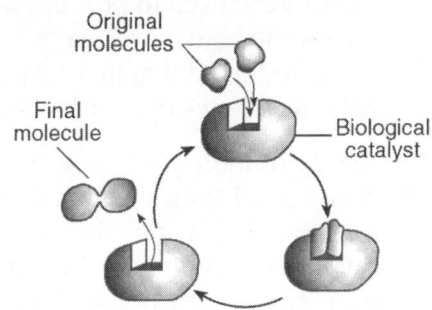

    A. enzymes and synthesis
    B. amino acids and glucose
    C. antigens and immunity
    D. ribosomes and sugars

27. A forest is cut down and replaced by a cornfield. A *negative* consequence of this practice would be a (an)
    A. increase in the oxygen that is released
    B. increase in the size of local predators
    C. decrease in the biodiversity of the area
    D. decrease in the amount of soil washed away

28. An investigation was carried out and the results are shown below. Substance X resulted from a metabolic process that produces ATP in yeast (a single-celled fungus). Which statement best describes substance X?

    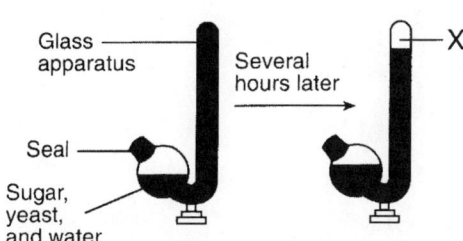

    A. It is oxygen released by protein synthesis.
    B. It is glucose that was produced in photosynthesis.
    C. It is starch that was produced during digestion.
    D. It is carbon dioxide released by respiration.

29. Which statement best describes the relationship between cells, DNA, and proteins?
    A. Cells contain DNA, which controls the production of proteins.

B. DNA is composed of proteins, which carry coded information about cell functions.
C. Proteins are used to produce cells, which link amino acids together into DNA.
D. Cells are linked together by proteins to make different kinds of DNA molecules.

30. The theory of biological evolution by natural selection includes the concept that
   A. species of organisms found on Earth today have adaptations that are not always found in earlier species
   B. fossils are the remains of present-day species and were all formed at the same time
   C. individuals may acquire physical characteristics after birth and pass these acquired characteristics on to their offspring
   D. the smallest organisms are always eliminated by the larger organisms within the ecosystem

End of Part 2

You may check your work on this part only.

Do not go to the next page.

PART 3

Directions for Questions 31 through 45 For each of the questions or incomplete statements below, choose the best of the answer choices given. Write your answers on a separate sheet of paper.

Base your answer to question 31 on the information below and on your knowledge of biology.

Analysis of a sample taken from a pond showed variety in both the number and types of organisms present. The data collected are in the table below.

Types of Organisms	Number Present in Sample
Bass (fish)	Two
Frogs (amphibian)	Forty
Insect larvae	Hundreds
Phytoplankton	Thousands

31. If the frogs feed on insect larvae, what is the role of the frogs in this pond ecosystem?
    A. herbivore   C. parasite
    B. consumer   D. host

32. The graph below shows the number of birds in a population over time. Which statement best explains section *X* of the graph?

    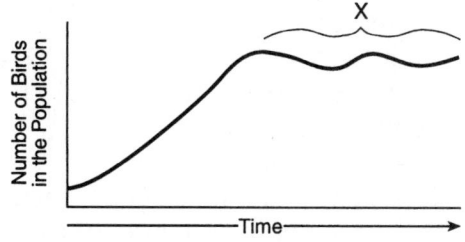

    A. Interbreeding between members of the population increased their mutation rate.
    B. An increase in the bird population caused a decrease in the predator population.
    C. The population reached the environment's carrying capacity and its growth rate stabilized.
    D. Another species came to the area and provided food for the birds.

33. Meiosis and fertilization are important for the survival of many species because these two processes result in
    A. a large number of gametes
    B. increasingly complex organisms
    C. genetic variability among offspring
    D. the cloning of superior offspring

34. Researchers performing a well-designed experiment should base their conclusions on
    A. the original hypothesis of the experiment
    B. data from repeated trials of the experiment
    C. a small sample size to ensure a reliable outcome
    D. results predicted before performing the experiment

35. The graph below shows the relative concentrations of different ions inside and outside of an animal cell. Which process is directly responsible for the net movement of K^+ and Mg^{++} *into* the animal cell?

    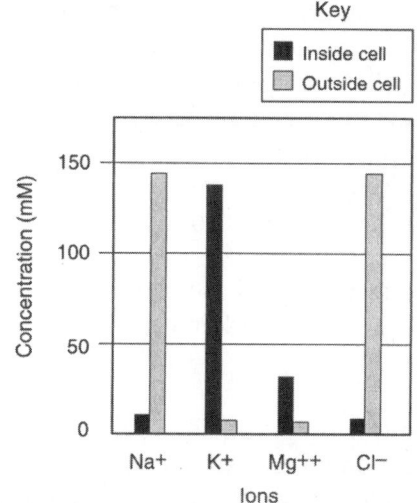

    A. electrophoresis   C. diffusion
    B. active transport   D. circulation

36. The information that controls the production of proteins in a cell must pass from the nucleus to the
    A. cell membrane
    B. mitochondria
    C. chloroplasts
    D. ribosomes

37. When a white insect lands on the bark of a white birch tree, its light color has a high adaptive value. If the birch trees become covered with black soot, then the white color of this type of insect would most likely
    A. retain its adaptive value
    B. increase in adaptive value
    C. mutate to a darker color
    D. decrease in adaptive value

38. What is the volume of the liquid in the graduated cylinder shown below?

    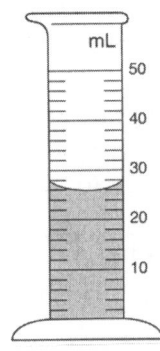

    A. 22 mL   C. 26 mL
    B. 24 mL   D. 28 mL

39. Strawberries can reproduce by means of runners, which are stems that grow horizontally above the ground. At the point where a runner touches the ground, a new plant develops. The new plant is genetically identical to the parent plant because
    A. it was produced sexually
    B. nuclei traveled through the runner to fertilize it
    C. it was produced asexually
    D. there were no other strawberry plants to fertilize it

40. The following graph shows the effect of temperature on an enzyme's action on a protein. Which change would *not* affect the enzyme's rate of action?

    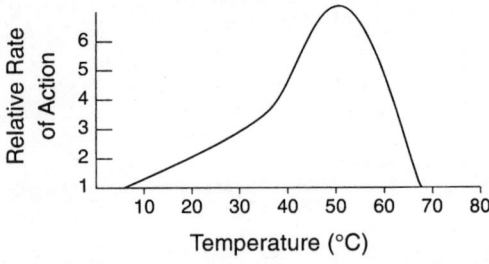

    A. the addition of cold water when the reaction is at 50°C
    B. an increase in temperature from 70°C to 80°C
    C. the removal of the protein when the reaction is at 30°C
    D. a decrease in temperature from 40°C to 10°C

41. Which statement is true of both mitosis and meiosis?
    A. Both are involved in asexual reproduction.
    B. Both occur only in the reproductive cells.
    C. The final number of chromosomes is reduced by half.
    D. DNA replication occurs before division of the nucleus.

42. The diagram below represents an energy pyramid. At each successive level, going from *A* to *D*, the amount of available energy

    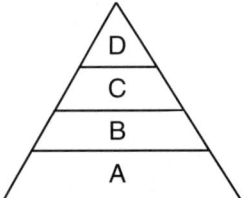

    A. increases, only
    B. decreases, only
    C. increases, then decreases
    D. remains the same

43. A scientist is planning an experiment on the effect of heat on the function of a certain enzyme. Which would *not* be an appropriate first step?
    A. doing research about enzymes in a library
    B. having discussions with other scientists
    C. completing a data table of expected results
    D. using what is already known about the enzyme

44. Leaves contain openings known as stomates, which are opened and closed by specialized cells to allow gas exchange with the outside environment. Which phrase best represents the net flow of gases involved in photosynthesis that would go through a leaf's stomates on a sunny day?
    A. carbon dioxide moves in; oxygen moves out
    B. oxygen moves in; nitrogen moves out
    C. carbon dioxide moves in; ozone moves out
    D. ozone moves in; carbon dioxide moves out

45. Which statement describes the ecosystem represented in the following diagram?

    A. This ecosystem is the first stage in an ecological succession.
    B. This ecosystem would most likely lack decomposers.
    C. All the organisms in this ecosystem are heterotrophs.
    D. This ecosystem includes both producers and consumers.

Directions for Question 46 Respond fully to the open-ended question that follows. Use one or more complete sentences to answer the question. Write your answer on a separate sheet of paper.

Base your answer to question 46 on the following chemical equation and on your knowledge of biology.

$$CO_2 + H_2O \xrightarrow[\text{CHLOROPHYLL}]{\text{LIGHT ENERGY}} C_6H_{12}O_6 + O_2$$

46. An important life process, carried out by producers, is described by this equation.

- Identify the process described and the two vital products of this reaction.
- Explain why cellular respiration is basically the opposite of the process shown in the equation.
- Identify the two waste products of cellular (aerobic) respiration.

End of Part 3

You may check your work on this part only.

DIAGNOSTIC CHECKLIST

Use this checklist to evaluate your Diagnostic Test answers. The checklist is designed so that you can determine which skills you need to improve in your preparation for the New Jersey Biology End-of-Course Test. Answer columns are provided so that you can compare your answers to those given by your teacher. If you miss an answer, check the box next to the number of that answer. Once you have checked all your answers, note which sections you need to review by referring to the sections listed in the reference column. Page numbers in the last column provide the location of those sections in this book.

Answers		Check if missed	References to standards and text	Chapter	Pages
Yours	Key				
1.			5.5.12 C1 Reproduction and Heredity	13	165–174
2.			5.10.12 A1 Environmental Systems	19	235–251
3.			5.2, 5.3, 5.4 Science, Math, & Technology	2	24–37
4.			5.10.12 A1 Environmental Systems	19	235–251
5.			5.10.12 A1 Environmental Systems	19	235–251
6.			5.5.12 C1 Reproduction and Heredity	14	175–184
7.			5.5.12 A1 Matter, Energy, & Organization	5	63–75
8.			5.5.12 A2 Matter, Energy, & Organization	6	76–86
9.			5.5.12 A2 Matter, Energy, & Organization	6	76–86
10.			5.5.12 B1/B2 Diversity & Biol. Evolution	11	144–152
11.			5.5.12 A2 Matter, Energy, & Organization	6	76–86
12.			5.5.12 B2 Diversity & Biol. Evolution	12	153–161
13.			5.5.12 A3 Matter, Energy, & Organization	7	87–101
14.			5.5.12 C2 Reproduction and Heredity	16	195–205
15.			5.1 Scientific Processes	1	13–23
16.			5.5.12 A2 Matter, Energy, & Organization	6	76–86
17.			5.5.12 C1 Reproduction and Heredity	15	185–194
18.			5.5.12 A2 Matter, Energy, & Organization	6	76–86
19.			5.5.12 C2 Reproduction and Heredity	17	206–220
20.			5.5.12 A1 Matter, Energy, & Organization	3	41–50
21.			5.5.12 B1/B2 Diversity & Biol. Evolution	11	144–152
22.			5.5.12 A1 Matter, Energy, & Organization	4	51–62
23.			5.10.12 B1 Environmental Systems	20	252–264
24.			5.5.12 C1 Reproduction and Heredity	14	175–184
25.			5.1 Scientific Processes	1	13–23
26.			5.5.12 A1 Matter, Energy, & Organization	5	63–75
27.			5.10.12 B1 Environmental Systems	20	252–264
28.			5.5.12 A2 Matter, Energy, & Organization	6	76–86
29.			5.5.12 C1 Reproduction and Heredity	14	175–184
30.			5.5.12 B1 Diversity & Biol. Evolution	10	134–143
31.			5.5.12 A3 Matter, Energy, & Organization	7	87–101
32.			5.10.12 A1 Environmental Systems	19	235–251
33.			5.5.12 C2 Reproduction and Heredity	16	195–205
34.			5.1 Scientific Processes	1	13–23
35.			5.5.12 A1 Matter, Energy, & Organization	4	51–62
36.			5.5.12 C1 Reproduction and Heredity	14	175–184
37.			5.5.12 B2 Diversity & Biol. Evolution	12	153–161
38.			5.2, 5.3, 5.4 Science, Math, & Technology	2	24–37
39.			5.5.12 C1 Reproduction and Heredity	15	185–194
40.			5.5.12 A1 Matter, Energy, & Organization	5	63–75
41.			5.5.12 C1/C2 Reproduction and Heredity	16	195–205
42.			5.5.12 A3 Matter, Energy, & Organization	7	87–101
43.			5.1 Scientific Processes	1	13–23
44.			5.5.12 A2 Matter, Energy, & Organization	6	76–86
45.			5.10.12 A1 Environmental Systems	19	235–251
46.			5.5.12 A2 Matter, Energy, & Organization	6	76–86

Unit I
Scientific and Laboratory Procedures

STANDARDS 5.1 through 5.4
The Process of Science

Enduring Understandings In order for students to be considered proficient at science, they should:

- know, use, and interpret scientific explanations of the natural world;
- generate and evaluate scientific evidence and explanations;
- understand the nature and development of scientific knowledge;
- participate productively in scientific practices and discourse.

Chapter 1

The Scientific Method and Inquiry

Standard 5.1 (Scientific Processes) All students will develop problem-solving, decision-making, and inquiry skills, reflected by formulating usable questions and hypotheses, planning experiments, conducting systematic observations, interpreting and analyzing data, drawing conclusions, and communicating results.

SCIENCE: LEARNING ABOUT THE WORLD

Science is a body of knowledge about our natural world. To make sense of all this knowledge, scientists have organized the information into major areas of understanding, such as chemistry, physics, biology, and earth science. **Biology** is the study of all living things, which includes humans. The study of biology also includes more specific areas, or *branches,* of study, such as anatomy, ecology, and genetics, among others. (See Table 1-1 on page 14.)

Science is also a way of learning about the world. Everything that we now know in biology and every other area of science is a result of this way of learning. While it is important for us to understand how our bodies work and how all living things exist, it is also important for us to understand how we *learned* what we know and how we continue to expand our knowledge. In other words, we need to understand the nature of scientific *thinking.* This is called **scientific inquiry**, and it is where we begin our study of biology.

The thinking process known as science has been devised by people in order to learn about how the world works. Based on observations and evidence collected from experimentation, and using what is already known, scientific explanations are developed about the world. These explanations are always subject to change as new evidence and observations are presented. Scientific methods exist to constantly test and re-test what we know in relation to existing explanations. In this way, scientific knowledge advances toward a more complete understanding of the world around us. A **theory** is a general statement that is supported by many scientific observations and experiments and represents the most logical explanation of the evidence. Theories become stronger as more supporting evidence and experimental data are gathered.

Natural phenomena and events are understood by people on the basis of existing explanations. To develop scientific explanations, people use evidence

Table 1-1 The Main Branches of Biology

Branch	Specific Field of Study
Anatomy	Body structures (human and animal)
Botany	Plants (characteristics and classification)
Biochemistry	Chemistry of life (internal)
Cytology	Cell structure and function
Ecology	Interaction of organisms with their environment
Embryology	Early developmental stages in an organism's life
Genetics	How characteristics are expressed and passed from parent to offspring
Physiology	Internal functions of organisms
Marine biology	Living things in the ocean
Microbiology	Microscopic life-forms
Taxonomy	Classification of living things
Zoology	Animals (behavior and classification)

that can be observed as well as information that is already known. Researchers think about these explanations by visualizing them; that is, they create mental pictures and develop mathematical models. It is also important to learn about the history of science and the particular individuals who have contributed to scientific understanding. While science provides knowledge about the world, it also challenges people to develop the values to use this knowledge ethically and effectively.

SCIENCE BEGINS WITH QUESTIONS

To develop scientific ideas, one needs to be curious, think, do library research, and discuss one's ideas with others, including experts. Scientific inquiry involves asking questions and locating information from a variety of sources. It also involves making wise judgments about how reliable and relevant the information is.

The questions that are studied often occur to us as we observe things around us. For example, you might have noticed that, over time, a plant near a window bends toward the light. (See Figure 1-1.) When you notice this, you are making an **observation**. Seeing many different plants do the same thing near many different windows is a discovery of a pattern or regularity. A scientific mind is very observant and also curious about why patterns exist. Science attempts to do much more than simply observe and describe. It seeks to find general explanations for why things are the way they are and why things behave the way they do. Thus, the question about plants might be posed as "What causes the tendency of plants to bend toward a light source?"

Figure 1-1 You can observe that, over time, a plant near a window bends toward the sunlight.

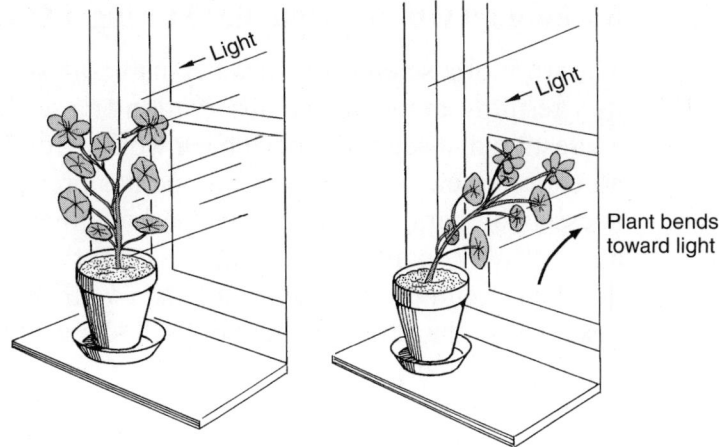

There are many questions in biology that science has studied and continues to study. Examples of these questions include:

- How is oxygen transported throughout the body by our blood?
- Why do birds migrate?
- Why are most plants green?
- Can skin tissue be grown in the laboratory and used to heal burn victims?
- How does DNA store genetic information?

For every biological question and problem studied by researchers, a similar approach is used to seek an answer. This approach, often called the *scientific method,* involves a way of conducting scientific investigations.

THE SCIENTIFIC METHOD: STEPS IN AN INVESTIGATION

Experiments conducted by researchers demonstrate how scientific knowledge can be gained by using the **scientific method**, which is an organized approach to problem solving. The main steps in this procedure are:

- State the problem (in question form).
- Collect information about the problem.
- Form a hypothesis (a possible answer).
- Design and conduct an experiment (use an experimental group with a variable; use a control group without the variable).
- Record observations and data.
- Check results; redo experiment (if necessary).
- Draw your conclusions (accept or reject your hypothesis).
- Communicate your results.

Making an Observation and Stating a Question

The work of a scientist always begins with asking a question. The question may result from making an observation or a series of observations. It may be a question that scientists have been working on for a long time; or it may be an entirely new question that no one has previously studied. In all cases, it is essential that the question be clearly defined and carefully stated. For example, "Why do plants like light?" is a poorly defined, vague question. "Why do plants bend toward the light?" is a well-defined, specific question. The next step in the scientific method will show why a scientific question must be clearly stated.

Forming a Testable Hypothesis

Once a specific question has been asked, a scientist can think about possible answers to the question. A possible answer—actually an educated guess at this stage—is called a *hypothesis*. The scientist also begins to collect information that may help answer the question. The information may include results from other researchers' observations or experiments. Often, one question may lead to more than one hypothesis. For example, perhaps plants bend toward the light because of differences in temperature inside and outside the window. Perhaps the plant's roots push the stem toward the window. Or, perhaps structures inside the plant's stems are affected by the light, thus causing it to bend. Any one of these hypotheses might be the answer to the question. This is why the question must be very clear.

To conduct a scientific investigation, a hypothesis must be testable. There must be a possible answer that can be studied, tested, and proved to be true or not. Most often, once a hypothesis has been tested—and proved either true or false—the result leads to another question, more hypotheses, and even more testing. This is the process of scientific research.

Designing a Scientific Experiment

A hypothesis is tested by conducting an experiment. The most common type of experiment in science is called a *controlled experiment*. It actually consists of two or more experiments carried out at the same time. In each experiment, all conditions are kept the same except for one *factor* that is controlled by the researcher. This factor, called the **independent variable**, is the one difference (according to hypothesis) that might explain the observation and answer the question. The **dependent variable** is the change that occurs (such as bending toward the light) because of the independent variable. The experimental setup without the independent variable is the *control*.

For example, to test our hypothesis that temperature differences cause the bending of plants toward light, we might design an experiment as follows: (a) Grow plants near a window in the winter, keeping the temperature inside the room as cold as the outside; and (b) grow plants near a window in the winter, making the temperature inside the room warmer than the outside. To test the hypothesis that structures in the stems cause the bending toward light, we might design this experiment: (a) Grow plants near the window

Temperature (°C)	Grams of starch hydrolyzed per minute
0	0.0
10	0.2
20	0.4
30	0.8
40	1.0
50	0.3
60	0.2

Figure 1-2 Data from an experiment about the effect of temperature on enzyme activity can be organized and displayed in a table.

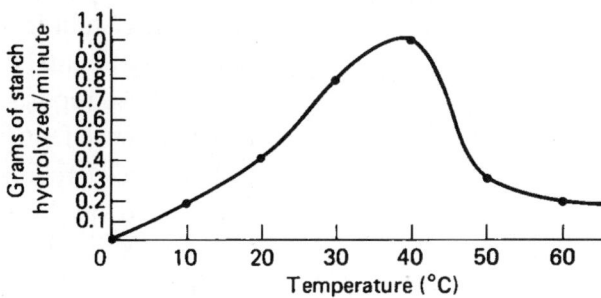

Figure 1-3 Data from an experiment about the effect of temperature on enzyme activity can be organized in the form of a line graph, too.

with their stems covered by paper that blocks the light; and (b) grow plants near the window without any coverings on the stems.

Conducting the Experiment

Once an experiment has been designed, the scientist must, of course, carry out the experiment. To conduct an experiment, the researcher must consider the following:

- establish the number of organisms to be tested, called the *sample size* or *group size;* a larger group usually ensures a successful experiment, with more valid results, although the experiment may take more time and expense to complete;
- determine what measurements are to be made, what tools are to be used to make the measurements, and how often these data should be collected;
- maintain careful records of the data that have been collected and select the best methods to organize the data into charts and graphs; Figure 1-2 shows data organized in a table from an experiment designed to investigate the effect of temperature (the variable) on the action of an enzyme that breaks down starch into sugars; the same data also can be shown in the form of a line graph, as shown in Figure 1-3;
- finally, and perhaps most importantly, review and analyze the data in order to determine if the results of the experiment prove or disprove the hypothesis; and if it is possible to draw a conclusion from this experiment, determine what that conclusion is.

COMMUNICATING SCIENTIFIC RESULTS

No scientist works alone. The process of scientific inquiry builds upon previous knowledge and extends our understanding. For example, when scientists first investigated why plants bend toward light, they would have relied on information already known about the nature of light, the cellular structure of plants, and how cells divide to make plants grow. Any new findings that scientists obtain from their experiments are shared with other scientists. For hundreds of years, this has been done through the publication of research reports. These reports are printed in scientific journals and mailed around the world. Now scientific findings are posted on the Internet as well, to be read by other researchers who work on similar topics. Experimental results are considered valid if other scientists can obtain the same results when they repeat the experiment under the same conditions.

It is through this community of scientific study that explanations about how the world works are constantly revised and expanded. As scientists study each other's work, new ideas develop. A conclusion that is based upon the results of an experiment is called an *inference*. In the plant investigation, the conclusion that light changes the growth of a plant stem is an example of an inference. In scientific inquiry, an inference always needs to be tested again (by conducting more experiments) in order to be confirmed. One way to do this is to make a new prediction from the inference. If the new prediction is tested and turns out to be true, it is strong evidence that the inference was valid. For example, based on the inference that light affects the growth of plant stems (causing them to bend), the prediction might be made that only plants with soft green stems can bend, whereas plants with hard wooden stems cannot. This would require a new experiment.

The study of science never ends because each new answer to a research problem leads to another question. This is *scientific inquiry*—the ongoing process of learning about the world around us.

Chapter 1 Review

Multiple Choice

1. Dr. Jones met a man who had been bitten by venomous snakes several times over many years, yet he was not sick. The doctor thinks that something produced in the man's blood protects him from harmful effects of the poison. This idea, or possible answer to a scientific question, is called a (an)
   A. hypothesis   C. observation
   B. inference   D. theory

2. A student placed slices of moist bread in a closed cupboard and noticed that mold grew faster on them than on slices of moist bread left out on a counter. The one difference in this experiment—the presence or absence of light—is called the
   A. theory   C. control
   B. hypothesis   D. variable

3. A logical explanation of natural phenomena that is supported by scientific observations and experiments is called a (an)
   A. hypothesis   C. factor
   B. inference   D. theory

4. A student wants to test how much water is necessary to produce the most bread mold. She kept one slice of bread dry while using varying amounts of water on other slices. The dry bread in this experiment is called a
   A. hypothesis   C. observation
   B. control   D. theory

5. The number of organisms that are tested in an experiment is called the
   A. variable size   C. sample size
   B. controlled size D. experimental factor

6. The scientific method is
   A. a way of posing a research question only
   B. used to organize data that is already known
   C. an organized approach to problem solving
   D. used by all scientists in an identical way

7. The best way to be sure that your experimental results are valid is to
   A. ignore information from other sources
   B. conduct your experiment one time only
   C. use more than one variable in the experiment
   D. test as large a sample size as possible

8. A researcher is reviewing another scientist's experiment and conclusions. The reviewer would most likely consider the experiment *invalid* if
   A. the sample size produced a great deal of data
   B. other individuals are able to duplicate the results
   C. it has conclusions that are not explained by the evidence
   D. the hypothesis was not supported by the data obtained

9. An experimental setup is shown in the diagram below. Which hypothesis would most likely be tested using this setup?

   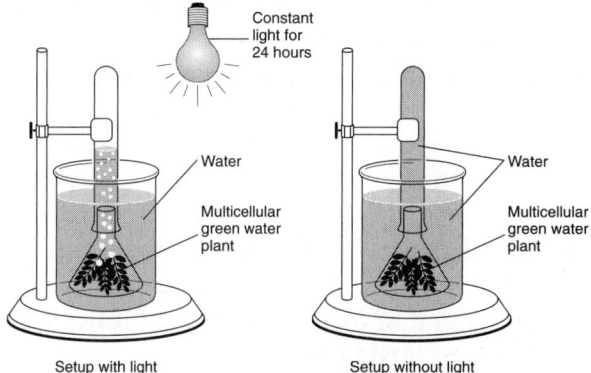

   A. Water plants release a gas in the presence of light.
   B. Roots of plants absorb minerals in the absence of light.
   C. Green plants need light for cell division.
   D. Plants grow best in the absence of light.

10. Which statement best describes a scientific theory?
    A. It is a collection of data designed to provide support for a prediction.
    B. It is an educated guess that can be tested by experimentation.
    C. It is a scientific fact that no longer requires any evidence to support it.
    D. It is a general statement that is supported by many scientific observations.

11. An experiment was carried out to determine which mouthwash was most effective against bacteria commonly found in the mouth. Four paper discs (*A* through *D*) were each dipped

into a different brand of mouthwash. The discs were then placed onto the surface of a culture plate that contained food, moisture, and bacteria commonly found in the mouth. The diagram below shows the growth of bacteria on the plate after 24 hours. Which change in procedure would have improved the experiment?

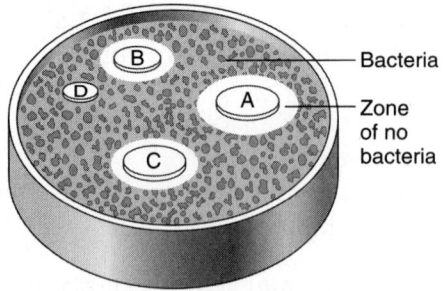

A. using a smaller plate with less food and moisture
B. using bacteria from many habitats other than the mouth
C. using the same size paper discs for each mouthwash
D. using the same type of mouthwash on each disc

12. Which source would provide the most reliable information for use in a research project investigating the effects of antibiotics on disease-causing bacteria?
A. the local news section of a newspaper from 1993
B. a news program on national television about antigens produced by various plants
C. a current professional science journal article on the control of pathogens
D. an article in a weekly news magazine about reproduction in pathogens

13. Researchers performing a well-designed experiment should base their conclusions on
A. the original hypothesis of the experiment
B. data from repeated trials of the experiment
C. a small sample size to ensure valid results of the experiment
D. results predicted before performing the experiment

14. A scientist is planning to carry out an experiment on the effect of heat on the function of a certain enzyme. Which would be the *least* appropriate first step to take?
A. doing research in a library or on the Internet
B. having discussions with other scientists
C. completing a data table of expected results
D. using what is already known about the enzyme

15. Which statement best describes the term *theory* as used when describing a scientific theory?
A. A theory is never revised even when new scientific evidence is presented.
B. A theory is an assumption made by scientists and implies a lack of certainty.
C. A theory is a scientific explanation that is supported by a variety of experimental data.
D. A theory is a hypothesis that has been supported by one experiment only.

16. The analysis of data gathered during a particular experiment is necessary in order to
A. formulate a hypothesis for that experiment
B. develop a research plan for that experiment
C. design a control group for that experiment
D. draw a valid conclusion for that experiment

17. When starch is digested, it is broken down into smaller sugar units. The graph below represents data obtained from an experiment on starch digestion. Which statement best describes point *A* and point *B* on the graph?

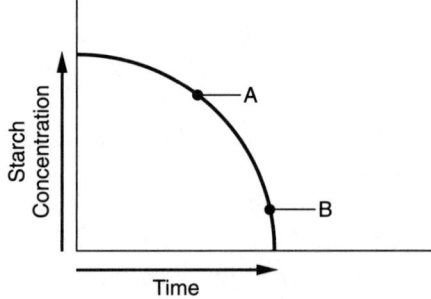

A. The concentration of sugars is greater at point *A* than it is at point *B*.
B. The concentration of sugars is greater at point *B* than it is at point *A*.

C. The starch concentration is the same at point *A* as it is at point *B*.
D. The starch concentration is greater at point *B* than it is at point *A*.

18. A great deal of information can now be obtained about the future health of people by examining the genetic makeup of their cells. There are concerns that this information could be used to deny an individual health insurance or employment. These concerns best illustrate that
    A. scientific explanations depend upon evidence collected from a single source
    B. scientific inquiry involves the collection of information from a large number of sources
    C. acquiring too much knowledge in human genetics will discourage future research in that area
    D. while science provides knowledge, values are essential to making ethical decisions when using this knowledge

19. A biologist reported success in breeding a tiger with a lion, producing healthy fertile offspring. Other biologists will accept this report as fact only if
    A. research shows that other animals can be crossbred
    B. the offspring are given a new scientific name
    C. the biologist included a control in the experiment
    D. they can repeat the experiment and get the same results

20. Why do scientists consider any hypothesis valuable to research?
    A. A hypothesis requires no further investigation.
    B. A hypothesis may lead to further research even if it is disproved by the experiment.
    C. A hypothesis requires no further investigation if it is proved by the experiment.
    D. A hypothesis can be used to explain a conclusion even if it is disproved by the experiment.

Analysis and Open Ended

21. Describe the basic steps of the scientific method in the sequence that they should be used.

22. Define and distinguish between the terms *hypothesis* and *theory*.

23. Why is the development of a research plan necessary before testing a hypothesis?

24. Why is a clear statement of a hypothesis very important to the experiment that will follow?

25. Describe three ways in which a scientist can collect and organize data.

26. Why is it so difficult for the results of one experiment to become a theory? Explain.

27. Compounds containing phosphorus that are dumped into the environment can upset local ecosystems because phosphorus acts as a fertilizer. The graph below shows measurements of phosphorus concentrations at two sites taken during the month of June, from 1991 to 1997.

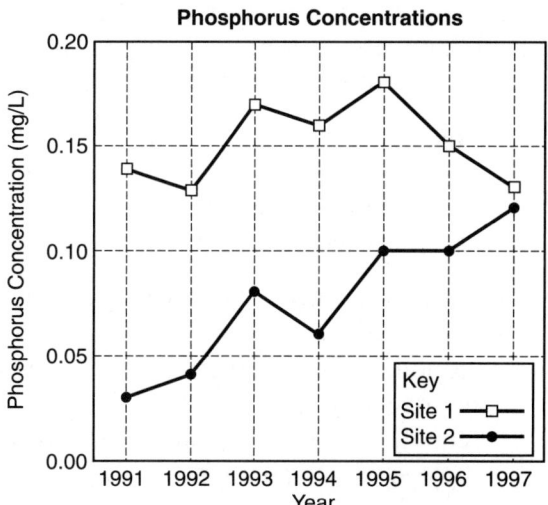

Which statement represents a valid inference based on information in the graph?
    A. There was a decrease in the amount of phosphorus dumped at Site 2 from 1991 to 1997.
    B. Pollution controls may have been put into effect at Site 1 starting in 1995.
    C. There was most likely no vegetation present near Site 2 from 1993 to 1994.
    D. There was a greater variation in phosphorus concentration at Site 1 than at Site 2.

28. The concentration of salt in water affects the hatching of brine shrimp eggs. Brine shrimp eggs will develop and hatch at room temperature in glass containers of salt solution. Describe a controlled experiment using three separate experimental groups that could be

used to determine the best concentration of salt solution in which to hatch brine shrimp eggs. Your answer must include:
- a description of how the control group and each of the three experimental groups will be different;
- *two* conditions that must be kept constant in the control group and the experimental groups;
- what data should be collected, and how often;
- *one* sample of experimental results that would indicate the best concentration of salt solution in which to hatch brine shrimp eggs.

29. A student hypothesized that lettuce seeds would not sprout (germinate) unless they were placed in the dark. The student planted 10 lettuce seeds in soil that was kept in the dark and planted another 10 lettuce seeds in soil that was exposed to the light. The data collected are shown in the table below.

Experimental Conditions	Number of Seeds Germinated
Seeds in soil kept in the dark	9
Seeds in soil exposed to light	8

One way to improve the reliability of these results would be to
A. conclude that darkness is necessary for lettuce seed germination
B. conclude that light is necessary for lettuce seed germination
C. revise the hypothesis
D. repeat the experiment

30. To determine which colors of light are best used by plants for photosynthesis, three types of underwater plants of similar mass were subjected to the same intensity of light of different colors for the same period of time. All other environmental conditions were kept the same. After 15 minutes of exposure to light, a video camera was used to record the number of gas bubbles each plant gave off (due to photosynthesis) in a 30-second time period. Each type of plant was tested six times. The average of the data for each plant type is shown in the following table.

Average Number of Bubbles in 30 Seconds

Plant Type	Red Light	Yellow Light	Green Light	Blue Light
Elodea	35	11	5	47
Potamogeton	48	8	2	63
Utricularia	28	9	6	39

Which statement is a valid inference based on the data?
A. Each plant carried on photosynthesis best in a different color of light.
B. Red light is better for carrying on photosynthesis than blue light is.
C. Each plant carried on photosynthesis best in red light and blue light.
D. Water probably filters out the red light and the green light.

31. A certain plant that usually grows in slightly basic soil produces flowers with white petals. Sometimes the same plant species produces flowers with red petals. A company that sells the plant wants to know if soil pH affects the color of the petals in this plant. Design a controlled experiment to determine if soil pH affects petal color. In your experimental design be sure to:
- state the hypothesis to be tested in the experiment;
- state *one* way the control group will be treated differently from the experimental group;
- identify *two* factors that must be kept the same in both the control and the experimental group;
- state *one* piece of evidence that would support your hypothesis.

32. A researcher conducts an experiment on corn seedlings and uses the results to make predictions and conclusions about all plants. Explain the researcher's error in using the results in this way.

33. Why is it important for a scientist to include all the steps that he or she used to conduct an experiment when writing the research report for other scientists to read?

Reading Comprehension

Base your answers to questions 34 to 36 on the information below and on your knowledge of biology. Use one or more complete sentences to answer each question.

> The main purpose of science is to look at events, occurrences, and patterns in nature and develop explanations for them. These explanations can always be changed as new observations are made and new evidence is found. A possible explanation of a natural event or pattern is called a hypothesis. Charles Darwin, in his own words, showed why he was a true scientist:
>
> "From my early youth I have had the strongest desire to understand or explain whatever I observed—that is, to group all facts under some general laws. I have steadily endeavored to keep my mind free, so as to give up any hypothesis, however much beloved (and I cannot resist forming one on every subject), as soon as facts are shown to be opposed to it. I followed a golden rule that whenever a published fact, a new observation or thought came across me, which was opposed to my general results, to make a memorandum of it without fail and at once. During some part of the day I wrote my Journal, and took much pains in describing carefully and vividly all that I had seen; and this was good practice . . . and this habit of mind was continued during the five years of the voyage. I feel sure that it was this training which has enabled me to do whatever I have done in science."
>
> On the voyage of the *Beagle,* Darwin saw seeds that had washed ashore on a small island near Australia. He wondered whether seeds could travel long distances in the ocean and still be able to grow. Back in England, he enthusiastically filled his home with bottles of seawater. Darwin soaked many different kinds of seeds in the salty water, for short and long periods of time. He used a variety of crop seeds—such as cabbage seeds, lettuce seeds, and celery seeds—23 kinds in all. He then tried to grow them in soil. Darwin's experiments showed exactly what the scientific method is: State the problem; collect information; form a hypothesis; perform an experiment; record observations and data; draw a conclusion; and share your results.

34. When does a scientist find it necessary to change a hypothesis?

35. Why did Charles Darwin keep a journal?

36. Describe how Darwin followed each of the steps of the scientific method in his experiment with seeds and salt water.

Chapter 2

Laboratory Tools and Techniques

Standard 5.2 (Science and Society) **All students will develop an understanding of how people of various cultures have contributed to the advancement of science and technology, and how major discoveries and events have advanced science and technology.**

Standard 5.3 (Mathematical Applications) **All students will integrate mathematics as a tool for problem solving in science, and as a means of expressing and/or modeling scientific theories.**

Standard 5.4 (Nature and Process of Technology) **All students will understand the interrelationships between science and technology and develop a conceptual understanding of the nature and process of technology.**

THE BIOLOGY LABORATORY: TOOLS AND DISCOVERIES

Scientific investigations are based on observations and measurements. Biologists also use a wide variety of tools and procedures to increase the range and accuracy of their natural senses. For example, we use eyeglasses in everyday life to aid our sense of sight. Similarly, a biologist uses microscopes, thermometers, weighing scales, and many other instruments to make detailed and precise observations and measurements. Careful attention and proper safety procedures are also essential in the laboratory at all times. Therefore, safety precautions related to laboratory tools will be examined in this chapter.

Early Discoveries in Biology

The invention of the *microscope* had an enormous influence on the development of biology as a science. Anton van Leeuwenhoek, a Dutch lens maker who lived more than 300 years ago, is considered the first person to have built a simple microscope. With his invention, the first single-lens microscope, van Leeuwenhoek was able to magnify the image of tiny objects more than 270 times their normal size. This allowed him to see things that had never been seen before, such as bacteria, protozoa, yeast cells, and blood cells. In 1665, Robert Hooke, an English physicist, made a compound light

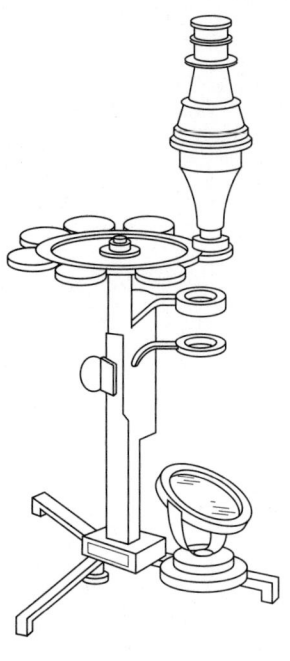

Early Microscope

Figure 2-1 An early microscope, used to observe tiny organisms and cells.

microscope using two lenses. With this device, he observed the inside structure of a thin piece of cork and called the spaces he saw "cells." In the 1800s, scientists in other countries used microscopes to learn much more about cells. Robert Brown in Scotland discovered the central structure in plant cells: the nucleus. Two German researchers—Matthias Schleiden studying plants and Theodor Schwann studying animals—concluded that the tissues of all living things are composed of cells. This is the basis for the *cell theory,* which states that all living things are made of cells (which come from other cells) and that the cell is the basic unit of structure and function of all living things. (See Figure 2-1.)

The biological studies continued. In the mid-1800s, an Austrian monk, Gregor Mendel, investigated the simple garden pea plant to develop his fundamental principles of heredity. In 1859, English naturalist Charles Darwin presented his theory of evolution by natural selection, based on years of observations and detailed plant and animal studies. In the early 1900s, Thomas Hunt Morgan studied the common fruit fly to learn about chromosomes and genes. Then, in 1953, scientists James Watson and Francis Crick, using precise measurements and molecular models, discovered the double-helix structure of DNA, allowing us to understand how this most important molecule serves as the basis for all life on Earth.

SCIENCE AND TECHNOLOGY

Science and technology affect the lives of people all over the world. As previously noted, *science* is a body of knowledge, discovered through the process of seeking answers to questions about our natural world. **Technology** is the process of using that scientific knowledge and other resources to develop new products and processes. These products and processes help solve problems and meet the needs of both the individual and society. While the emphasis in science is on gaining knowledge of the natural world, the emphasis in technology is on finding practical ways to apply that knowledge to solve problems. The fields of science and technology frequently help to advance each other. Scientific discoveries often lead to the development of even better technological devices, such as high-powered microscopes. These technologies may, in turn, lead to new discoveries or to a better understanding of scientific principles.

MICROSCOPES AND MEASUREMENTS

Using a Microscope

The **compound light microscope** uses two lens types, the *objective lens* and the *ocular lens,* or *eyepiece,* to magnify the image of a specimen. The magnified image seen through the objective lenses is further enlarged by the

Figure 2-2 Diagram of a compound light microscope, which uses two types of lenses.

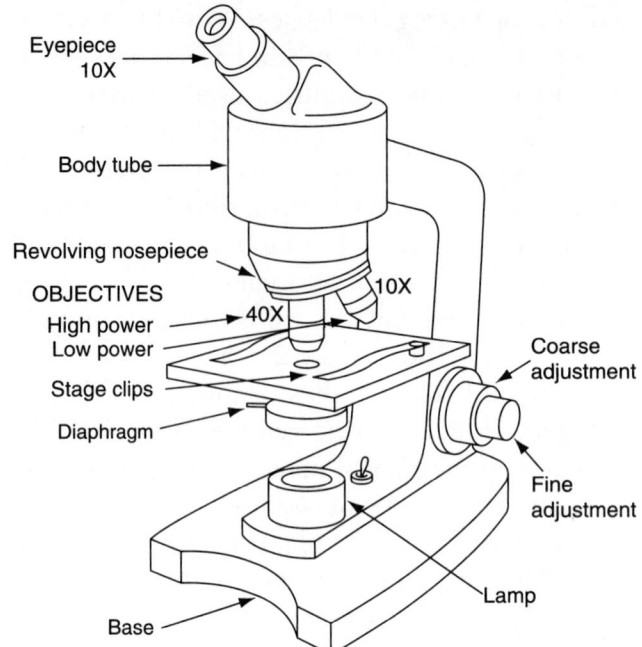

ocular lens. (See Figure 2-2.) When viewing a specimen, one should begin with the low-power objective, first using the coarse-adjustment knob to focus and then the fine-adjustment knob. Further study of specimen details can then be done by using the high-power objective and the fine-adjustment knob only. The area seen while looking through a microscope is called the *field* of view. The actual diameter of the field being viewed decreases as you switch from low power to high power. For example, if the magnification is 100x under low power and 400x under high power, then the field diameter under high power will be one-fourth of what it is under low power. In this example, if four cells in a row could be observed under low power, then only one of those cells could be observed under high power, but with more detail. The field also appears dimmer under high power than under low power, because the objective is closer to the slide. By opening the diaphragm (under the stage), you can allow more light in to reach the specimen. (See Table 2-1.)

Many specimens, such as tissue samples or tiny organisms, need to be studied with a microscope by first preparing a **wet mount** and applying a stain. (See Figure 2-3.) The following steps are used to make a wet mount:

1 Use a medicine dropper to place a drop of water onto the center of a clean slide.

2 Use the medicine dropper to place the specimen into the water on the slide.

3 Gently lower a coverslip (placing one edge down first, at a 45° angle) over the drop of water to avoid trapping air bubbles.

4 Add a drop of a stain such as iodine or methylene blue at one edge of the coverslip (to see more detail in the specimen). Draw the stain (dye) out with a piece of paper toweling.

Table 2-1 Parts of the Compound Light Microscope and Their Functions

Part	Function
Base	Supports the microscope
Arm	Used to carry microscope; attaches to the base, stage, and body tube
Body tube	Holds the objective lens and eyepiece
Stage	Platform on which the glass slide with the specimen is placed (over the hole in the stage through which light passes)
Clips	Hold the slide in position on the stage
Nosepiece	Holds the objective lenses; rotates so that the different objective lenses can be moved in line with the specimen and eyepiece
Coarse adjustment	Larger knob used for rough-focusing with the low-power objective
Fine adjustment	Smaller knob used for focusing with the high-power objective and for final-focusing with the low-power objective
Lamp or mirror	Directs light to the specimen (on the stage)
Diaphragm	Controls the amount of light reaching the specimen
Objective lenses	Lenses mounted on the nosepiece
Ocular lens	Lens at the top of the body tube; commonly called the eyepiece

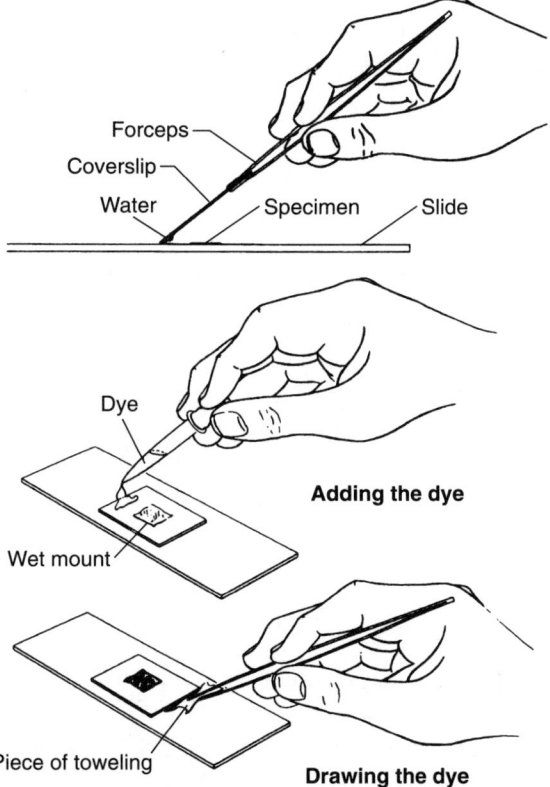

Figure 2-3 The wet mount technique is used to prepare a biological specimen for viewing under a microscope.

Chapter 2/Laboratory Tools and Techniques

Many cell parts and structures are not visible with a compound light microscope. University and professional laboratories use *electron microscopes* to magnify images more than 250,000 times. In place of light beams and optical lenses, the electron microscope has an electron beam and electromagnetic lenses. Beams of electrons passing through the specimen are focused on a television screen for viewing. This is called a *transmission electron microscope* (TEM). Another piece of equipment, just as advanced and expensive, is the *scanning electron microscope* (SEM), in which an electron beam is passed back and forth over the surface of a specimen, revealing very fine details about the surface structure of the object.

Taking Measurements

As part of their scientific investigations, researchers take a variety of measurements. One or more of the following factors may be measured while conducting an experiment: length, volume, temperature, and mass. The tools for measuring these factors are described below.

- *Length:* The *metric ruler* is used to determine the length of a specimen. A one-meter ruler, or meterstick, is divided into 100 centimeters, with each centimeter further divided into 10 millimeters. One meter equals 1000 millimeters. (See Figure 2-4.)

- *Volume:* A clear numbered column called a *graduated cylinder* is used to measure the volume of a liquid in liters and milliliters. One liter equals 1000 milliliters. (See Figure 2-5.) The surface of the liquid is compared to the measurement scale on the cylinder to determine the volume. To be most accurate, the measurement should be read at the bottom, or *meniscus,* of the curved surface of the liquid. (See Figure 2-6.)

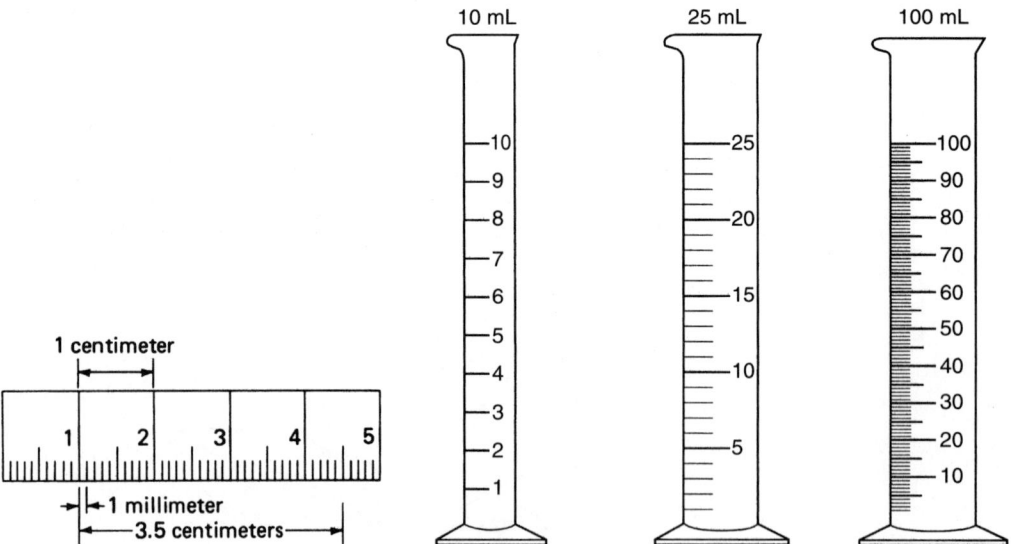

Figure 2-4 A centimeter ruler.

Figure 2-5 Graduated cylinders are used to measure the volume of a liquid; measurements are in milliliters (mL).

Figure 2-6 Always read the volume of a liquid at the bottom of the meniscus (not at the top of the curve).

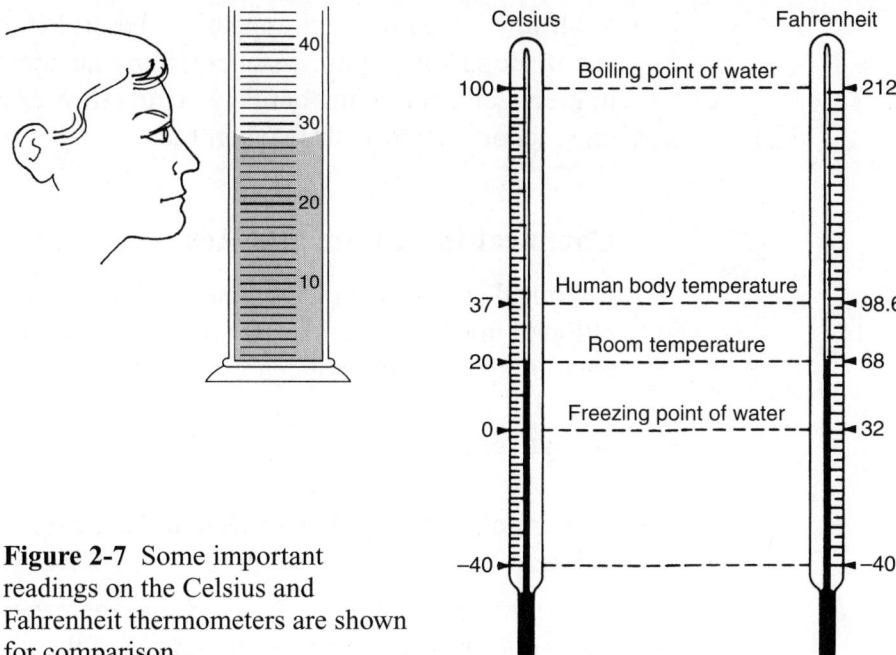

Figure 2-7 Some important readings on the Celsius and Fahrenheit thermometers are shown for comparison.

Figure 2-8 The triple-beam balance is used to measure mass; measurements are in grams.

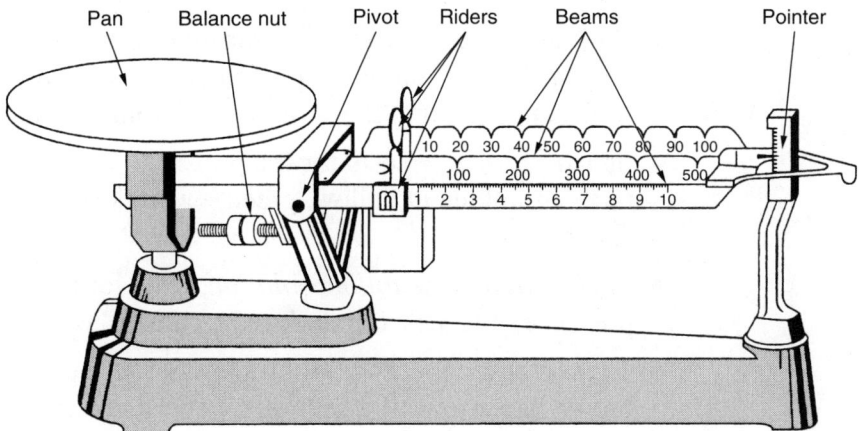

- *Temperature:* The temperature of a substance can be measured by using a *Celsius thermometer.* Zero degrees Celsius (0°C) is the freezing point of freshwater; 100 degrees Celsius (100°C) is the boiling point of freshwater. (See Figure 2-7.)
- *Mass:* A *triple-beam balance* or *electronic balance* is used to measure the mass of an object in kilograms and grams. One kilogram equals 1000 grams. (See Figure 2-8.)

CHEMICALS AND SAFETY PROCEDURES

One of the most important things to do *before* you go to work in a lab is: Be prepared. Study the assigned investigation before you come to class. When in the laboratory, maintain a clean, open work area that is free of everything except those materials needed for the investigation.

If any lab equipment appears to be broken or in poor condition, do not use it. Report it to your teacher. If any accident or injury occurs, report it immediately; and if any damage occurs to your clothing or personal belongings, report that to your teacher, too.

Chemical Indicators and Reagents

Chemical *stains* (such as iodine and methylene blue) are used in the preparation of microscope slides. Chemical *indicators* are used to test for the presence of specific substances or to determine chemical characteristics of a material. The following list describes typical chemical tests used in a biology laboratory.

- *Litmus paper* is an indicator used to determine whether a solution is acidic or basic. Blue litmus turns red in an acid. Red litmus turns blue in a base.
- *pH paper* has an indicator soaked into the paper that changes to one of many colors based upon the actual pH of the solution. The color of the pH paper is matched against a color chart to determine the pH of the solution tested.
- *Bromthymol blue* is an indicator that turns a solution yellow in the presence of carbon dioxide. If the carbon dioxide is removed from the solution containing the bromthymol yellow, it turns back to blue.
- *Benedict solution* tests for the presence of simple sugars. When heated, it turns from blue to yellow, green, or brick red, depending on the amount of sugar present.
- *Lugol solution,* or *iodine solution,* tests for the presence of starch. In its presence, it turns from dark red to blue-black.
- To test for the presence of fat, gently rub the food sample on a piece of brown paper toweling or brown paper (supermarket) bag. A translucent grease spot shows that fat is present.
- *Biuret solution,* which is colorless, turns violet in the presence of protein.

Handling Chemicals Safely

Laboratory investigations often require the use of different chemicals; they may also require the use of heat, such as from a Bunsen burner, and breakable glassware, such as beakers and test tubes. Spilled chemicals can cause burns; and heated chemicals can give off toxic vapors. Because of these risks, it is very important to keep alert when working in a lab and to follow the safety procedures outlined below.

- Do not handle chemicals or equipment unless you are told by your teacher to do so.
- Never eat, drink, chew gum, or apply makeup in the laboratory.
- Read the label on every container before you use it. Do not use chemicals from containers that are not clearly labeled.

Figure 2-9 Always wear safety goggles and use caution when heating chemicals in a test tube.

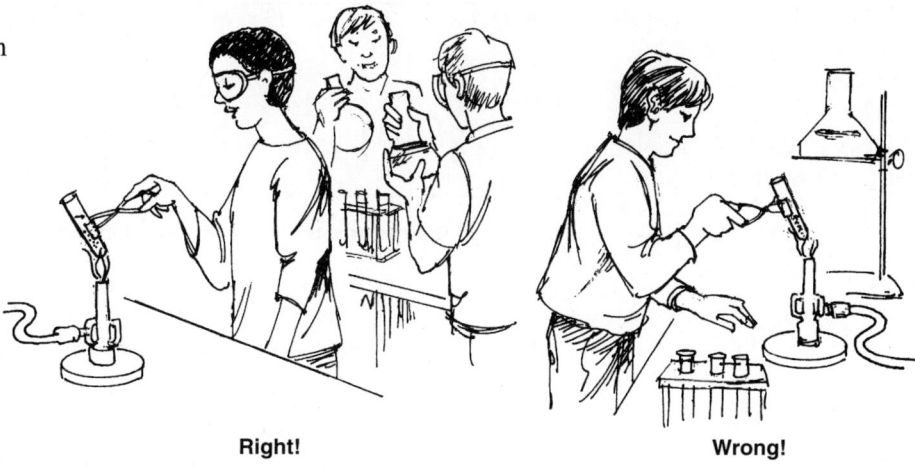

Right!   Wrong!

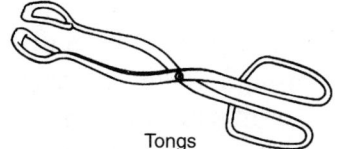

Tongs

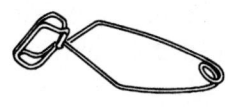

Test-tube holder

Figure 2-10 Lab equipment needed to handle hot objects.

- Wear protective clothing and keep chemical stains off counters and other materials. Do not touch stains with your fingers; use medicine droppers to transfer stains from container to slide.
- Tie back long hair and remove all dangling jewelry. Roll up long sleeves and do not wear loose-fitting sleeves.
- Never taste chemicals or inhale the vapors from a chemical, since they can be toxic and dangerous; gently wave hand over open container to smell non-toxic chemicals.
- When you heat a liquid in a test tube, be certain that the opening of the test tube is pointed away from you and away from anyone else nearby. (See Figure 2-9.)
- Never pour excess reagents back into stock bottles. (Place the excess reagents in a designated waste container.) Never exchange the stoppers between different bottles.
- Wear safety goggles and a laboratory apron, especially when heating substances in the laboratory.
- Always handle hot test tubes with a test-tube holder and hot beakers with tongs or an oven mitt. (See Figure 2-10.)
- Keep all flammable substances, such as alcohol, far away from an open flame.
- Know the locations of the fire extinguisher, fire blanket, first aid kit, safety shower, eyewash station, fire exit, and fire alarm.
- Clear counters carefully after use and then wash your hands thoroughly.

OTHER LABORATORY TECHNIQUES AND PROCEDURES

Chromatography is a method used for separating and analyzing mixtures of complex chemical substances. The mixture to be separated is placed on a material such as chromatography or filter paper. A solvent is added, which begins to move through the material. Different substances in the mixture will

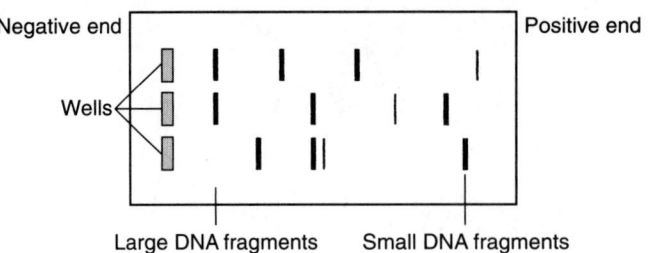

Figure 2-11 Gel electrophoresis is used to separate molecules that have different charges; the smaller DNA fragments move farthest from the wells.

move along with the solvent at different rates, causing them to be separated and allowing them to be studied.

Electrophoresis is a technique used to separate molecules that have different electrical charges. An electric current is run through a material, usually a gel, in which the mixture has been placed. Different substances move at different rates in the electrical field. The resulting pattern of bands shows the different substances that were originally mixed together. Gel electrophoresis is the process used, for example, to analyze one's DNA and compare it to the DNA of another person. (See Figure 2-11.)

Centrifugation is the method used to separate materials of different densities from one another. The original liquid mixture is placed in a test tube, which is spun around in a **centrifuge**. The heaviest particles settle to the bottom; the least-dense material forms a layer on the top. The *ultracentrifuge*, the most powerful of these machines, can spin at rates of 100,000 revolutions per minute, allowing the lightest-weight particles in cells to be separated from one another.

SCIENTIFIC INVESTIGATIONS: PUTTING IT ALL TOGETHER

Observations made during scientific research usually need to be analyzed to see what they mean. The data may be organized and represented in a variety of ways. For example, diagrams, tables, charts, graphs, equations, and matrices can represent data. When the data are interpreted, the result may be the statement of a new hypothesis. Another result may be the conclusion that a general understanding or explanation of a natural phenomenon is, in fact, correct.

The mathematical processes of **statistical analysis** are used to determine if the results obtained are valid or if they might have been simply due to chance. Statistics also allows a researcher to find out the degree to which the predicted results (based on the hypothesis) match the actual results. Based on this "matching," the scientist can conclude whether the proposed explanation is, or is not, supported by the data. Data and observations are *valid* if they measure what they are supposed to be measuring. They are *reliable* if repeated trials give the same results. Statistical procedures help indicate the reliability of a set of observations. *Accuracy* is a general term that includes both the validity and reliability of a set of observations.

The analysis of the data, followed by public discussion, can lead to a revision of the original explanation, the development of new hypotheses, and the design of new research plans. When claims are made based on the collected evidence, the reliability of the claims should be questioned if the design of the experiment was faulty. For example, if there were small sample sizes, incomplete or misleading use of data, or a lack of controlled conditions, the results would not be reliable. Also, great care should be taken not to confuse facts with opinions.

When all the research and data analysis are concluded, a written report is prepared for the public to study. This report includes a literature review, the research, the results, and suggestions for further research. One purpose of making the results public is to allow the research to be repeated by someone else. Science assumes that, through the collection of similar evidence, different researchers will come to the same conclusions and explanations of natural phenomena. *Peer review*—the study of research reports by fellow scientists—is important as a check on the quality of the research. It also results, at times, in the suggestion of alternative explanations for the same observations.

Chapter 2 Review

Multiple Choice

1. A scientist wants to study the internal structure of a chloroplast (part of a plant cell) in great detail. The best instrument for this detailed examination would be a (an)
   A. compound microscope
   B. simple light microscope
   C. electron microscope
   D. ultracentrifuge

2. Blood plasma can be separated from red blood cells because they have different densities. To separate these parts of blood tissue, a biologist would use a (an)
   A. micro-dissection instrument
   B. centrifuge or ultracentrifuge
   C. compound light microscope
   D. electron microscope

3. To view cells under the high power of a compound light microscope, a student places a slide of the cells on the stage and moves the stage clips to secure the slide. She then moves the high-power objective into place and focuses on the slide with the coarse adjustment. Two steps in this procedure are incorrect. For this procedure to be correct, she should have focused under
   A. high power first, using the fine adjustment, then low power using the fine adjustment
   B. low power using the coarse and fine adjustments, then high power using only the fine adjustment
   C. low power using the coarse and fine adjustments, then high power using only the coarse adjustment
   D. low power using only the fine adjustment, then high power using the fine adjustment

4. A student used iodine solution to test a material, which turned very dark blue after the iodine was added. This proved the presence of
   A. proteins
   B. simple sugars
   C. oxygen
   D. starch

5. When a student adds bromthymol blue to a solution containing carbon dioxide, the solution should
   A. show no color change
   B. turn yellow
   C. turn blue-black
   D. turn red-orange

6. An unknown solution may contain glucose. Which reagent could be used to test for the presence of this simple sugar?
   A. litmus paper
   B. bromthymol blue
   C. Lugol solution
   D. Benedict solution

7. Which piece of equipment should be used to transfer a one-celled organism onto a microscope slide?
   A. pair of scissors
   B. dissecting needles
   C. medicine dropper
   D. test tube

8. While a student is heating a liquid in a test tube, the mouth of the tube should always be
   A. corked with a rubber stopper
   B. pointed toward the student
   C. pointed straight upward
   D. aimed away from everybody

9. During the observation of a microscope slide, a sharp crack is heard as the objective presses against the slide and breaks it. Which of the following actions might have caused this result?
   A. changing the diaphragm opening under high power
   B. using the fine adjustment to focus under low power
   C. using the coarse adjustment to focus under high power
   D. switching back to low power after focusing under high power

10. You are observing a wet mount of a paramecium culture under the low power of a compound microscope. After a time you notice that the range of movement of the organisms is shrinking toward the center of the mount. What should you do to restore the activity of the organisms?
    A. Add a drop of iodine stain.
    B. Switch to high power and refocus.
    C. Use a brighter light source.
    D. Add a drop of water to the edge of the coverslip.

11. An acid-base indicator that changes to either red or blue only, and is easy to use, is called
    A. methylene blue
    B. bromthymol blue
    C. Lugol solution
    D. litmus paper

12. When a person exhales into a solution of bromthymol blue, the solution changes color. It changes to what color and why?
    A. dark blue due to presence of oxygen
    B. yellow due to presence of oxygen
    C. dark blue due to presence of carbon dioxide
    D. yellow due to presence of carbon dioxide

13. When you heat a test tube that contains liquid, which of the following is a dangerous action?
    A. wearing safety goggles
    B. using a metal test-tube holder
    C. wearing a laboratory apron
    D. looking into the tube opening to observe a reaction

14. In the laboratory, which of the following actions is *not* recommended?
    A. wearing loose-fitting clothing and dangling jewelry
    B. having your setup checked by the teacher before starting a procedure
    C. reading the instructions through to the end before starting an investigation
    D. being familiar with the location and use of the emergency equipment

15. Safety goggles must be worn in the laboratory, and most especially when
    A. testing pH levels
    B. heating solids
    C. heating liquids
    D. cleaning up after an experiment

16. A student performed the following experiment: a dry piece of white bread was tested with iodine solution; the bread turned blue-black. Another piece of white bread was chewed and then tested with Benedict solution; the mixture turned red. The student therefore concluded that when a piece of bread is chewed, starch is changed to sugar. One error in the student's procedure was that he did *not*
    A. test the chewed piece of bread for starch
    B. test the dry piece of bread for sugar
    C. consider the age of the piece of bread
    D. use Biuret solution to test for starch

17. An unknown solution was tested with Benedict solution and the liquid remained blue. The most reasonable conclusion from this evidence is that the solution contained no
    A. sugars
    B. protein
    C. starch
    D. fats

18. In order to measure and pour 10 milliliters of Benedict solution into a test tube, which is the best procedure to ensure accuracy of the measurement?
    A. Use a 10-ml graduated cylinder.
    B. Use a 100-ml graduated cylinder.
    C. Weigh out 10 milliliters on a metric balance.
    D. Fill a 1-milliliter medicine dropper ten times.

19. A slide of human blood cells was observed under the low-power objective of a compound light microscope that had clean lenses. When the microscope was switched to high power, the image was dark and fuzzy. Which two parts of the microscope should be used to correct this situation?
    A. nosepiece and coarse adjustment
    B. diaphragm and ocular lens
    C. objective and fine adjustment
    D. diaphragm and fine adjustment

20. Which structure is best seen by using a compound light microscope?
    A. a cell's nucleus
    B. a paramecium
    C. a DNA sequence
    D. a mitochondrion

Analysis and Open Ended

21. The following diagrams show four different one-celled organisms (shaded) in the field of view of the same microscope using different magnifications. Which illustration shows the largest one-celled organism?

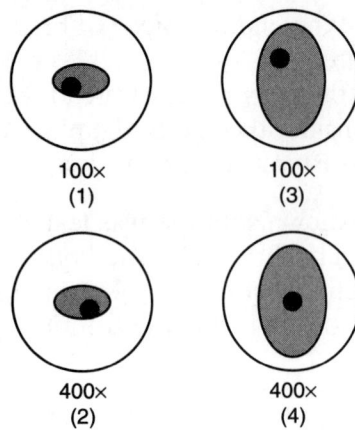

A. diagram 1   C. diagram 3
B. diagram 2   D. diagram 4

22. A peppered moth and part of a metric ruler are represented in the diagram below.

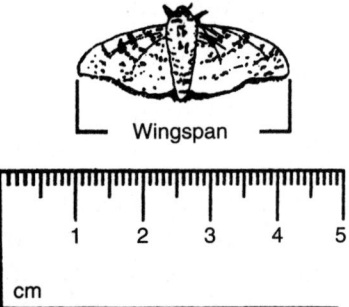

Which row in the chart below best represents the ratio of body length to wingspan of the peppered moth?

Row	Body Length:Wingspan
(1)	1:1
(2)	2:1
(3)	1:2
(4)	2:2

A. row 1   C. row 3
B. row 2   D. row 4

23. How much water should be removed from the following graduated cylinder in order to leave exactly 5 milliliters of water in the cylinder?

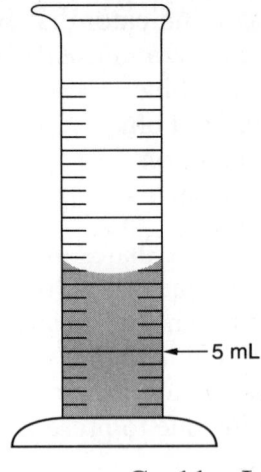

A. 6 mL   C. 11 mL
B. 7 mL   D. 12 mL

24. The diagram below shows how a coverslip should be lowered onto some single-celled organisms during the preparation of a wet mount. Why is this the preferred procedure?

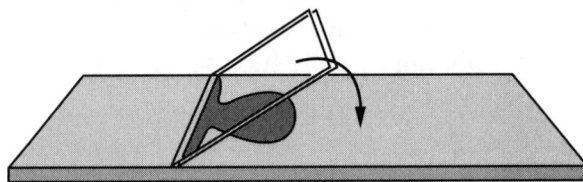

A. The coverslip will prevent the slide from breaking.
B. The organisms will be more evenly distributed.
C. The possibility of breaking the coverslip is reduced.
D. The possibility of trapping air bubbles is reduced.

25. Identify a piece of laboratory equipment that normally would be used to measure 5 milliliters of glucose solution for an experiment. Explain how the reading should be done to get an accurate measurement.

26. Why do scientists take measurements when recording data?

27. Explain why scientists use tools and instruments during experiments.

28. Describe the procedure for preparing a wet mount.

29. Explain the similarities and differences between the transmission electron microscope (TEM) and the scanning electron microscope (SEM).

30. Identify the metric units used for each of the following: length, volume of a liquid, temperature, and mass.

31. Suppose that a new biology laboratory room has been opened in your school. Prepare a poster listing the safety rules and procedures that should be followed by students working in the lab.

32. A person is determined to follow a fat-free diet. This person has a piece of cheddar cheese and a slice of apple. Explain how this person could test these two food items for the presence of fat. Conduct the test yourself at home and describe your results.

Reading Comprehension

Base your answers to questions 33 to 35 on the information below and on your knowledge of biology. Use one or more complete sentences to answer each question.

> Light microscopes are used by scientists to view extremely small objects, such as cells. However, due to the physical properties of light, objects below a certain size cannot be seen in a sharp, focused image—no matter how well made and powerful the light microscope is. Bacteria with a diameter of 0.5 micrometer are about the smallest living things that can be observed with a high-quality light microscope. (A micrometer is 1/1,000,000 of a meter.) During the twentieth century, scientists learned to use a beam of electrons (similar to what is used in a television set), instead of a beam of light, to produce images of very small objects. Electron microscopes have now been used to observe and produce photographs of the detailed, internal structures of plant and animal cells, as well as of the smallest living things, such as bacteria and viruses.
>
> Modern electron microscopes are at least 1000 times more powerful than light microscopes, allowing clear observations of objects that are as small as 0.5 nanometer. (A nanometer is 1/1,000,000,000 of a meter.) Two types of electron microscopes are the transmission electron microscope (TEM) and the scanning electron microscope (SEM). In a TEM, a beam of electrons is passed through an object to show the details within it. In an SEM, a beam of electrons is passed over an object, producing a detailed, three-dimensional image of the surface of its cells. Electrons would bounce off the gas molecules that are present in the air. So there must be a vacuum where the specimen is placed inside an electron microscope. In addition, specimens are usually treated with stains that interact with the electron beams in order to produce the images. As a result, cells cannot be viewed with an electron microscope when they are still alive.

33. How have scientists been able to overcome the limitations of light microscopes?

34. In a few sentences, describe the similarities and differences between light microscopes and electron microscopes.

35. If scientists were to invent a new kind of microscope, how might it combine the best of a light microscope with the best of an electron microscope?

Unit II
Matter, Energy, and Organization of Living Things

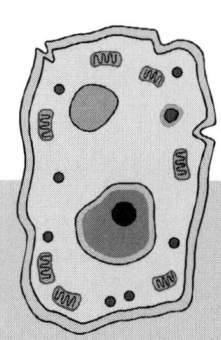

STANDARD 5.5.12 A
Characteristics of Life

All students will gain an understanding of the structure, characteristics, and the basic needs of organisms and will investigate the diversity of life.

Enduring Understanding I The cell is the basic unit of life. Living cells are composed of elements that form large, complex molecules. The primary source of energy to sustain most life is derived from a conversion of light energy to chemical energy through the process of photosynthesis.

Chapter 3

Organic Molecules: From Atoms to Cells

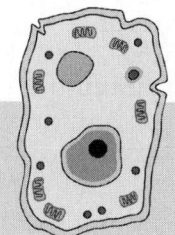

Standard 5.5.12 A1 **Relate the structure of molecules to their function in cellular structure and metabolism.**

LEVELS OF ORGANIZATION IN LIVING THINGS

The study of how living things are put together begins with *atoms*. The next level of organization above atoms is *molecules,* then the families of *organic compounds,* and then *organelles* and *cells.* Beyond that, the organization of living things continues. Plants and animals, which are composed of enormous numbers of **cells**, have groups of similar cells called **tissues** that work together to perform specific functions. For example, the nervous tissue in your brain consists of billions of nerve cells, or *neurons,* functioning together.

A group of tissues, in turn, works together as an **organ**. The brain is an organ made up of different types of tissue, including nervous, blood, and connective tissue. Organs that work together make up **organ systems**. The nervous system, for example, includes the brain, spinal cord, and sensory organs. Finally, different organ systems work together as a functioning living thing, or **organism**. (See Figure 3-1 on page 42.)

To study life is to learn about all the levels of life's organization, sometimes one at a time, sometimes all together. At each level there are characteristics that were not present on the previous level. While we may not be able to define life, we can say what living things do. Properties of life include order, reproduction, growth and development, digestion, nutrition, excretion, respiration, movement, coordination, use of energy, response to the environment, regulation, immunity, and adaptation. We can see that the structures found in single-celled, or **unicellular**, organisms also function much like the tissues and systems found in many-celled, or **multicellular**, organisms, thus allowing them to perform all the life processes needed to stay alive.

Figure 3-1 The levels of organization in a living organism start with the cell.

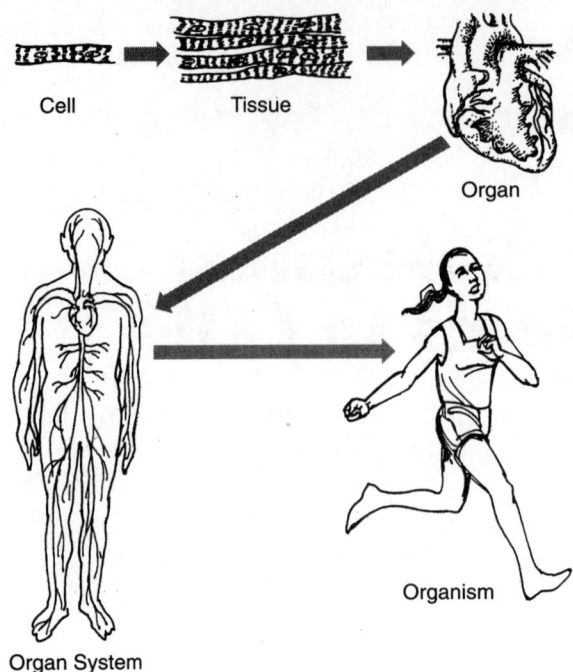

Energy, Matter, and Organization

Everything in the world tends to get more disorganized as time passes. Physicists recognize that natural events tend to increase disorder in the universe; or, put more simply, the universe is running down. It becomes more disordered over time, a trend known as *entropy*.

In contrast, living things are highly organized. How do they keep themselves in order and functioning properly in a universe where things are constantly running down?

To maintain the state of organization necessary for life to exist, all living things require **energy**. Energy exists in different forms, such as heat, motion, light, and electricity. And living things need a continuous input of energy to stay organized and remain alive.

MATTER, ATOMS, AND LIFE

All matter is made up of **atoms**, particles far too small to be seen with the unaided eye or even through an ordinary microscope. Each atom has an extremely dense nucleus in its center. Distributed in the mostly empty space around the nucleus are *electrons*. Electrons have energy. The orbitals in which electrons move make up shells, or energy levels. The electron shells closest to the nucleus have the least energy; the electron shells farthest from the nucleus have the most energy. It is possible for electrons to move from one shell to another. A ball rolling down from the top of a hill loses energy as it moves down the hill. So, too, does an electron when it moves from an outer, higher-energy level to an inner, lower-energy level. The movement of electrons between energy levels in atoms is what produces all the changes, or *transformations,* of energy.

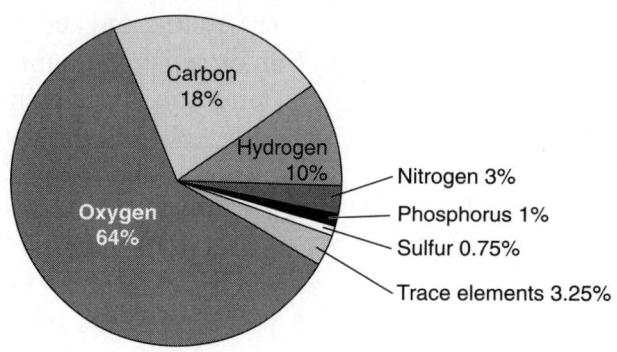

Figure 3-2 The graph shows the percentages of the main elements that make up an organism's body.

An organism is actually a very complex system for transforming energy. A **system** is an organized group of structures that works together to perform a task. Through natural selection, evolution has produced countless species of organisms that are efficient energy transformers. Where does the energy come from that keeps organisms as different as grass, ants, and elephants alive? Sunlight is the main source of energy for most life on Earth. Plants capture this light energy; and then plants, animals, and other organisms use it to live, grow, and reproduce. (Exceptions include deep-sea communities around hydrothermal vents, which obtain chemical energy from the mineral-rich water.)

Atoms Bond Together: Molecules

There are only about 100 different types of atoms. Of these basic types, called *elements,* 92 are found naturally on Earth. Carbon, hydrogen, nitrogen, and oxygen are the four most important elements for organisms. Also important for living things are *phosphorus, sulfur,* and a few other elements. (See Figure 3-2.)

Atoms of most elements combine with other atoms to form larger structures called **molecules**. The atoms in a molecule are kept together by a kind of realtionship called a *chemical bond.* One type of chemical bond is the *covalent* bond. In such a bond, atoms share electrons with each other, making each atom more stable. (The other type of bond, called *ionic,* occurs when electrons are transferred from one atom to another.)

Elements combine to form thousands of different *compounds,* but only a small number of compounds are important for living things. Many of the compounds found in living things are called **organic compounds**. All organic compounds contain the elements carbon and hydrogen. Carbon is of special importance; organic compounds usually have a skeleton of carbon atoms bonded to each other.

ORGANIC COMPOUNDS

Organic compounds accomplish many complex processes that keep us and all other organisms alive. These biochemical processes include capturing and transforming energy, building new structures, storing materials, repairing structures, and keeping all chemical activities in the body working properly. In organisms, chemical reactions are always putting things together (by condensation, or dehydration, synthesis) or taking things apart (by hydrolysis). These chemical activities in an organism are called its *metabolic activities,* or **metabolism**.

Organic molecules are often very large. A large molecule, such as the blood protein hemoglobin, is called a **polymer**. Polymers are formed by the linking together of smaller molecules, or *subunits,* which are monomers. In polymers, these subunits are held together by covalent bonds.

THE FAMILIES OF ORGANIC COMPOUNDS

Even though organisms contain many different molecules, there are relatively few different types of subunits that make up these molecules. The subunits located in the organic compounds found in living organisms, from bacteria to whales, are almost identical. However, the different arrangements of the subunits create an enormous variety of polymers. Some polymers have more than 100 subunits; others have thousands of subunits. As a result, there are many different kinds of polymers.

Organic compounds are grouped into four major families: carbohydrates, proteins, lipids, and nucleic acids. These are the families of compounds found in all living things.

Carbohydrates

The **carbohydrates** include simple sugars as well as polymers, which are made up of sugar subunits. Carbohydrates are formed from the elements hydrogen, oxygen, and carbon. In living things, the main functions of carbohydrates are storing energy and providing strong building materials for certain types of cells.

Polysaccharides (*poly* means "many"; *saccharides* means "sugars") are large molecules made up of many sugar subunits, or *monosaccharides,* joined together. Energy is stored in these molecules. Our muscles must contain stored energy to allow them to work at a moment's notice. In humans, this energy is stored in muscles and in the liver in the form of *glycogen.* The polysaccharide glycogen is made up of many subunits of **glucose**, a simple sugar. (See Figures 3-3a and 3-3b.)

Plants store energy in the form of **starch**, a polysaccharide contained in such foods as corn and rice. *Cellulose,* another important polysaccharide found in plants, helps build up tough structures such as wood.

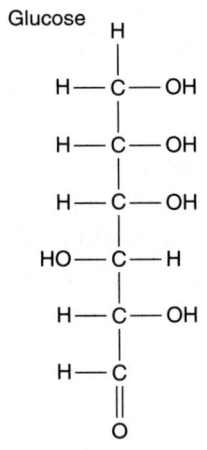

Figure 3-3a Glucose is a simple sugar. Like most other organic compounds, it has a skeleton of carbon atoms to which other atoms form bonds.

Proteins

Hundreds of thousands of different **proteins** exist. No two people other than identical twins have exactly the same proteins. Proteins are responsible for a wide variety of functions in organisms. They are used to build materials,

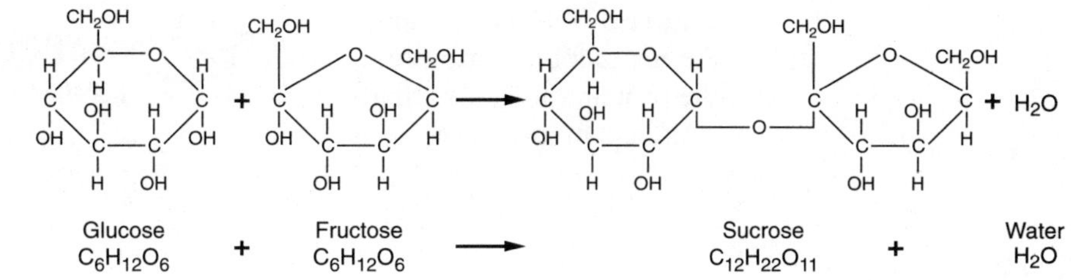

Figure 3-3b Glucose and fructose are monosaccharides. Sucrose is a disaccharide. Two monosaccharides linked together form one disaccharide and water.

Glucose $C_6H_{12}O_6$ + Fructose $C_6H_{12}O_6$ → Sucrose $C_{12}H_{22}O_{11}$ + Water H_2O

transport other substances, send signals, provide defense, and control chemical and metabolic activities.

The building plan for proteins is the same as for other organic compounds. Proteins are large polymers that are made up of smaller subunits called **amino acids.** There are 20 different types of amino acid molecules. By combining these amino acids in a row, in different sequences, many types of proteins are made. Proteins can easily have more than 100 amino acid subunits. (See Figure 3-4.)

The order in which its amino acids are linked determines the characteristics of a protein molecule. Every different sequence produces a different protein. However, the sequence does not behave like a string of letters; instead, the protein chains twist, turn, and bend into specific three-dimensional shapes. The shape of a protein molecule is called its *conformation*. Every protein molecule has a very specific conformation, and that is what determines its function.

Lipids

Oils and fats make up the second major family of organic compounds, the **lipids**. One type of lipid makes up the basic structure of the cell membrane. However, the main purpose of lipids is energy storage, which they do more efficiently than carbohydrates. Energy stored in lipids is for long-term use. During physical activity, the carbohydrate glycogen gets used up quickly. However, lipids, or fat deposits, do not disappear quickly.

Nucleic Acids

The **nucleic acids** consist of *deoxyribonucleic acid (DNA)* and *ribonucleic acid (RNA)*. **DNA** and **RNA** molecules are responsible for storing the genetic information that contains directions for building every molecule that makes up an organism. (See Figure 3-5 on page 46.)

The pattern for building nucleic acids is similar to that of the other organic molecules. Individual subunits, called **nucleotides**, are combined in a

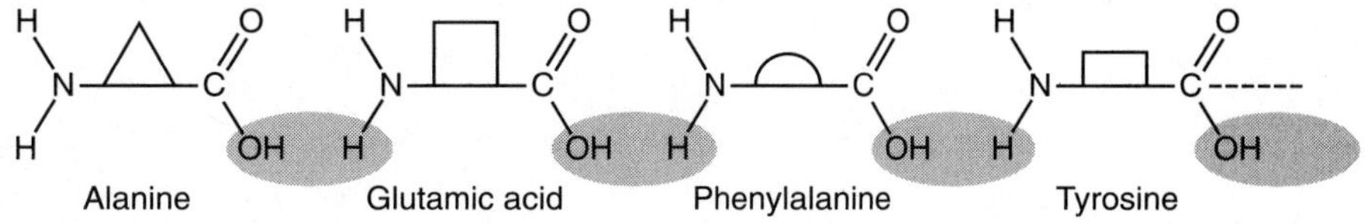

Figure 3-4 Amino acids combine in linear sequences to form a great variety of proteins.

Alanine Glutamic acid Phenylalanine Tyrosine

Chapter 3 / Organic Molecules: From Atoms to Cells

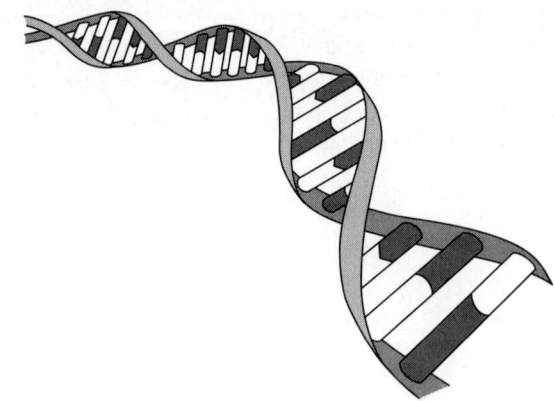

Figure 3-5 DNA contains the instructions for building the protein molecules that make up an organism.

linear sequence to build the polymers DNA and RNA. DNA molecules usually contain thousands of nucleotides linked together in a specific sequence. In a ribosome, the sequence in the DNA, which has been copied into RNA, is used as a building plan to construct a protein molecule. Specific portions of the DNA molecule are used to make specific proteins. Because proteins make us who we are—and because DNA makes our proteins—it is really our DNA that makes us, physically, who we are.

The numerous organic compounds work together to support life by maintaining the structures and functions of living cells.

Chapter 3 Review

Multiple Choice

1. As time passes, the universe tends to become
   A. more organized
   B. less organized
   C. much simpler
   D. a little smaller

2. To maintain organization, all living things need a source of
   A. time   C. brains
   B. patience   D. energy

3. The densest region of an atom is the
   A. space around its nucleus
   B. nucleus at its center
   C. inner electron shells
   D. outer electron shells

4. Which event is most like an electron moving from an outer shell to an inner shell?
   A. a ball rolling down a hill
   B. a fish swimming upstream
   C. a soccer ball rolling across a field
   D. a baseball player running to a base

5. All transformations of energy involve
   A. a change in height
   B. the direct use of sunlight
   C. the organization of atoms
   D. the movement of electrons

6. How do molecules differ from atoms?
   A. Atoms occur naturally, whereas molecules do not.
   B. Atoms are individual particles, whereas molecules are combinations of atoms.
   C. There are many different types of atoms but only a few different types of molecules.
   D. Atoms are found in living things, but molecules are found in nonliving things.

7. The two elements found in every organic compound are
   A. nitrogen and oxygen
   B. oxygen and hydrogen
   C. carbon and hydrogen
   D. carbon and oxygen

8. The chemical reactions that form or break bonds in an organism are called its
   A. transport   C. transformation
   B. conformation   D. metabolism

9. Which statement concerning simple sugars and amino acids is correct?
   A. They are both wastes resulting from protein synthesis.
   B. They are both needed for the synthesis of larger molecules.
   C. They are both building blocks of starch.
   D. They are both stored as fat molecules in the liver.

Refer to the diagram below to answer question 10.

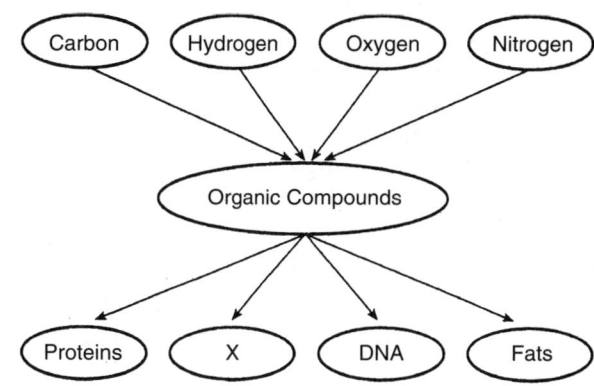

10. What substance could the letter *X* represent?
    A. carbohydrates
    B. carbon dioxide
    C. ozone
    D. water

11. Which family of organic compounds is used mainly to store energy for the body?
    A. lipids
    B. carbohydrates
    C. proteins
    D. nucleic acids

12. An iodine test of a tomato plant leaf revealed that starch was present at 5:00 P.M. on a sunny afternoon in July. When a similar leaf from the same tomato plant was tested with iodine at 6:00 A.M. the next morning, the test indicated that less starch was present. This reduction in starch content most likely occurred because the starch was changed
    A. directly into proteins
    B. into molecules of fat
    C. into polysaccharides
    D. back to sugars for energy

13. The subunits that make up proteins are called
    A. amino acids   C. fats and lipids
    B. single atoms   D. nucleic acids

Refer to the diagram below, which provides some information about proteins, to answer question 14.

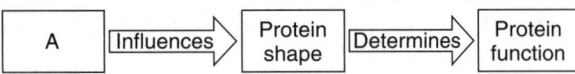

14. Which phrase does the letter *A* most likely represent?
    A. sequence of amino acids
    B. sequence of starch molecules
    C. sequence of simple sugars
    D. sequence of ATP molecules

15. The subunits of DNA are called
    A. amino acids
    B. nucleotides
    C. polysaccharides
    D. cell units

16. How is RNA related to proteins?
    A. Proteins are made up of RNA molecules.
    B. RNA determines which proteins are made.
    C. RNA is copied into DNA to build a protein.
    D. DNA is copied into RNA to build a protein.

17. DNA molecules are important because they store
    A. fats for energy
    B. genetic information
    C. carbohydrates
    D. polysaccharides

18. Every single-celled organism is able to survive because it carries out
    A. metabolic activities
    B. heterotrophic nutrition
    C. autotrophic nutrition
    D. sexual reproduction

19. Which sequence represents the correct order of levels of organization found in a complex organism?
    A. cells → organelles → organs → organ systems → tissues
    B. organelles → cells → tissues → organs → organ systems
    C. tissues → organs → organ systems → organelles → cells
    D. organs → organ systems → cells → tissues → organelles

20. Large molecules, such as starches or proteins, are known as
    A. tissues   C. organelles
    B. subunits   D. polymers

Refer to the diagrams of the organisms shown below to answer question 21.

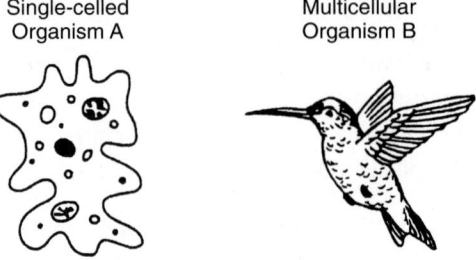

21. Which statement concerning organism *A* and organism *B* is correct?
    A. Organism *A* contains tissues and organs, while organism *B* lacks these structures.
    B. Organism *A* and organism *B* both have structures that help them to stay alive.
    C. Organism *A* and organism *B* have the same organs to perform their life functions.
    D. Organism *A* lacks the structures that aid survival, while organism *B* has them.

Analysis and Open Ended

22. How do living things maintain the high level of organization that they need to stay alive?

23. Why is carbon particularly important for the existence of life (as we know it)?

24. List four important functions of the organic compounds found in living things.

25. Explain why athletes need to eat lots of complex carbohydrates during training.

Refer to the diagram at right to answer questions 26 and 27. (Molecule is shown as a straight chain.)

26. Based on the elements and molecular structure that make up glucose, you could determine that it is an example of a (an)
    A. carbon molecule
    B. hydrogen molecule

C. organic compound
D. inorganic compound

27. When many of these glucose subunits join together, they make up a
    A. protein molecule
    B. polysaccharide
    C. lipid molecule
    D. DNA molecule

28. Hemoglobin and hair are both proteins, yet they each have different structures. Explain.

29. The diagram below illustrates a reaction in which
    A. several amino acids join to form a protein molecule
    B. inorganic compounds form an organic compound
    C. simple sugars join to form a larger sugar molecule
    D. polysaccharides are broken down into simple sugars

30. Identify three important characteristics of proteins. Your answer should include the following:
    - what the subunits are that make up proteins
    - four main functions of proteins in living things
    - what determines structure and function of a protein

31. In what way do the particular proteins in our bodies depend on our DNA?

32. In terms of levels of organization, what is the difference between a *tissue* and an *organ*? Give one example of each.

33. List the main levels of organization of living things, going from the level of atoms up to that of an organism.

Glucose $C_6H_{12}O_6$ + Fructose $C_6H_{12}O_6$ → Sucrose $C_{12}H_{22}O_{11}$ + Water H_2O

Reading Comprehension

Base your answers to questions 34 to 36 on the information below and on your knowledge of biology. Use one or more complete sentences to answer each question.

People try to lose weight by dieting. Researchers now know that dieting may reduce the amount of fat in the body, but it also changes how the body functions. When people diet, they usually limit the amount of food calories they take in. A dieter's body reacts to protect itself. Because of our evolutionary history, the body thinks that there is an actual shortage of food—that starvation is imminent. The body does not know that the dieting person is intentionally limiting the amount of food taken in, and it reacts by slowing down to survive the food shortage. The result? Fewer calories are burned during normal activities, and not much weight loss occurs. The fat cells that normally store fat are being emptied, but they still remain in the body. The person feels hungry and may end the diet. The "starved" fat cells quickly refill their reserves and a type of on-again, off-again dieting may result. The fluctuating weight loss and gain that occurs can be dangerous.

Most researchers now realize that the best way to avoid becoming overweight is to reduce the amount you eat somewhat and to increase physical activity. Exercise increases the amount of energy used by the body. It also increases the amount of muscle tissue, which even when resting burns more calories than other types of body tissues.

There are other serious health risks involved in severe weight loss that is caused by a refusal to eat. The disorder called *anorexia nervosa* is most common in young women. Abnormal fears of being overweight, as well as other fears, may lead to anorexia nervosa. An anorexic person appears unhealthy. This disorder can be fatal.

Bulimia is another eating disorder. Unlike most anorexics, a person with bulimia might appear healthy. However, this person swings between overeating and getting rid of the food, often by taking laxatives or inducing vomiting. Some studies show that as many as 20 percent of college-age women suffer from some form of bulimia. This disorder can be dangerous. It can damage the heart, kidneys, or digestive system. Counseling to help a person understand the reasons behind these eating disorders is important. It is also important to learn how to make wise choices about what one eats. In some severe cases, hospital treatment may be necessary.

34. In a paragraph, describe how evolution explains the unintended effects of dieting.

35. Explain why exercising is a healthier way to lose weight than dieting.

36. Compare and contrast the eating disorders *anorexia nervosa* and *bulimia*.

Chapter 4

Cell Structure and Function

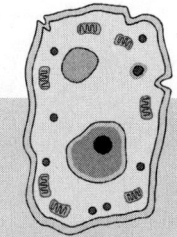

Standard 5.5.12 A1 Relate the structure of molecules to their function in cellular structure and metabolism.

THE CELL: THE BASIC UNIT OF LIFE

The idea that organisms are made up of *cells* is one of the central ideas of modern biology. Referred to as the **cell theory**, its main points are:

- All organisms are made up of one or more cells.
- The cell is the basic unit of structure and function of living things.
- All cells arise from previously existing cells.

A TOUR OF THE CELL: ORGANELLES

Some living things are made up of a single cell; but most have many cells that work together on behalf of the organism. Yet, almost everything an organism does to stay alive is accomplished by each individual cell: getting food, using food for energy, transporting substances, growing, reproducing, and eliminating wastes. Each of these activities involves a large number of chemical reactions. Organization is needed for all of these reactions to take place correctly. In many cells, these reactions take place inside special internal structures called **organelles**.

Organelles are dispersed throughout the **cytoplasm**, which fills up the cell and transports materials within it. Cytoplasm is a thin gel, made up mostly of water, with many other chemicals dissolved in it. The **cell membrane** encloses the cell's cytoplasm. In a typical animal cell, there are many **ribosomes**, the organelles that are attached to the *endoplasmic reticulum* and at which proteins are built. The **lysosomes**, which are scattered throughout the cell, are organelles that are involved in breaking down food. The

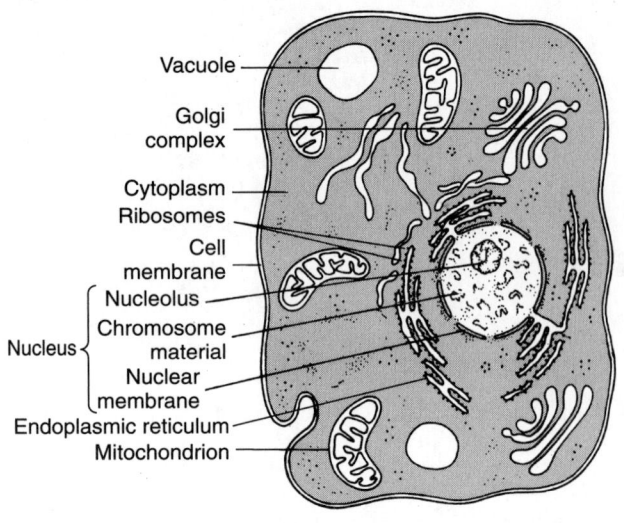

Figure 4-1 The organelles of a typical animal cell.

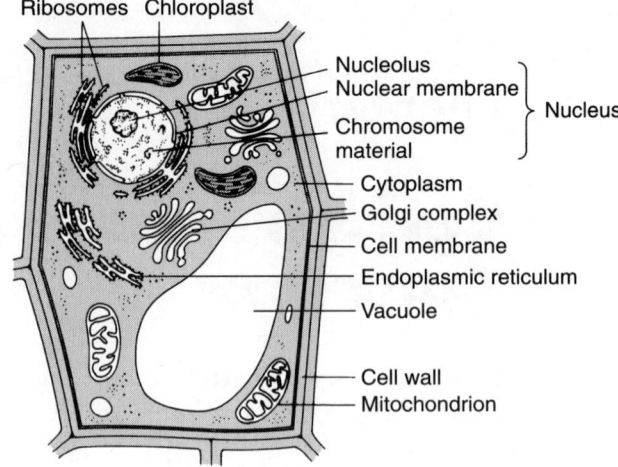

Figure 4-2 The organelles of a typical plant cell.

mitochondria (singular, *mitochondrion*), shaped like tiny kidney beans, are the organelles in which the cell's energy is released. (See Figure 4-1.)

By far the largest structure in a *eukaryotic* cell is the **nucleus**, which, as its control center, is responsible for information storage. The nucleus often fills the entire central portion of the cell. In addition, other specialized structures may be present to handle a variety of functions, such as: the *Golgi complex* (or *Golgi apparatus*), which packages many materials; **vacuoles**, which store materials such as food and wastes; and, in plants and algae, **chloroplasts**, the organelles in which cells convert energy from the sun into food, and the *cell wall,* a rigid, nonliving structure that surrounds their cell membranes. (See Figure 4-2.)

INTRODUCTION TO THE CELL MEMBRANE

The inside of a single-celled organism is very much alive. However, the physical environment outside the cell is the opposite—a nonliving place where many changes occur. What stands between a cell and the potentially hostile environment that surrounds it? An ultrathin, extremely important layer separates the living world inside a cell (or single-celled organism) from the nonliving world outside. This is the *cell membrane,* or **plasma membrane**.

Without a cell membrane, there could be no cell. The cell membrane performs two primary, yet very different, functions: (1) It separates the cell from its outside environment; and (2) it enables *communication* and *movement* of materials between the cell and its environment. In multicellular organisms, protein molecules float within the lipid molecules of the cell membrane. This arrangement of molecules in constant motion is described as the *fluid mosaic model* of the cell membrane. The protein molecules enable much of the transport, or **movement**, of materials across the membrane; as such, they are referred to as *carrier* (or *transport*) *proteins*. These proteins often extend right

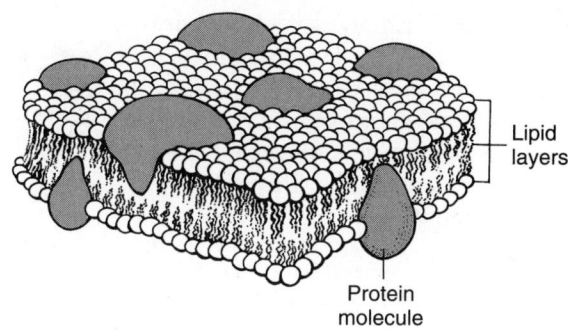

Figure 4-3 The cell membrane acts as a barrier, separating the inside of the cell from the outside. Protein molecules, floating in the lipid layers, enable much of the transport of materials across the membrane.

through the plasma membrane, from one side to the other. In addition, some proteins in the cell membrane function as *receptor molecules;* chemical signals can attach to these **receptors**, enabling communication between cells. (See Figure 4-3.)

TRANSPORT ACROSS THE CELL MEMBRANE

For a cell to remain alive, it must have a very special collection of chemicals inside it. These chemicals may be quite different from the chemicals located in the outside environment. Some substances that are abundant outside the cell are not found inside the cell. Other substances that are scarce outside the cell are present in larger quantities inside the cell. The cell membrane creates and maintains this special environment inside the cell. How does it do this?

The cell membrane allows some substances—that is, molecules—to pass through but keeps other substances out. This ability to determine which molecules can pass through is called **selective permeability**. Thus the plasma membrane is said to be *selectively permeable,* or *semipermeable.* It controls which molecules move through it, as well as whether the molecules move into or out of the cell. The membrane also determines whether the transport of these specific molecules is rapid or slow.

Passive Transport

Molecules are constantly in motion; and they naturally move from where they are more concentrated to where they are less concentrated. Typically, there is an overall, or *net,* movement of molecules from an area of high concentration, where the molecules are more abundant (called *hypertonic*), to an area of lower concentration (called *hypotonic*). This molecular flow results in a balance in concentration (called *isotonic*). The difference in concentration between the two areas is called a *concentration gradient*. The random movement of molecules down the gradient (that is, from a higher to a lower concentration) is called **diffusion**. (See Figure 4-4.)

Diffusion happens automatically with a cell if its membrane is permeable to the molecules and if there is a difference in concentration of the molecules on either side of the cell membrane. This movement is called **passive**

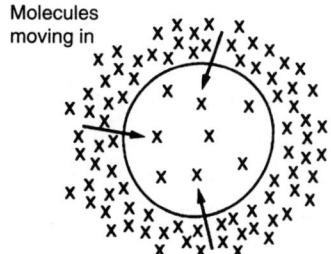

Figure 4-4 Diffusion is the movement of molecules from an area of higher concentration to an area of lower concentration.

Figure 4-5 Plasmolysis is observed when a plant cell is placed within a concentrated sugar or salt solution. Water will move by osmosis from the inside to the outside of the cell. Note how the cell contents shrink away from the cell wall.

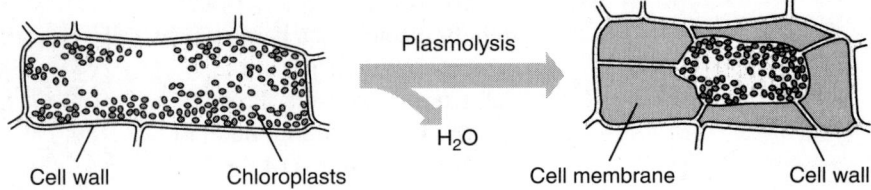

transport, because no energy is used by the cell and no work is done. For example, one of the basic needs of most cells is oxygen. There are few oxygen molecules inside a cell, but there is usually an abundance of oxygen molecules in the water or other liquid that surrounds the cell. Thus, oxygen molecules diffuse across the cell membrane into the cell by passive transport.

The diffusion of water molecules across a cell membrane, which is so important for living cells, is given a special name: **osmosis**. When plant cells are put in a strong salt or sugar solution, the abundant freshwater inside the plant cells automatically moves out of the cells to where there are more salt or sugar (solute) molecules and relatively few water molecules. Plant cell membranes can be seen pulling away from the cell walls, in a process called *plasmolysis,* as their cells lose water. (See Figure 4-5.) The reverse happens when limp celery stems are put in freshwater. The celery stems are limp because their cells have too little water in them. When the celery stem is put in water, osmosis occurs and water molecules move into the cells. As a result, the cells expand, the cell membranes push against the cell walls, and the plant cells—and thus the celery stems—become firm, or *turgid,* again.

Facilitated Diffusion

Facilitated diffusion is the form of transport in which a carrier protein in the cell membrane is needed to help move a specific molecule across it. (*Facilitate* means "to help" or "to make easier.") The carrier protein, acting like a gate, increases the rate of transport of certain substances across the membrane. Like diffusion, this process does *not* require energy because the molecules are moving down a concentration gradient. Thus, it is a type of passive transport. Glucose molecules, for example, move across the cell membrane by means of facilitated diffusion. (See Figure 4-6.)

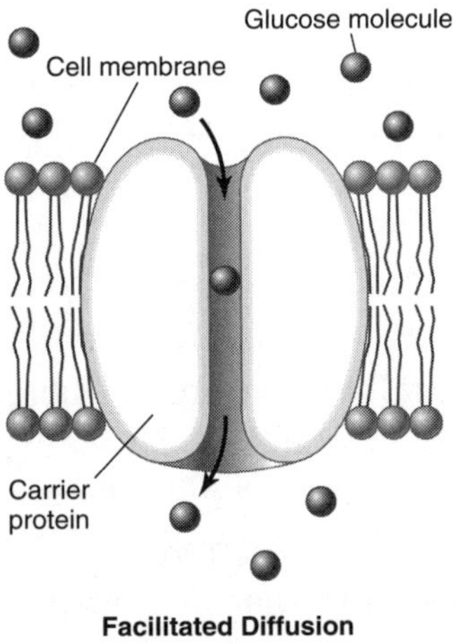

Active Transport

The movement of a substance against a concentration gradient (from an area of low concentration to an area of higher concentration) is known as **active transport**. When substances are moved from an area of low concentration to an area of high concentration, energy is needed and work is done. This kind of transport of materials across the cell membrane is one of the most important activities of cells. As mentioned previously, protein molecules within the cell membrane function as *carriers* to enable the transport of materials across the membrane. (See Figure 4-7.) Other than using energy from your food to keep you warm, the most important use of energy in your body is to

Figure 4-6 Facilitated diffusion, a type of passive transport, requires the presence of a carrier protein in the cell membrane; substances move down a concentration gradient.

Figure 4-7 Active transport across a cell membrane requires the presence of a carrier protein and the energy supplied by ATP; substances move against a concentration gradient.

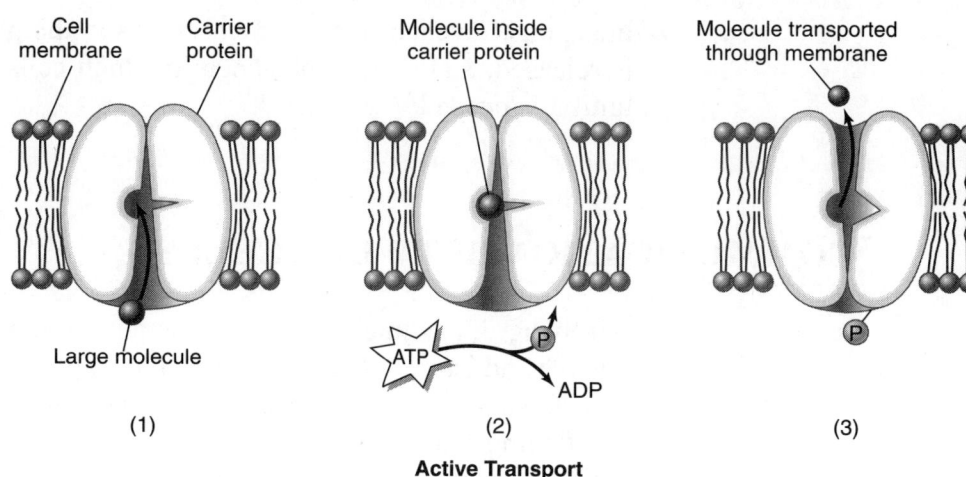

help pump substances across your cell membranes by active transport—a process that goes on all the time. Cells get the energy for carrying out active transport from *ATP* (adenosine triphosphate) molecules.

ENERGY TRANSFORMATIONS INSIDE THE CELL

The energy used in the chemical reactions that take place inside cells is associated with the electrons of atoms. The greater an electron's distance from its nucleus, the more stored, or *potential,* energy it has. When some atoms join, the electrons shared between them form the covalent bonds of a new chemical compound. Each type of covalent bond has a specific amount of energy, depending on which atoms are involved. Whenever covalent bonds are formed or broken, the amount of stored energy changes. Chemical reactions are mainly energy transformations in which the energy stored in chemical bonds is transferred to other newly formed chemical bonds or released as heat or light. (See Figure 4-8.)

For example, the glucose in the food you eat is used by your cells after it is digested. In a chemical reaction that requires oxygen, the high-energy chemical bonds in the glucose molecules are broken. When glucose molecules are broken apart, **carbon dioxide** and water are formed. The energy levels of the chemical bonds in the carbon dioxide and water are lower than

Figure 4-8 When covalent bonds are formed or broken, the amount of stored energy changes. Here two hydrogen atoms share their single electrons to form a stable covalent bond.

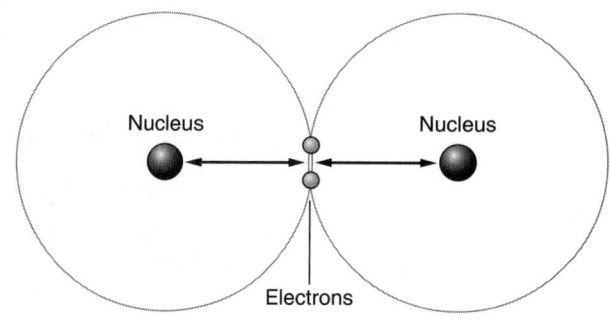

the energy levels of the chemical bonds in the glucose. Thus, some energy is released; ATP is the substance in which cells store this released energy until it is needed.

COMMUNICATION BETWEEN CELLS

All living things interact with their environment in many ways. Conditions outside and inside an organism are constantly being checked. When needed, adjustments are made to maintain internal stability. Whatever the interaction is—finding food, maintaining body temperature, or protecting oneself from disease—communication is required. Information must be received from the environment, processed, and responded to. Organisms—particularly complex, multicellular ones—must organize the information they receive and respond to it. This makes it necessary for all parts of an organism to work in a coordinated fashion. Therefore, to maintain stability, an organism must have a means for *integration*—making all of its body parts work together—and a means for *control*—acting in an organized and appropriate fashion.

Every function of an organism must involve its cells. This includes the communication of an organism with its environment and the communication within the organism among all of its parts. Importantly, the only way that cells communicate is chemically. Communication for a cell means having chemicals moving into and out of it. The work of the two organ systems responsible for integration and control—the *nervous system* and the *endocrine system*—is based on this chemical communication between cells. And, as stated above, special proteins in the plasma membrane act as receptors for these chemical signals.

The Nerve Cell: A Cell for Rapid Communication

How does a message travel through the nervous system? The cell theory tells us that the messages must travel along pathways composed of cells. The specialized cells that make up these pathways are the **nerve cells**, or *neurons*. The message itself is a *nerve impulse*. Every nerve cell does three things: it receives, conducts, and sends impulses; and the cell membrane is the most important part of a nerve cell involved in conducting an impulse. Through the rapid movement of positive ions across the cell membrane, an *electrical voltage* is created. Electrical voltage is a form of energy. In a nerve cell, the voltage changes that occur at one place on the membrane trigger the same kind of changes at the next spot on the membrane. This movement of voltage changes along the length of a neuron *is* the nerve impulse. (See Figure 4-9.)

Crossing the Gap Between Nerve Cells

If you accidentally touch a hot pan on a stove, you immediately pull your hand away. The nerve pathway from your finger to your spinal cord and brain and then back to your finger consists of many nerve cells. Close examination shows that nerve cells do not touch each other. They are separated by a gap.

So how does the impulse get from one nerve cell to another? Extremely important chemicals are released by the ends, or *end brush,* of one nerve cell. The chemicals are released as the impulse arrives at the nerve cell endings, in an area called the *synapse.* These chemicals diffuse across the gap to the next nerve cell. Once received by the next neuron, the chemicals make a new nerve impulse possible. In this way, the message continues along the entire nerve pathway, moving from one nerve cell to another. (See Figure 4-10.)

The endocrine system (the other system involved in chemical communication between cells) will be discussed in the following chapter.

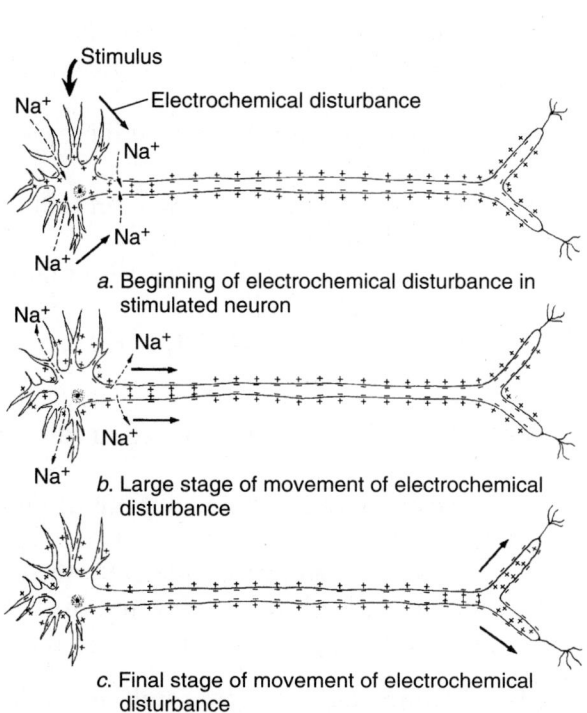

Figure 4-9 A nerve impulse is the movement of cell membrane voltage changes along the length of a nerve cell.

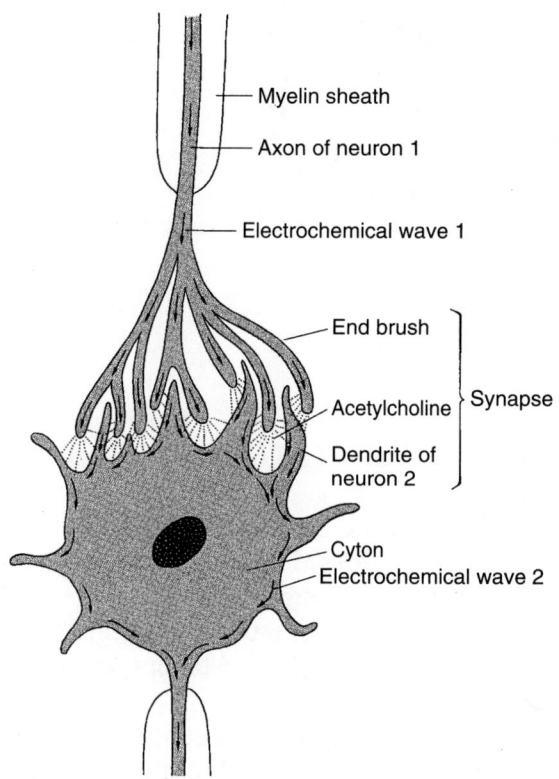

Figure 4-10 Nerve impulses travel from one nerve cell (neuron) to the next by way of chemicals that diffuse across the gap (at the synapse).

Chapter 4 Review

Multiple Choice

1. Which letter indicates the cell structure that directly controls the movement of molecules into and out of the cell?

   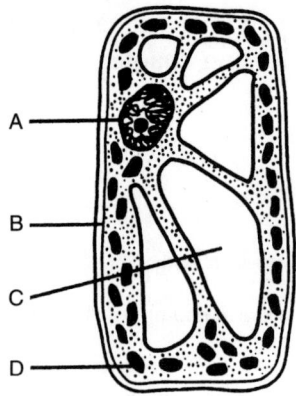

   A. letter *A*
   B. letter *B*
   C. letter *C*
   D. letter *D*

2. Which statement about the functioning of the cell membrane of all organisms is *not* correct?
   A. The cell membrane forms a boundary that separates the cell's contents from the outside environment.
   B. The cell membrane forms a solid barrier that keeps out all substances, either good or bad.
   C. The cell membrane is capable of receiving and recognizing chemical signals.
   D. The cell membrane controls the movement of molecules into and out of the cell.

3. What happens during diffusion?
   A. Molecules move automatically from an area of higher concentration to an area of lower concentration.
   B. Molecules are pumped from an area of lower concentration to an area of higher concentration.
   C. Small molecules form bonds to build large molecules.
   D. Large polymers break down into smaller molecules.

Base your answer to question 4 on the following diagram, which represents a cell in water.

Formulas of molecules that can move freely across the cell membrane are shown. Some molecules are located inside the cell and others are in the water outside the cell.

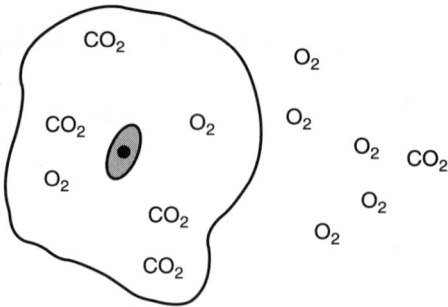

4. Based on the distribution of these molecules, what would most likely happen to them after a period of time has passed?
   A. The concentration of O_2 will increase inside the cell.
   B. The concentration of O_2 will remain the same outside the cell.
   C. The concentration of CO_2 will remain the same inside the cell.
   D. The concentration of CO_2 will decrease outside the cell.

5. A plant cell shrinks when placed in salt water due to the
   A. osmosis of water molecules out of the cell
   B. osmosis of water molecules into the cell
   C. diffusion of salt molecules into the cell
   D. diffusion of salt molecules out of the cell

6. Placing limp celery in water will make the celery stalk firm again due to
   A. carrier proteins
   B. osmosis
   C. active transport
   D. a catalyst

7. A high "concentration gradient" means that the concentration of a substance is
   A. low on both sides of the cell membrane
   B. high on both sides of the cell membrane
   C. about the same on both sides of the cell membrane
   D. high on one side of the cell membrane and low on the other side

Base your answer to question 8 on the following diagram, in which the dark dots represent small molecules. These molecules are moving out of the

58 Preparing for the New Jersey Biology Competency Test

cells, as indicated by the arrows. The number of dots represents the relative concentrations of the molecules inside and outside of the two cells.

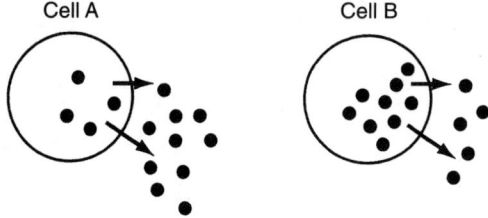

8. ATP is required to move the molecules out of
   A. cell *A* only
   B. cell *B* only
   C. both cell *A* and cell *B*
   D. neither cell *A* nor cell *B*

Refer to the set of diagrams below, which shows the movement of a large molecule across a cell membrane, to answer question 9.

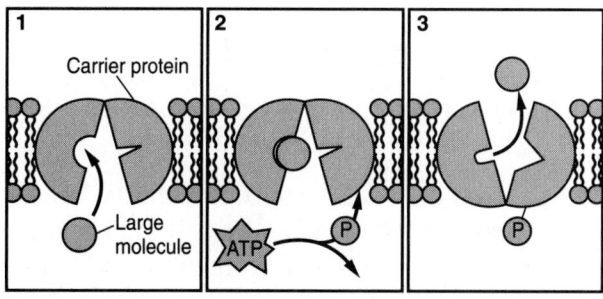

9. Which process is best represented by this set of diagrams?
   A. active transport
   B. protein synthesis
   C. passive transport
   D. energy transfer

10. When covalent bonds are formed or broken,
    A. the amount of stored energy changes
    B. some energy is always lost
    C. energy is always gained
    D. new atoms are formed

11. Which of the following is *not* an idea of the cell theory?
    A. Organisms are made up of one or more cells.
    B. Cells bond together just like atoms do.
    C. The cell is the basic unit of structure in living things.
    D. All cells arise from previously existing cells.

12. While viewing a slide of rapidly moving sperm cells, a student concludes that these cells require a large amount of energy to maintain their activity. The organelles that are most directly involved in releasing this energy are the
    A. vacuoles
    B. chloroplasts
    C. ribosomes
    D. mitochondria

13. The organelle that stores wastes for a cell is the
    A. vacuole
    B. chloroplast
    C. ribosome
    D. Golgi complex

14. Chloroplasts are important organelles because they
    A. are necessary to release stored energy
    B. store wastes in both plants and animals
    C. use energy from the sun to make food
    D. are in the nuclei of plant and animal cells

15. Nerve cells are essential to an animal because they carry out
    A. chemical communication between cells
    B. regulation of reproductive rates in cells
    C. transport of nutrients to various organs
    D. an exchange of gases within the body

16. A nerve impulse results from
    A. the removal of fluid from between cells
    B. electrical voltage changes on the cells
    C. the movement of cells through the body
    D. a direct collision between two cells

17. A nerve impulse is transmitted along the length of a cell's
    A. plasma membrane
    B. mitochondrion
    C. endoplasmic reticulum
    D. Golgi apparatus

18. How does a nerve impulse cross the gap from one nerve cell to the next?
    A. The impulse pulls the nerve cells together to close the gap.
    B. The impulse forms a bridge of proteins across the gap.
    C. The first nerve cell releases chemicals that diffuse across the gap.
    D. The nerve cell membranes open up and connect at the gap.

Refer to the diagrams below to answer questions 19 and 20.

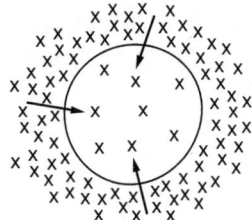
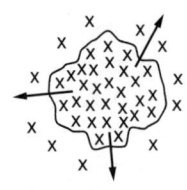

19. Both diagrams represent the movement of molecules from an area of
    A. low concentration to an area of high concentration
    B. high concentration to an area of low concentration
    C. low concentration to an area of equal concentration
    D. high concentration to an area of equal concentration

20. The diagrams could be used to illustrate all of the following types of transport *except*
    A. diffusion   C. active
    B. osmosis   D. passive

Refer to the following diagram to answer questions 21 and 22.

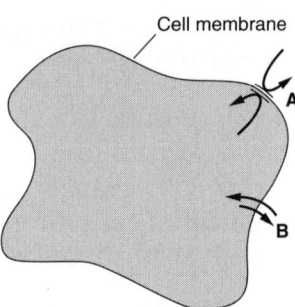

21. Which part of the diagram shows the cell membrane acting to allow materials into and out of the cell?
    A. *A* only   C. both *A* and *B*
    B. *B* only   D. neither *A* nor *B*

22. Which part of the diagram shows the cell membrane acting to separate the inside of the cell from the outside environment?
    A. *A* only   C. both *A* and *B*
    B. *B* only   D. neither *A* nor *B*

Base your answer to question 23 on the following diagram, which illustrates one type of cell communication, and on your knowledge of biology.

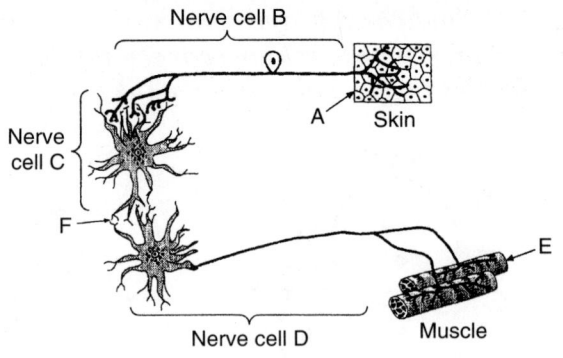

23. In region *F*, there is a gap between nerve cells *C* and *D*. Nerve cell *D* is stimulated to respond by
    A. the flow of blood proteins from cell *C* to cell *D*
    B. the movement of a virus from cell *C* to cell *D*
    C. a chemical message diffused from cell *C* to cell *D*
    D. a blood vessel that forms between cell *C* and cell *D*

Analysis and Open Ended

24. Name three important cell organelles and describe the main function of each.

25. Discuss the "selective permeability" of cells by explaining (a) why the cell membrane is said to be selectively permeable and (b) why this characteristic is important for the health and functioning of a cell.

Base your answers to questions 26 and 27 on the diagram below, which represents a unicellular organism in a watery environment. The small triangles represent molecules of a specific substance.

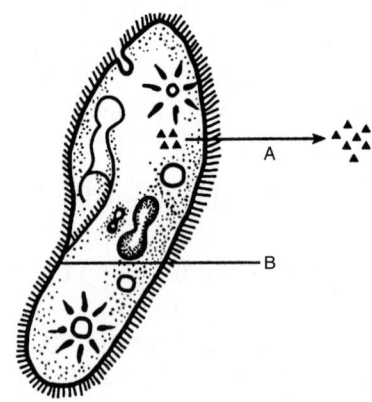

26. What kind of transport is represented by arrow *A*? State two ways in which active

transport is different from diffusion (passive transport).

27. In the cells of multicellular organisms, structure *B* often contains special proteins known as *receptor molecules*. What specific function do these protein molecules carry out for the cell?

28. Briefly explain the importance of "integration" and "control" for maintaining stability in an organism. What two systems accomplish this in animals?

29. Describe the means by which cells communicate with each other.

30. What are the three main functions of every nerve cell?

31. Describe how a nerve impulse travels along the length of a neuron. Your answer should explain the following: (a) what a nerve impulse is; and (b) the role of the cell membrane in this process.

32. What happens to a nerve impulse when it reaches the end of a nerve cell?

33. Use the following terms to construct a concept map that shows how a cell regulates its internal environment: *active transport; diffusion; uses energy; does not use energy; cell membrane; osmosis.*

34. What are the two main things that can happen—as a result of chemical reactions—to the energy stored in chemical bonds? For what important activity do cells use this energy?

35. What, specifically, would happen to a cell if its mitochondria were removed? Explain why.

Reading Comprehension

Base your answers to questions 36 to 38 on the information below and on your knowledge of biology. Use one or more complete sentences to answer each question.

The spleen is an organ that helps your body fight disease. It also helps break down old red blood cells. For John Moore, removal of his spleen cured the leukemia, a type of cancer of the blood that he was found to have in 1976. Removing the spleen is a standard treatment for this disease. The leukemia did not return and Mr. Moore was very happy.

But his attitude was soon to change. For what Mr. Moore did not know was that the physicians at the University of California who removed his spleen kept some of the cells from the spleen alive. The physicians put the cells in a special nutrient-rich solution. As the cells continued to reproduce, the physicians studied them. They discovered that cells from Mr. Moore's spleen produced an interesting blood protein. The physicians received a patent on the cells, which made them the legal owners of the cells they removed from Mr. Moore. In turn, the cells and the blood protein they made were being sold to a company that planned to develop a new medicine from the cells—a medicine that would be sold for a large profit. Mr. Moore did not think this was right. He told the physicians, "Don't use my cells; I own them." Then he went to court.

The lawsuit was heard by the Supreme Court of California, which ruled in July 1990 in favor of the physicians. Later, the United States Supreme Court also ruled in the physicians' favor. The courts felt that scientific research would be threatened if researchers did not have the freedom to work with human cells. Besides, the courts said that John Moore never expected to get his cells returned to him when he gave permission for his spleen to be removed in the medical procedure. More than a

> decade later, Mr. Moore still remained healthy. And by then the drug company stock owned by Mr. Moore's original doctor—in exchange for the rights to use the Moore cell line—was worth over five million dollars.

36. Explain why the removal of his spleen was a necessary procedure for the patient John Moore.

37. Why was the tissue from Mr. Moore's spleen of great interest to a company that produces medicines?

38. How was the legal system involved in the scientific research that arose from the removal of Mr. Moore's spleen?

Chapter 5

Homeostasis and Enzymes

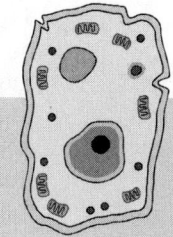

Standard 5.5.12 A1 Relate the structure of molecules to their function in cellular structure and metabolism.

HOMEOSTASIS

Organisms live in a world of changing outside conditions. But, to remain alive, every organism needs to keep the conditions inside its body fairly constant. An organism must have ways to keep its internal conditions from changing when its external environment changes. This ability of all living things to detect external changes and to maintain a constant internal environment is known as **homeostasis**.

Over the course of evolution, there has been a trend toward the development of larger multicellular organisms from earlier, microscopic unicellular ones. So, is there an advantage to being multicellular? Although unicellular organisms perform all the necessary life functions, being single-celled makes it more difficult for an organism to maintain homeostasis. Being multicellular makes it possible to have many types of protection against changes in the environment. In other words, an organism with many cells is able to have structures and systems that protect its individual cells from external changes, thus helping it to stay alive. To maintain homeostasis, organisms actually must make constant changes. That is why homeostasis is often referred to as maintaining a **dynamic equilibrium**. *Dynamic* means "active," and *equilibrium* means "balanced." Homeostasis requires active balancing.

The Cell and Homeostasis

One of the most fascinating facts about our bodies is that each of our many, many cells is surrounded by liquid. The smallest blood vessels in our bodies, the *capillaries,* are close to every cell. There is a small amount of space between the capillaries and the body cells. This space is filled with fluid. The fluid that surrounds all cells is made up mostly of water, with many substances dissolved in it. This **intercellular fluid** (or *ICF,* meaning "between

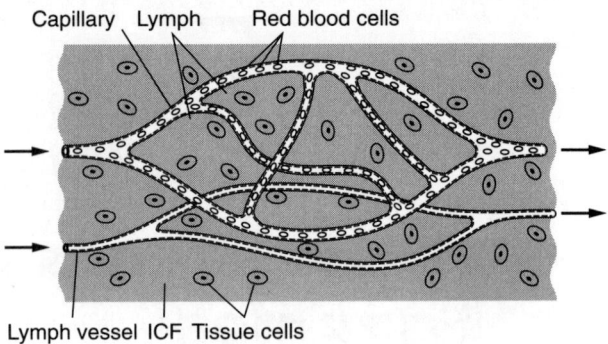

Figure 5-1 All the cells in our body are surrounded by intercellular fluid (ICF). Materials are exchanged between the cells and the fluid, which helps to maintain stable conditions inside each of the cells.

cells") is important for helping to maintain stable conditions inside each cell. Many materials are exchanged between the cells and the fluid. In turn, materials may be exchanged between the fluid and the blood in the capillaries. All of this is done to make sure that each and every body cell is able to maintain homeostasis and remain healthy. (See Figure 5-1.)

SYSTEMS FOR MAINTAINING HOMEOSTASIS

Multicellular animals have evolved highly organized, complex organ systems especially suited to maintaining a relatively constant internal environment. These organ systems include the *excretory system,* which regulates the chemistry of the body's fluids while removing harmful wastes; the *nervous system,* which uses electrochemical impulses to regulate body functions; the *endocrine system,* which produces hormones—chemical messengers that help regulate the functions of the body; and, finally, the *immune system,* which uses a set of defenses to protect the body from dangerous substances and harmful microorganisms that could upset its internal balance. (See Figure 5-2.)

Maintaining Homeostasis During Exercise

Exercise involves increased muscle activity. This activity creates changes within the body. To maintain homeostasis, the body needs to be able to

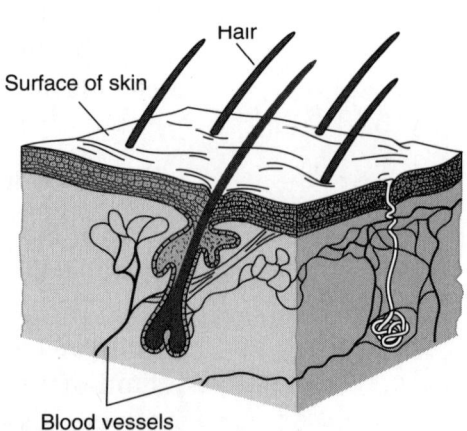

Figure 5-2 Multicellular organisms have systems and structures that help them maintain homeostasis. For example, our skin has features that detect and respond to changes in external temperature; it is also part of the excretory system, which regulates body chemistry and removes harmful wastes.

Figure 5-3 A structure in the brain (the medulla) monitors the amount of CO_2 in the body, adjusting the breathing rate to maintain proper levels.

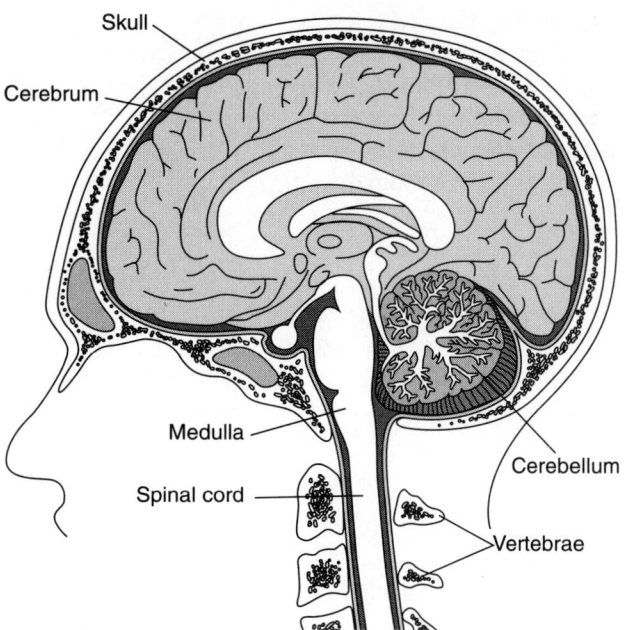

respond to these changes. Any event or change that causes an organism to react is called a **stimulus** (plural, *stimuli*). The resulting reaction of the organism is called a **response**.

An example of a change that occurs when we exercise is the increase of carbon dioxide (CO_2) in the body, produced by muscle cells as a result of cellular respiration. The level of CO_2 increases in both the intercellular fluid and the blood. To maintain homeostasis, the body first must be able to detect this change and then respond to it.

A structure in the brain detects the increased CO_2 level in the blood passing through the brain and in the fluid around the brain cells. In response, this part of the brain sends signals to the chest to increase the rate of breathing and the amount of air taken in on each breath. These changes in breathing increase the exchange of gases in the lungs, lowering the CO_2 levels in the body. These lower levels are then detected in the brain, which in turn sends a signal to reduce the breathing rate. This process is an example of a (negative) **feedback mechanism**, a system that reverses the original response to a stimulus when the desired condition is reached. Such feedback mechanisms are important in maintaining homeostasis. (See Figure 5-3.)

HOMEOSTASIS AND FEEDBACK MECHANISMS

Carbon dioxide levels in your body are regulated in much the same way that a thermostat regulates the air temperature in your home. The following are the parts of a (negative) feedback mechanism used in maintaining homeostasis:

- *Sensor.* Something must be able to detect a change. For example, a thermometer attached to a thermostat is a sensor. In the body, specific structures in the brain act as a sensor to detect changes in CO_2 levels.

Figure 5-4 Both CO_2 levels and body temperature are regulated by feedback mechanisms, much as a thermostat controls the temperature in a room.

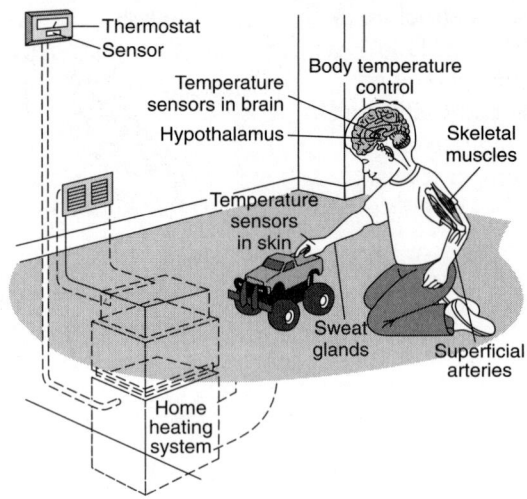

- *Control unit.* Something must know what the correct, or healthy, level should be. A thermostat in a house is set to a particular comfort level. Information in the brain is "preset" to recognize the correct CO_2 level.
- *Effector.* Something must take instructions from the control unit and make the necessary changes. In a house, the effector would be a furnace or an air conditioner. In the body, the effector for CO_2 levels would be the muscles in the chest that are used for breathing. (See Figure 5-4.)

Plants and Homeostasis

Maintaining water balance is essential to homeostasis in all living things. Plants, as well as animals, must maintain water balance. Openings, or *stomata,* in the surface of a leaf are adapted to control the loss of water. (*Note:* The singular term *stomate,* or *stoma,* means "mouth.") Each opening is surrounded by two *guard cells.* These guard cells, like any other cells, allow water to diffuse through their cell membranes. When water (in a plant) is abundant, it moves into the guard cells. The increased quantity of water increases the pressure within the guard cells, which are somewhat curved in shape. When they are filled with water, the guard cells become even more curved. As a result, the space between them expands, the stomata widen, and excess water evaporates out of the leaf and into the air around the plant. (See Figure 5-5.)

When water is scarce, the guard cells contain less water; they become less curved in shape and the space between them closes. In this way, water loss is reduced and the plant is able to maintain its water balance.

Figure 5-5 Special openings in the surface of a leaf function to maintain water balance in plants. These openings, or *stomata,* control water loss through the actions of the guard cells that surround them.

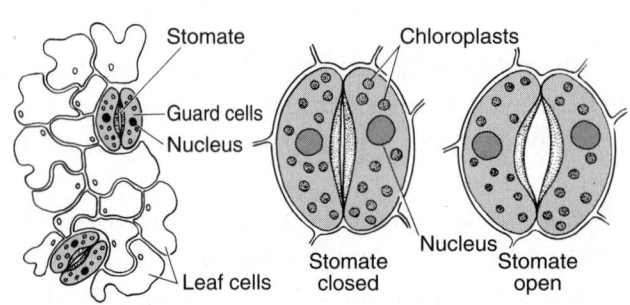

THE ENDOCRINE SYSTEM: HORMONES AND HOMEOSTASIS

You have learned that nerve cells use chemicals for communicating with each other. An organism's nervous system controls its responses to both internal and external stimuli. The other body system that depends on chemical communication between cells is the endocrine system. This system is made up of special *endocrine glands* whose main function is to help the body maintain homeostasis. The glands produce **hormones**, the chemical "messengers" that are released directly into the blood and carried throughout the body by the *circulatory system*. When a hormone arrives at its special *target cells* (often in a body part far from the gland that made it), it causes specific changes. For example, your pancreas releases the hormone insulin when your blood sugar level is too high, thus lowering it to a safe level. (See Figure 5-6.)

How do hormones do their work? Some hormones bind to specific *receptor molecules* (proteins) found in the cell membrane. The binding of a hormone with a receptor protein then causes a change inside that cell, usually involving the cell's enzymes. Other hormones pass right through the plasma membrane and bind to receptor proteins in the cytoplasm. The hormone-receptor complex may then move to the nucleus and interact with the cell's DNA, affecting gene activity.

Both the nervous system and the endocrine system are communication networks. However, there are important differences in the two systems. Impulses sent by the nervous system usually produce rapid responses, frequently produced by the actions of muscles. Hormones usually produce slower, longer-lasting changes, which often involve metabolic activity within the target cells.

An important characteristic of hormones is that only small amounts are usually needed to produce the required effect. The group of target cells for a hormone is usually very sensitive to that particular hormone. Feedback

Figure 5-6 The human endocrine system: The pituitary gland controls the release of hormones by many other glands in the body. The actions of the hypothalamus connect the nervous and endocrine systems, thereby controlling hormones that are released from the pituitary gland.

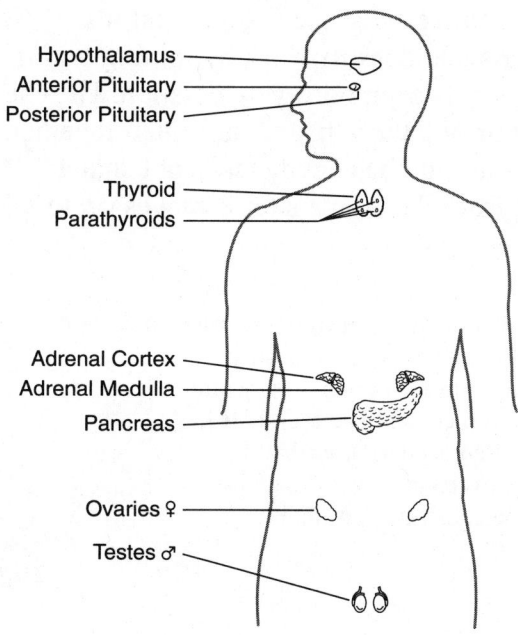

mechanisms work to control the amounts of many hormones that are released. In this way, the release of a specific hormone has the effect of stopping any further release of it until the hormone is needed once again.

ENZYMES: THE CELL'S MIRACLE WORKERS

To be safe, chemical reactions in the cell must take place in a series of small steps, rather than in a single large burst of activity, such as in a heated test tube. The steps must be very precise, too; and they must occur in the correct order, one after the other. One problem for the cell is that the substances, or *reactants,* in a reaction, must get changed into exactly the right products and not into something else. Another problem is that the reaction must occur at a relatively low temperature so that the cell is not harmed.

These problems are solved by substances called **enzymes**. Because a particular enzyme is needed for each type of chemical reaction (enzymes are specific to their reactions), cells have thousands of different kinds of enzymes. The correct enzyme can make a reaction occur 10 times a second that otherwise might occur only once every 100 years. Substances that are responsible for changing the rate at which chemical reactions occur, without being chemically changed themselves, are called **catalysts**. Enzymes are called *organic catalysts,* because they are made up of either proteins or nucleic acids (and are in living things). Enzymes are very accurate in their work because they are, usually, proteins that have a very specific shape.

The shape of an enzyme molecule includes a spot somewhere on its surface that is like a pocket. This "pocket" is called the *active site,* and it is exactly the right size and shape to fit a particular substance. The substance that fits an enzyme, and which is acted on by the enzyme, is called a *substrate.* If two different substrates are involved in a chemical reaction, there is a precise fit in the active site for each one. The *lock-and-key model* of enzyme action is often used to describe this exact fit between an enzyme's active site ("lock") and the substrate ("key") that fits it. (See Figure 5-7.)

An enzyme does its work by joining with the substrate in this close fit. But this is only a temporary association. The chemical reaction occurs while the enzyme and the substrate are fitted together. The substrate changes in a specific way, but the enzyme does not change. The product of the reaction is released from the active site; it moves on to take part in another chemical

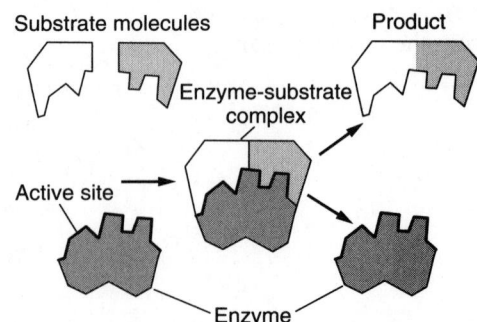

Figure 5-7 Enzymes speed up the rate of chemical reactions in the body by temporarily joining, at their active site, with other molecules. The lock-and-key model is often used to describe this exact fit between an enzyme's active site and the substrate that fits it.

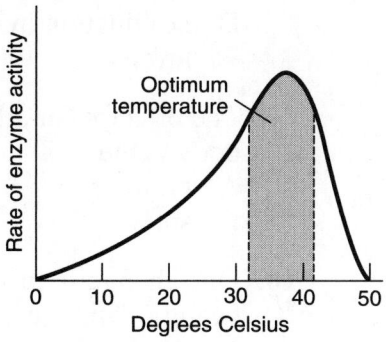

Figure 5-8 A typical enzyme in the body works best at temperatures between 35°C and 40°C.

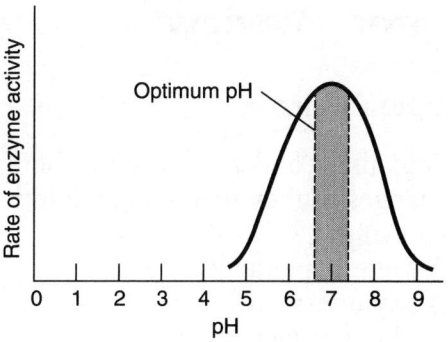

Figure 5-9 Most enzymes function best at about pH 7, which is neutral.

process. However, the unchanged enzyme, with its active site once again available, is reusable; it gets to catalyze the same reaction on yet another molecule of the same substance. In effect, the enzyme is recycled, getting to do its job over and over again without itself being changed. *Note:* The name of an enzyme usually ends in "ase" and relates to its substrate; for example, the enzyme *lactase* helps to digest the sugar *lactose*.

Helping Enzymes Do Their Jobs

For an enzyme to work well, the protein molecule must maintain its correct shape. Two conditions in the cell, temperature and pH, are very important for maintaining the shape of an enzyme molecule. In fact, the main reason animals need to maintain a constant body temperature is to allow cellular enzymes to function properly. (See Figure 5-8.) The **pH** scale is used to measure how acidic or basic a solution is. Slight changes in pH quickly change an enzyme's shape and, thus, its ability to affect a substance. Most enzymes function best at about pH 7, which is neutral. Our cells must maintain that pH in order to survive. (See Figure 5-9.)

Chapter 5 Review

Multiple Choice

1. Organisms undergo constant chemical changes as they maintain an internal balance known as
   A. interdependence
   B. synthesis
   C. homeostasis
   D. recombination

2. What characteristic has evolved that helps to maintain homeostasis?
   A. taller bodies with larger cells
   B. shorter bodies with fewer cells
   C. multicellular bodies with many cells
   D. multicellular bodies with fewer cells

3. A system in dynamic equilibrium
   A. makes constant changes
   B. changes in intervals or steps
   C. changes very infrequently
   D. never changes at all

4. Intercellular fluid is made up mostly of
   A. blood   C. mineral salts
   B. water   D. cytoplasm

5. Intercellular fluid is important for the exchange of materials between
   A. body cells and arteries
   B. body cells and veins
   C. veins and capillaries
   D. body cells and capillaries

6. As a result of exercise, CO_2 levels increase in the
   A. blood only
   B. intercellular fluid only
   C. blood and intercellular fluid
   D. muscles only

7. The brain sends a signal to increase the breathing rate when the CO_2 level has
   A. not changed for a while
   B. decreased too much
   C. increased too much
   D. increased, then decreased

8. The increased breathing rate signaled by the brain serves
   A. to increase the CO_2 level in the body
   B. to decrease the CO_2 level in the body
   C. to decrease the O_2 level in the body
   D. no function in changing O_2 and CO_2 levels

9. The effector for adjusting CO_2 levels in the body is the
   A. brain   C. capillaries
   B. spine   D. chest muscles

10. One change in the body results in another change. This second change reverses the first change in order to maintain homeostasis. This sequence describes a type of
    A. control mechanism
    B. feedback controller
    C. feedback mechanism
    D. effector mechanism

11. Why might a blood clot be important to maintaining homeostasis?
    A. It slows the flow of blood through the body.
    B. It prevents the loss of blood from the body.
    C. It increases the amount of water in the blood.
    D. It adds more cells to the blood supply.

Base your answer to question 12 on the data in the graph below.

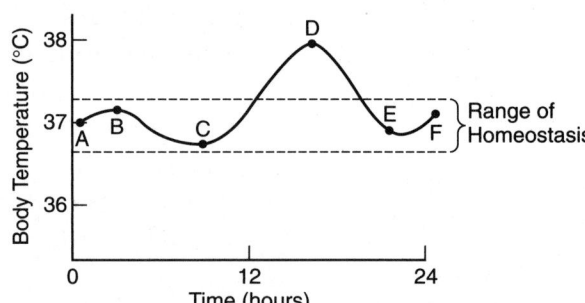

12. The graph shows evidence of disease in the human body. A disruption in the dynamic equilibrium is indicated by the temperature change that occurs between points
    A. *A* and *B*   C. *C* and *D*
    B. *B* and *C*   D. *E* and *F*

13. Which homeostatic adjustment would the human body make in response to an increase in environmental temperatures?
    A. a decrease in glucose levels
    B. an increase in perspiration
    C. a decrease in fat storage
    D. an increase in urine production

14. Chemical reactions within a cell usually take place
    A. over extremely long periods of time
    B. in a series of small steps
    C. all at once in a single burst
    D. over a period of several days

15. When a person's brain sends signals to the chest to increase the breathing rate after heavy exercise, it is an example of the body
    A. regulating its internal water balance
    B. maintaining a dynamic equilibrium
    C. disrupting its own homeostasis
    D. increasing enzyme reaction rates

Base your answer to question 16 on the figure below, which represents a view of the underside (lower surface) of a leaf.

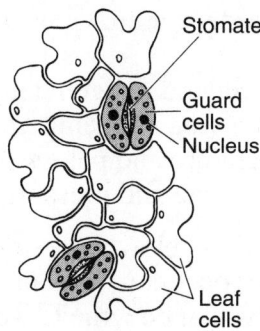

16. What is one main function of the stomata and guard cells?
    A. to store food for the cold season
    B. to help control the loss of water
    C. to undergo mitotic cell division
    D. to give support to the leaf's veins

17. The changing shape of a leaf's guard cells helps to
    A. allow the plant to grow stronger
    B. prevent the plant from losing sap
    C. regulate the temperature of the plant
    D. maintain the plant's water balance

Base your answer to question 18 on the table below, which shows the rate of water loss in three different plants.

Plant	Liters of Water Lost Per Day
Cactus	0.02
Potato plant	1.00
Apple tree	19.00

18. One reason (besides size differences) that each plant loses a different amount of water than the other plants is that each has
    A. its particular guard cells that are adapted to maintain homeostasis
    B. the same number of chloroplasts but different rates of photosynthesis
    C. different types of insulin-secreting cells that regulate water levels
    D. the same rate of photosynthesis but different numbers of chloroplasts

19. The endocrine system maintains homeostasis by
    A. regulating physical coordination
    B. controlling the size of blood vessels
    C. sending nerve impulses to the brain
    D. releasing hormones into the blood

20. Hormones carry out their work by
    A. replacing the nucleus of a cell
    B. breaking down the membrane of a cell
    C. binding to receptor proteins in a cell's membrane or cytoplasm
    D. changing the shape of a cell and then becoming part of the cell

21. The endocrine system differs from the nervous system in that the endocrine system
    A. produces faster, short-term changes
    B. produces slower, long-term changes
    C. operates in isolated regions of the body
    D. produces changes in the brain only

22. Hormones and secretions of the nervous system are chemical messengers that
    A. store genetic information
    B. extract energy from nutrients
    C. carry out the circulation of materials
    D. coordinate system interactions

23. Which of the following statements is true?
    A. Hormones are carried by the respiratory system.
    B. Hormones are produced by the nervous system.
    C. Hormones are usually needed in small amounts.
    D. Hormones are usually needed in large amounts.

24. The body does not produce too much of most hormones because
    A. the excretory system controls the amount released

B. feedback mechanisms control the amount released
C. cells can store any excess hormones they receive
D. the body can produce only fixed amounts of each hormone

25. Which statement describes a feedback mechanism involving the human pancreas?
    A. The production of estrogen stimulates the formation of gametes for sexual reproduction.
    B. The level of sugar in the blood is affected by the amount of insulin in the blood.
    C. The level of oxygen in the blood is related to heart rate.
    D. The production of urine allows for excretion of cell waste.

26. The pancreas produces one hormone that lowers blood sugar level and another that increases blood sugar level. The interaction of these two hormones most directly helps humans to
    A. maintain a balanced internal environment
    B. dispose of wastes formed in other body organs
    C. digest needed substances for other body organs
    D. increase the rate of cellular communication

27. What process is represented by the boxed sequence below?

    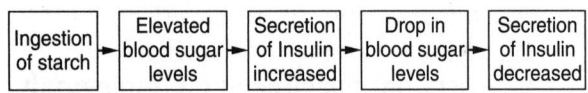

    A. a feedback mechanism in multicellular organisms
    B. the differentiation of organic molecules
    C. an immune response by cells of the pancreas
    D. the disruption of cellular communication

28. A hormone causes specific changes in a body when it
    A. delivers chemicals to nerve cells
    B. makes the body's fluids more acidic
    C. arrives at its special target cells
    D. attacks invading harmful microorganisms

29. To carry out its chemical reactions, each cell contains
    A. one specific type of enzyme for the cell
    B. fewer than twenty different enzymes
    C. thousands of different kinds of enzymes
    D. thousands of copies of the same enzyme

30. How do chemical reactions occur at the relatively low temperature found within cells?
    A. Some energy is destroyed before it heats up the cell.
    B. Some energy is stored temporarily in ATP molecules.
    C. Enzymes are used to slow (decrease) the rate of reactions.
    D. Enzymes are used to speed (increase) the rate of reactions.

31. Two conditions that must be kept constant in a cell in order for enzymes to work properly are the
    A. pH level and oxygen content
    B. surface area and temperature
    C. temperature and pH levels
    D. cell volume and pressure

32. The equation below represents a chemical reaction that occurs in humans. To support the hypothesis that enzyme C works best in a slightly basic environment, data should be collected about the

    $$\text{Substance X} + \text{Substance Y} \xrightarrow{\text{ENZYME C}} \text{Substance W}$$

    A. amino acid sequence of enzyme C
    B. shapes of molecules X and Y after the reaction occurs
    C. amount of substance W produced at various pH levels
    D. temperature of enzyme C before and after the reaction

Base your answer to question 33 on the diagrams below, which show an enzyme and four different molecules.

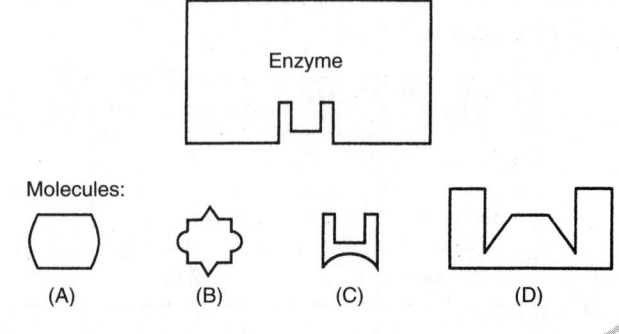

33. The enzyme would most likely affect reactions that involve
   A. molecule *A* only
   B. molecule *C* only
   C. molecules *B* and *D*
   D. molecules *A* and *C*

34. Which statement best describes the interaction between an enzyme and another substance?
   A. a temporary association in which the substance changes
   B. a temporary association in which the enzyme changes
   C. the final product in a series of slow chemical reactions
   D. the pocket into which the enzyme and the substance fit

Analysis and Open Ended

35. How does being multicellular increase an organism's ability to maintain homeostasis and survive?

36. Write a brief paragraph comparing the life of a cell in your body with that of an ameba in the soil. Why is it more likely that the body cell will survive for a long time, but the ameba will not?

Refer to the diagram below to answer question 37.

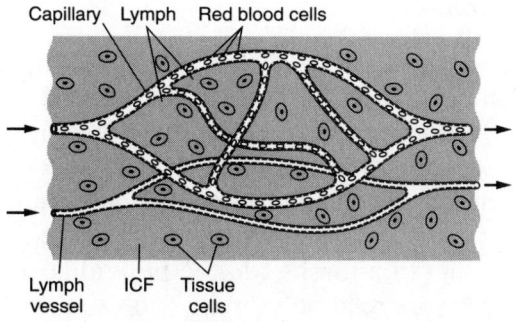

37. Use your knowledge of biology and the diagram to explain why intercellular fluid (ICF) is so important for homeostasis.

Base your answer to question 38 on the information and diagrams below.

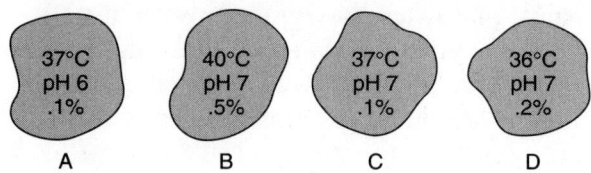

To survive, an organism must maintain the health of its cells. The normal internal environment of a human's cells would include a temperature of 37°C, a pH of 7, and a water/salt balance of 0.1 percent.

38. Which of the cells shown above would belong to someone who is *not* maintaining homeostasis?

39. List, and briefly describe the roles of, the three main parts of a feedback mechanism.

40. Use the diagram below to explain how feedback mechanisms help maintain homeostasis.

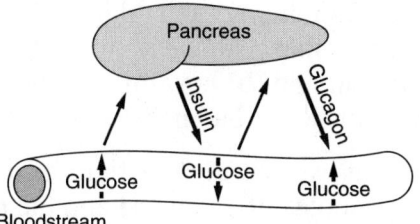

41. Briefly explain the way our bodies adjust breathing rates in order to maintain homeostasis.

Study the following graph to answer questions 42 and 43.

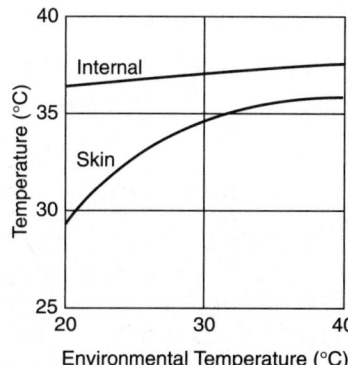

42. The graph shows the effect of external (environmental) temperatures on a student's skin and internal temperatures. Which statement best describes what happens as the environmental temperature increases?
   A. The skin temperature increases, then decreases to 20°C.
   B. The internal temperature increases abruptly to about 30°C.
   C. The skin temperature decreases, due to sweating, to 30°C.
   D. The skin temperature increases, then levels off at about 36°C.

43. What is the difference between the effects of rising external temperatures on the student's internal temperature and skin temperature? Explain how homeostasis is responsible for the effects seen in the graph.

44. In desert environments, organisms that cannot maintain a constant internal body temperature, such as snakes and lizards, rarely go out during the hottest daylight hours. Instead, they stay in the shade, under rocks, or in burrows. Explain how this behavior helps these organisms maintain homeostasis.

45. Describe how plants maintain their water balance. Your answer should include: (a) the structure that plants have to perform this function; and (b) how this structure works to maintain water balance.

46. In what way are the functions of the contractile vacuoles of an ameba and the guard cells of a plant similar?

47. Identify the four main organ systems that are involved in maintaining homeostasis. Briefly describe each of their roles in this process.

Base your answers to questions 48 and 49 on the diagram below, which represents an enzyme and four types of molecules present within a solution in a flask.

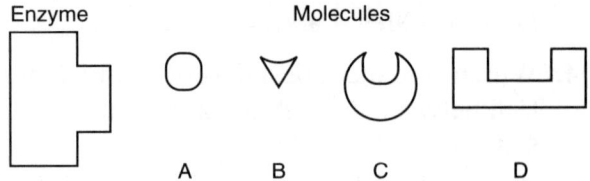

48. Which molecule would most likely react with the enzyme? Why?

49. What would most likely happen to the rate of reaction if the temperature of the solution in the flask were increased gradually from 10°C to 30°C?

50. Enzyme molecules are affected by changes in conditions within organisms. Explain how a long-term, extremely high body temperature during an illness could be fatal to a human.

51. Briefly explain why it is important for the cells of our bodies to maintain a neutral pH.

Reading Comprehension

Base your answers to questions 52 to 55 on the information below and on your knowledge of biology. Use one or more complete sentences to answer each question.

In 2002, flight engineers Carl Walz and Dan Bursch set the record for the longest United States space flight, with 196 days in space as members of Expedition 4 on the International Space Station (ISS). Typically, ISS crews have six or seven members who live on the station for 3 to 6 months. The crews live in a world of weightlessness—the station has no up or down, so there are no real ceilings or floors. While the total inside space of the station is about equal to that of a jumbo jet, the individual spaces in which the astronauts actually live and work are relatively small, each about the size of a school bus's interior. Crews sleep standing up or camping out where they feel comfortable by attaching their sleep restraints to the wall with Velcro.

Biomedical researchers are interested in studying the effects of weightlessness on humans. Being "weightless" is a brand-new challenge never experienced before in the millions of years humans have lived on Earth. And yet, time and again, space travel has demonstrated the marvelous, and often subtle, abilities of the human body to adapt. The body's reactions to weightlessness are teaching us a great deal about its normal responses to gravity. Astronauts report that when they grab the wall of a spacecraft and move their bodies back and forth, they feel as if they are staying in one place and that the spacecraft is moving. Being free of gravity's effects makes us aware of new things. Humans have evolved many automatic reactions to deal with the constant pressure of living in a downward-pulling world. Until we leave that world, we are usually not aware of such reactions.

These reactions include the use of signals from our eyes, from the fluid-filled tubes in our ears, from pressure receptors on the bottom of our feet, and from the distribution of liquids in our blood vessels. A sophisticated control system has evolved to keep gravity from pulling all the liquid in our body to our legs. Within minutes of being in a weightless environment, the veins in an astronaut's neck begin to bulge. The astronaut's face begins to fill out and become puffy. In this situation, the fluids in an astronaut's body are not being pulled down by gravity. The fluids spread throughout the body. Because the body seeks to maintain homeostasis, this new distribution of fluid causes other changes in the body in order to control fluid movement. Included in these are changes in hormone levels, kidney function, and red blood cell production.

Keeping body systems stable, even when external conditions change—that is, maintaining a *dynamic equilibrium*—is as necessary for life in space as it is on Earth. The unexpected result of "living" in space is a better understanding of how the human body works back here on Earth.

52. Describe three ways in which life on the ISS is very different from everyday life on Earth.

53. Why are the effects of weightlessness on humans of interest to researchers?

54. How do the body's responses to weightlessness help explain homeostasis?

55. Describe some adaptations of the body related to living in a world with gravity.

Chapter 6

Photosynthesis and Respiration

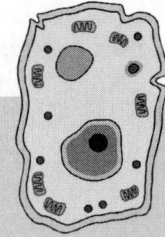

Standard 5.5.12 A2 Explain how plants convert light energy into chemical energy.

FOOD: MATTER AND ENERGY

An apple is a type of food. It contains complex organic compounds; its atoms are held together as molecules by chemical bonds that are rich in stored energy. When you eat an apple, you get both the matter and the energy you need to help build your body and to stay alive.

The apple tree that produced the fruit represents the group of organisms that are **autotrophic**, meaning "self-feeding." Like other plants, the apple tree makes its own food, taking in the **inorganic** substances carbon dioxide (CO_2) and water (H_2O) and changing them into **organic** compounds, such as sugars and starches. Animals (including humans) represent the other group of organisms, which are **heterotrophic**, meaning "other-feeding." Since they cannot make their own food, humans and other animals must get their complex organic compounds by eating other organisms. (See Figure 6-1.)

Figure 6-1 The grass, like all other plants, is autotrophic because it makes its own food. The cows, like all other animals, are heterotrophic because they have to eat other organisms in order to survive.

For the apple tree to combine the inorganic raw materials of CO_2 and H_2O into organic compounds such as sugar and starch, it needs a source of energy. The rays of sunlight that fall on the leaves of the apple tree provide that energy. The process of making this food, by using light as the source of energy, is called **photosynthesis**. All green plants are photosynthetic *autotrophs*. Without plants to capture the energy of sunlight and convert it into the chemical forms that are edible, most animals would have no constant source of food and could not exist.

PHOTOSYNTHESIS

Plants are able to make their own energy-rich carbon compounds. In particular, they make the simple sugar *glucose,* whose chemical formula is $C_6H_{12}O_6$. Plants get the carbon for these glucose molecules from inorganic CO_2 molecules in the air. In addition, plants release oxygen (O_2) molecules as a waste product into the air. (See Figure 6-2.) Scientists discovered that photosynthesis requires the green pigment *chlorophyll*. The chemical reactions of photosynthesis occur within the chlorophyll-containing organelles called *chloroplasts*. These are found inside cells in plant leaves and stems. Some scientists consider this process of photosynthesis the single most important chemical reaction that occurs on Earth. This all-important photosynthetic reaction can be summarized by the following chemical equation:

$$\underset{\text{Reactants}}{CO_2 + H_2O} \xrightarrow[\text{CHLOROPHYLL}]{\text{LIGHT ENERGY}} \underset{\text{Products}}{C_6H_{12}O_6 + O_2}$$

CARBON DIOXIDE   WATER   GLUCOSE   OXYGEN

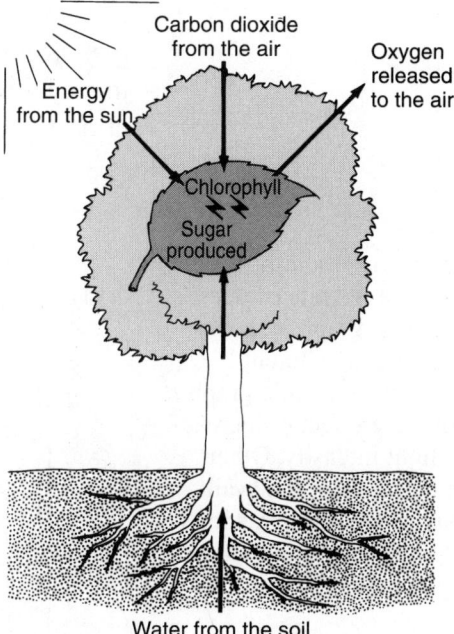

Figure 6-2 The diagram illustrates the basic process of photosynthesis—plants take in inorganic substances from the environment and produce organic substances such as glucose (a sugar).

Figure 6-3 Cells inside a leaf contain chloroplasts, which capture the sunlight that is used for photosynthesis. The structure of the leaf allows these cells to get maximum exposure to the light.

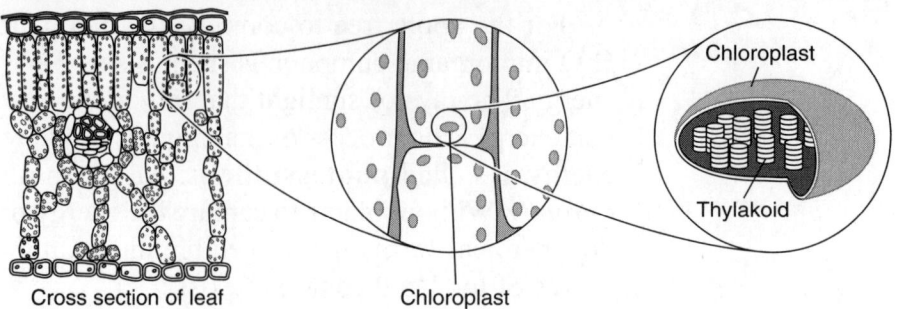

Leaves: Photosynthetic Factories

The structures that are present inside a leaf are well organized. Such organization allows the cells that contain chlorophyll to get maximum exposure to light. (See Figure 6-3.) Plant leaves also contain special openings called *stomates*. The stomates are opened and closed by guard cells, which controls the exchange of CO_2 and O_2 gases between the leaf and the outside environment. At the same time, the leaf controls the amount of water that is lost to the air.

As with any chemical reaction, the reactions of photosynthesis can occur at different rates. The factors that affect the rate at which photosynthesis occurs are temperature, light intensity, CO_2 concentration, availability of water, and the presence of certain minerals. (See Figure 6-4.)

The Purpose of Photosynthesis

A plant does not specifically go through the process of photosynthesis in order to make food for people and other animals. The apple tree, for example, is simply making food for itself so that it can live long enough to reproduce successfully. The apples contain seeds, which may get carried away to new places by animals that eat the apples. This makes it possible for the apple tree to produce more apple trees in other places. However, most of the glucose made in the leaves of the tree does not go into storage in the form of apples. Rather, it gets taken to different parts of the tree to be used by the tree to stay alive. In fact, the tree uses the same process as you do to get the energy it needs from the glucose it has made: *cellular respiration*.

Figure 6-4 Factors that influence the rate of photosynthesis: graph *A* shows the effect of different temperatures, while graph *B* shows the effect of increasing light intensity. Other factors, such as the availability of CO_2, water, and minerals, also affect the rate of photosynthesis.

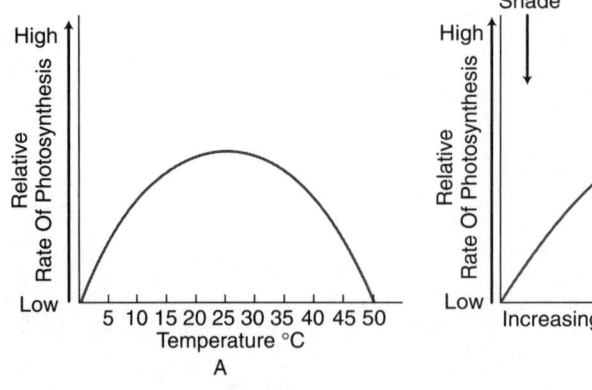

78   Preparing for the New Jersey Biology Competency Test

To summarize: All plants (and algae) are autotrophs that are able to make their own food by photosynthesis and use it for energy through cellular respiration. All animals (and fungi) are heterotrophs that must obtain energy by feeding on other organisms. They use cellular respiration, just as plants do, to obtain energy from the food they eat.

CELLULAR RESPIRATION: RELEASING THE ENERGY IN FOOD

Consider the relationship between the sunlight that falls on the leaves of an apple tree and the chemical process of photosynthesis. During photosynthesis, the light energy of the sun is converted into the stored chemical energy of glucose in the apple. After you eat the apple, your cells are ready to use that stored chemical energy. How does this happen?

The release of energy cannot occur all at once. Too much heat would be released inside your cells. Instead, the release of energy occurs in a series of enzyme-controlled small steps. The energy stored in glucose is converted into a usable form, the energy source of all cells, **adenosine triphosphate**, or **ATP**. This process is known as **cellular respiration**. Cells use the energy from ATP to perform many functions, such as obtaining materials and eliminating wastes. (See Figure 6-5.)

Cellular respiration is basically the opposite of the process of photosynthesis. Instead of being produced in the cells, the energy-rich glucose molecules are taken apart to release their stored energy. Cellular respiration can be summarized by the following chemical equation:

$$\underbrace{\underset{\text{GLUCOSE}}{C_6H_{12}O_6} + \underset{\text{OXYGEN}}{O_2}}_{\textit{Reactants}} \xrightarrow{\text{ENZYMES}} \underbrace{\underset{\substack{\text{CARBON} \\ \text{DIOXIDE}}}{CO_2} + \underset{\text{WATER}}{H_2O} + \underset{\text{ENERGY}}{ATP}}_{\textit{Products}}$$

Oxygen is needed to break down the glucose, and CO_2 and H_2O are released as waste products. This process is referred to as **aerobic** respiration because oxygen is used to help produce the ATP molecules. In contrast, during *anaerobic* respiration (such as the fermentation of yeast), energy is released without the use of oxygen. Anaerobic respiration begins with

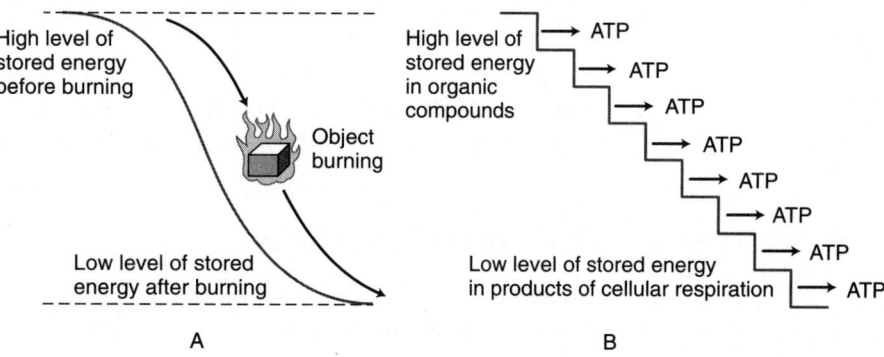

Figure 6-5 The burning of an object (diagram *A*) is due to the sudden release in one step of the energy stored in that object. By contrast, in cellular respiration (diagram *B*), energy is released from organic compounds as ATP in a series of small, enzyme-controlled steps.

Chapter 6/Photosynthesis and Respiration

glycolysis, the splitting of the glucose molecule (*glyco* means "sugar" and *lysis* means "breaking down"); this process yields a small amount of energy as ATP. Aerobic respiration continues after glycolysis through a complex series of chemical reactions in a process called the *Krebs cycle,* which in turn leads to the final stage of cellular respiration. At this stage, oxygen is used to make it possible for the efficient release of large amounts of stored energy from the glucose molecules.

ATP and ADP

When a cell needs energy, the third phosphate group of an ATP molecule is removed by the addition of water. This process, called *hydrolysis* (*hydro* means "water") yields *adenosine diphosphate,* or *ADP,* and inorganic phosphate (P_i). A large amount of energy is then released. This is a reversible reaction; when the cell has extra energy that it needs to store, the ADP combines with a phosphate group and reforms ATP. However, energy is needed for converting ADP back into ATP; the energy comes from the breakdown of glucose, which occurs during cellular respiration. Thus, ADP can be recharged to produce ATP by the addition of a phosphate group.

BREATHING AND RESPIRATION

Animals move O_2 and CO_2 into and out of their bodies by mean of a *respiratory system.* Although cellular respiration refers to the energy-releasing chemical reactions that occur in cells, the word *respiration* also means the process of exchanging gases. In many animals, air is physically pumped into and out of the body by the process of *breathing.* In multicellular animals, it is necessary to have breathing or respiration that involves a respiratory system in order to allow the life-sustaining activities of cellular respiration. However, the lungs themselves cannot move air into or out of the body, because lungs have no muscles.

All mammals, including humans, move air into their lungs by lowering the air pressure within them. Two sets of structures are involved: the ribs and the muscles between them. When the muscles move the ribs upward and outward, the rib cage expands. At the same time, the *diaphragm,* a large flat muscle that lies across the bottom of the chest cavity, contracts and moves down. This movement increases the size of the chest cavity and decreases the air pressure in the lungs, so they fill with air. To move air out, the diaphragm and the muscles between the ribs relax. This movement decreases the size of the chest cavity and increases the air pressure in the lungs, so air moves out. Breathing is the physical process of *inhalation* (when the muscles contract) and *exhalation* (when the muscles relax). (See Figure 6-6.)

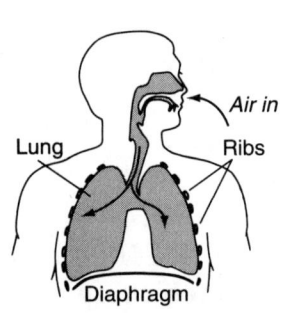

A. Inhaling

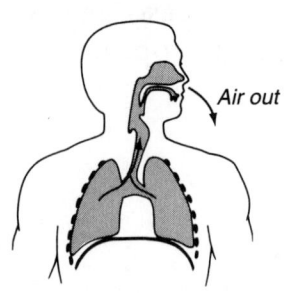

B. Exhaling

Figure 6-6 The process of breathing is controlled by the muscles of the ribs and the diaphragm.

Gas Exchange Surfaces

All aerobic organisms, both plants and animals, exchange oxygen and carbon dioxide with their environment. (Recall that the oxygen is used to produce

ATP molecules during cellular respiration.) Respiration in all organisms involves the diffusion of gases across cell membranes. This occurs for plants in the spongy layer of cells within the plants' leaves. Although gas exchange in animals may involve a complex respiratory system, the actual process of taking in and releasing gases is identical to that of a single-celled organism, such as the ameba. In other words, gases must cross a barrier to be moved into or out of the animal. This barrier, a part of the animal's body, is known as the *respiratory surface*. Respiratory surfaces in different animal species vary in size and shape. (See Figure 6-7.) However, they all share certain requirements:

- The respiratory surface must remain moist at all times so that gases can diffuse across the cell membranes.
- The respiratory surface must be very thin so that gases are able to pass through it.
- There must be a source of oxygen, either in the air or dissolved in the water.
- The respiratory surface must be closely connected to the transport system that delivers gases to and from cells.

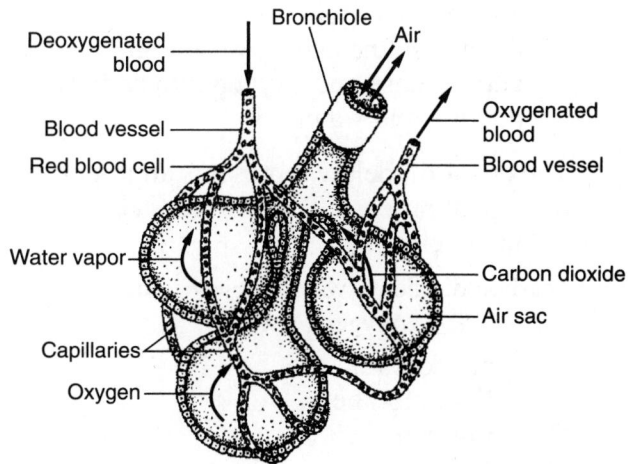

Figure 6-7 The lining of the alveoli acts as our respiratory surface—it is where gas exchange occurs in our bodies.

Chapter 6/Photosynthesis and Respiration

Chapter 6 Review

Multiple Choice

1. Which of the following is an autotroph?
   A. lizard
   B. cactus
   C. shark
   D. deer

2. In heterotrophs, energy for the life processes comes from the chemical energy stored in the bonds of
   A. water molecules
   B. organic compounds
   C. oxygen molecules
   D. inorganic compounds

3. During photosynthesis,
   A. animals use the energy of sunlight to convert starch in plants into food
   B. animals use the oxygen released by plants to produce carbon dioxide
   C. plants use the energy of sunlight to convert carbon dioxide and water into glucose and oxygen
   D. plants use the energy of sunlight to convert glucose and oxygen into carbon dioxide and water

4. The equation below represents an important biological process. This process is carried out within a cell's

   carbon dioxide + water $\longrightarrow$ glucose + oxygen (+ water)

   A. mitochondria
   B. cell membranes
   C. ribosomes
   D. chloroplasts

5. The source of energy for photosynthesis is
   A. oxygen
   B. sunlight
   C. carbon dioxide
   D. glucose

6. To occur, photosynthesis requires the presence of the green substance
   A. tree sap
   B. glucose
   C. chlorophyll
   D. copper

7. The approximate mass of a field of corn plants at the end of its growth period was 3 tons per hectare. Most of this mass was produced from
   A. water and organic compounds absorbed from the soil
   B. minerals and organic materials absorbed from the soil
   C. minerals from the soil and oxygen from the air
   D. water from the soil and carbon dioxide from the air

8. The diagram below represents part of the life process that occurs inside a leaf chloroplast. If the process were to be interrupted by a chemical at point X, there would be an immediate effect on the release of which substance?

   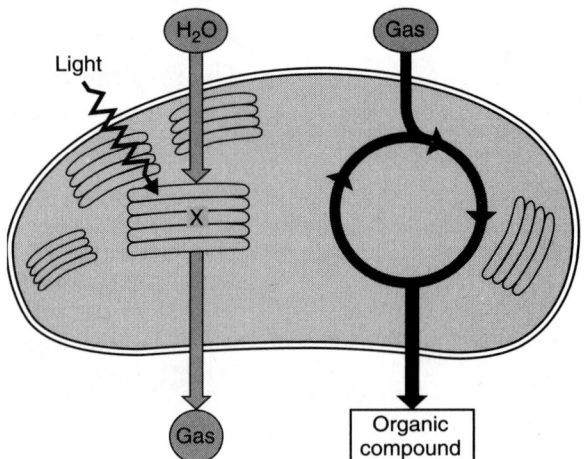

   A. chlorophyll
   B. carbon dioxide
   C. nitrogen
   D. oxygen

9. Look at the chart below to answer this question. Which phrase would you choose to fill in the missing title?

   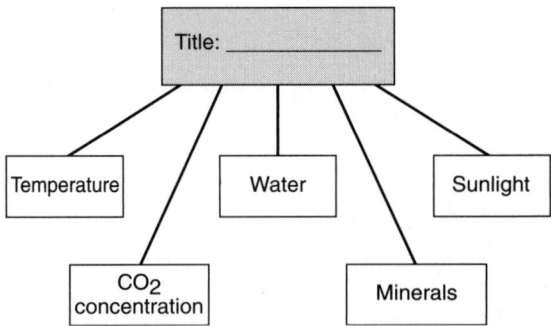

   A. Some living factors in the environment
   B. The chemical process of photosynthesis
   C. Factors that affect the rate of photosynthesis
   D. The nonliving things that make up a plant

82 Preparing for the New Jersey Biology Competency Test

10. The food produced by plants during photosynthesis is used
    A. by the plants themselves only
    B. by animals that eat them only
    C. by both the plants and the animals that eat them
    D. up at the end of the reaction

11. If stored energy were to be released too quickly, a cell would
    A. release too much heat
    B. produce ATP molecules
    C. become an autotroph
    D. become a heterotroph

12. How do animals and plants interact in terms of the two gases involved in photosynthesis?
    A. Animals take in the CO_2 released by plants and release O_2 to the plants.
    B. Animals take in the O_2 released by plants and release CO_2 to the plants.
    C. Plants and animals usually compete for the same O_2 available in the air.
    D. Plants and animals usually compete for the same CO_2 available in the air.

Answer question 13 based on the following information and graph.

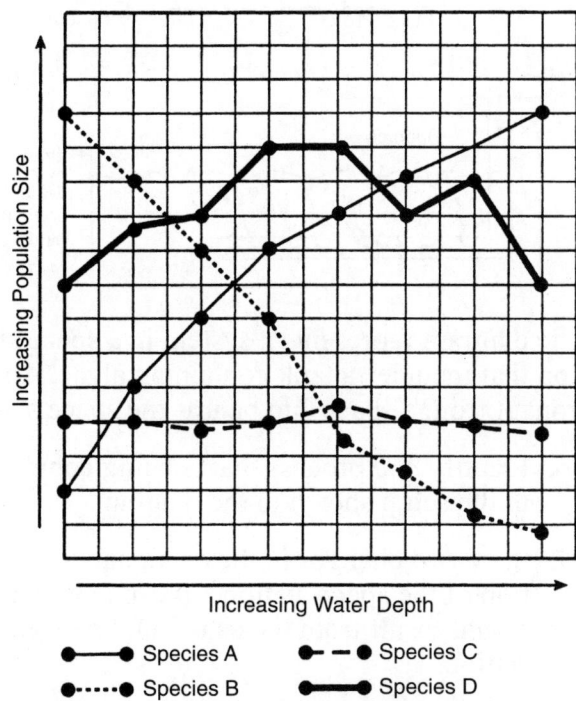

As the depth of the ocean increases, the amount of light that penetrates to that depth decreases. At about 200 meters, there is almost no light present.

The previous graph illustrates the population size of four different species at different water depths.

13. Which species most likely performs photosynthesis?
    A. species A
    B. species B
    C. species C
    D. species D

14. Cellular respiration occurs in
    A. autotrophs only
    B. heterotrophs only
    C. autotrophs and heterotrophs
    D. humans that do aerobics only

15. Eating a sweet potato provides energy for human metabolic processes. The original source of this energy was in the
    A. protein molecules stored within the potato
    B. sun's light captured during photosynthesis
    C. starch molecules absorbed by the potato plant
    D. vitamins and minerals absorbed from the soil

16. Plant leaves contain openings that are opened and closed by guard cells, allowing for gas exchange between the leaf and the outside air. Which phrase best describes the net flow of gases involved in photosynthesis into and out of the leaves on a sunny day?
    A. carbon dioxide moves in, oxygen moves out
    B. oxygen moves in, nitrogen moves out
    C. carbon dioxide and oxygen move in, ozone moves out
    D. water and ozone move in, carbon dioxide moves out

17. Refer to the following diagrams (*A* and *B*) to answer this question. The energy change in diagram *B* is different from the energy change in diagram *A* because, in diagram *B*,

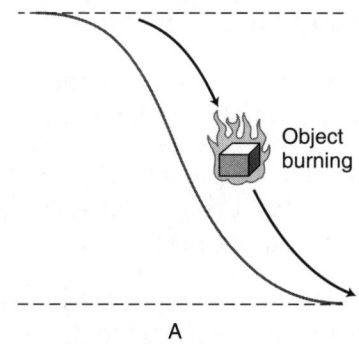

A

Chapter 6 / Photosynthesis and Respiration

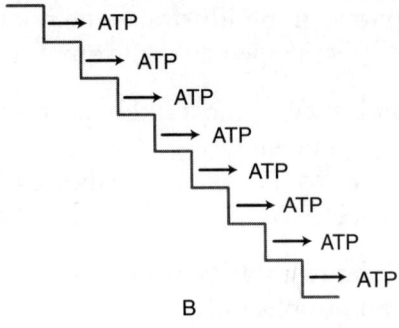

A. the energy is released suddenly in one big step
B. the energy is released in a series of small steps
C. there is less stored energy at the beginning
D. there is less stored energy remaining at the end

18. In animals, the process known as *breathing* refers to the actual
    A. exchange of O_2 and CO_2 gases at the respiratory surface
    B. movement of oxygen-carrying blood through the body
    C. physical pumping of the air into and out of the body
    D. energy-releasing chemical reactions that occur in cells

19. An animal's respiratory surface must be all of the following *except*
    A. extremely thin
    B. moist at all times
    C. tough and muscular
    D. near an oxygen source

20. The transport system must be closely connected to the respiratory system so that it can
    A. break down gases absorbed in the lungs
    B. move gases into and out of the chest
    C. absorb gases inhaled through the nose
    D. deliver O_2 and CO_2 to and from the cells

Analysis and Open Ended

21. Explain why plants are defined as *autotrophs* and why animals are defined as *heterotrophs*.

22. Why might the process of photosynthesis be considered a "bridge" between the living and nonliving parts of the world?

23. Briefly describe three ways in which the structures of a leaf enable the process of photosynthesis to occur. Your answer should include the following factors: (a) light; (b) water; and (c) gases.

Refer to the chemical equation below to answer questions 24 and 25.

$$CO_2 + H_2O \xrightarrow[\text{CHLOROPHYLL}]{\text{LIGHT ENERGY}} C_6H_{12}O_6 + O_2$$

24. What important life process is described by this equation? What are the two products of this reaction?

25. Explain why "cellular respiration is basically the opposite" of the process shown in the equation. What are the two (waste) products of cellular respiration?

Base your answers to questions 26 to 28 on the information and diagram below and on your knowledge of biology.

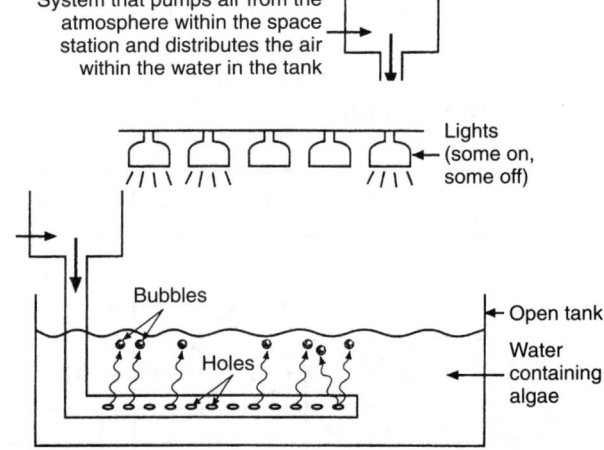

The diagram represents a system in a space station that includes a tank containing algae. An astronaut from a spaceship boards the space station.

26. Identify *one* process that is being controlled in the setup shown in the diagram.

27. State *two* changes in the chemical composition of the space station atmosphere that would result from the astronaut boarding the station.

28. State *two* changes in the chemical composition of the space station atmosphere that would result from turning on more lights in the algae setup.

Base your answers to questions 29 and 30 on the information and diagram below.

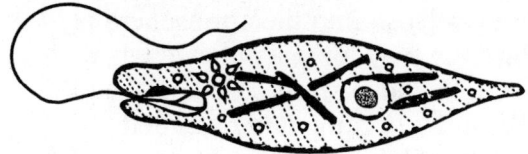

The diagram represents a single-celled organism known as *Euglena*. This organism is able to carry out both photosynthesis and cellular respiration.

29. Choose *one* of the two processes that *Euglena* carries out. Write down the word for it; then use words or chemical symbols to summarize the reaction for the process you chose.

30. State *one* reason why the process you chose is essential for the survival of the *Euglena*.

Base your answers to questions 31 and 32 on the following word equations of two biological processes and on your knowledge of biology.

Photosynthesis:

carbon dioxide + water $\xrightarrow{\text{ENZYMES}}$ glucose + oxygen (+ water)

Respiration:

glucose + oxygen $\xrightarrow{\text{ENZYMES}}$ carbon dioxide + water

31. State *one* reason why *each* of the processes shown is important for living things.

32. Choose *one* of the two processes shown and identify the following: (a) the source of the energy in the process you chose; and (b) where the energy ends up at the end of that process.

33. Briefly explain the difference between the processes of "breathing" and "respiration."

34. Refer to the diagram below to explain the process of breathing. Your answer should include the following:

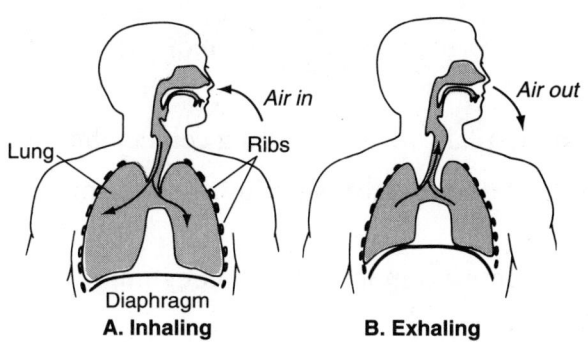

- what happens to the rib cage during inhalation and exhalation
- what happens to the diaphragm during inhalation and exhalation
- how these movements affect the size (volume) of the chest cavity
- how these movements affect the air pressure within the lungs

35. Describe the four characteristics required by an animal's respiratory surface.

Reading Comprehension

Base your answers to questions 36 to 38 on the information below and on your knowledge of biology. Use one or more complete sentences to answer each question.

> We walk on land. Even the very name Earth is used to mean land. But look at a world map and you will see a lot of blue space. In fact, more than 70 percent of Earth's surface is covered by water, mostly oceans. Unseen in these waters—drifting along with waves and currents—are countless numbers of tiny organisms. Photosynthetic bacteria, protists, and plants are included in these drifters. Some of these unicellular species are so small that if 12 million cells were lined up in a row, the line would be only about 1 centimeter long. In some places in the oceans, these microscopic organisms are so numerous that a cup of seawater may hold 24 million individuals of a single species, and that cup would contain other species as well!

These life-forms are very small, but their importance to the overall life on the planet is huge. Tiny sea-dwelling organisms are the beginning food source for almost all living things in the oceans; and the oxygen they release into the atmosphere is necessary for other organisms' cellular respiration. It is easy for us land dwellers to understand that many animals eat plants to get food. We have seen cattle and sheep grazing on grasses in a pasture. The drifting cells in the ocean could be called the *grass fields* or *pastures* of the sea. Just like grass on land, the sea drifters capture energy from the sun and convert inorganic CO_2 and water into organic molecules, which become important foods for other organisms. On land, plants bloom with wild displays of colorful flowers in spring. The photosynthetic drifters in the pastures of the seas are said to "bloom" in the spring, too, as the water warms and nutrients from ocean depths are brought to the surface by currents. A great deal has been learned recently about the seasonal explosive growth of these photosynthetic cells in the ocean from photographs taken by orbiting satellites.

36. Explain why the drifting cells in the ocean can be called the grass fields or pastures of the sea.

37. Describe three ways in which the microscopic drifting cells in the ocean are similar to plants on land.

38. How has modern technology improved our ability to study life in the ocean?

Chapter 7

Energy and Matter in Ecosystems

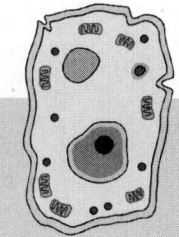

Standard 5.5.12 A3 Describe how plants produce substances high in energy content that becomes the primary source of energy for life.

THE BASIC CHARACTERISTICS OF ECOSYSTEMS

An ecosystem is made up of living and nonliving factors. In other words, *biotic factors,* such as plants and animals, and *abiotic factors,* such as water, air, and soil, function together in an ecosystem. For organisms to survive, there must be a continuous source of **energy**. The flow of energy between organisms and their environment is a basic characteristic of an ecosystem. Organisms are made up of matter. The flow of matter between organisms and their environment is another basic characteristic of an ecosystem.

What is the source of energy for almost all ecosystems on Earth? It is the sun. While energy is constantly reaching Earth from the sun, matter is not. The amount of matter on Earth remains constant. However, matter moves back and forth between organisms and the environment in all ecosystems.

ENERGY FLOW THROUGH ECOSYSTEMS

In most ecosystems, energy arrives as sunlight. Some organisms are able to use this energy directly. Other organisms use it indirectly—they get their energy by feeding on other organisms. Ecologists describe the flow of energy between organisms in a system of feeding levels, called **trophic levels**. (See Figure 7-1 on page 88.)

On the first trophic level are organisms that use energy, such as sunlight, directly from the environment. These first-level organisms are called **producers**. Green plants and algae are producers, or autotrophs, because they use the process of photosynthesis to make their own food with water, carbon dioxide, and sunlight. Organisms that feed on producers are in the next trophic level; they are called **consumers**, or heterotrophs. A caterpillar that

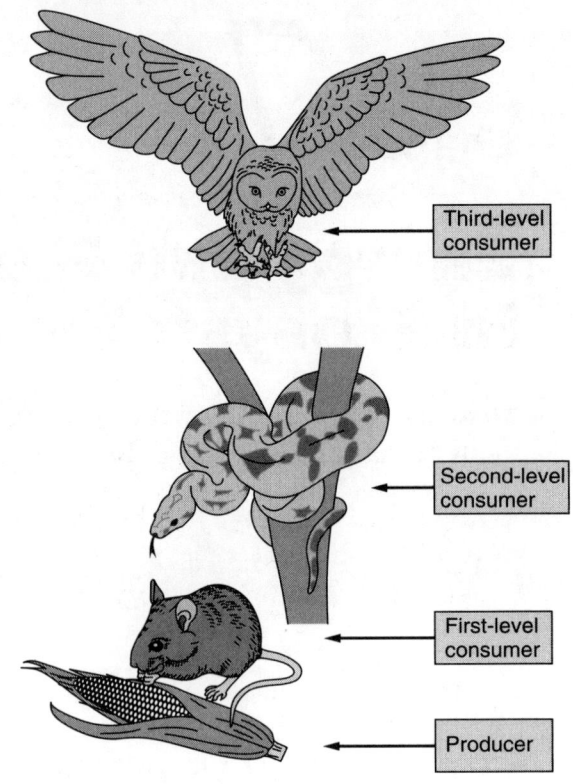

Figure 7-1 The trophic levels describe the flow of energy in an ecosystem, from the producer to the different levels of consumers.

eats oak leaves is a consumer. It is also a type of **herbivore**, because it feeds on plants. Additional levels exist in which consumers feed on consumers. For example, a small bird may eat a caterpillar. A large hawk may then eat the small bird. These animals are called **carnivores**, because they eat other animals. Each of these steps is called a trophic level because it describes the source of the organisms' food. We can describe the flow of energy in an ecosystem by using these trophic levels.

Food Chains and Food Webs

Energy enters an ecosystem at the producer level and is passed along from an organism in one trophic level to an organism in a higher trophic level. This transfer of food energy from one organism to the next is called a **food chain**. The path from oak leaf to caterpillar to small bird to hawk is a food chain. But a simple food chain like this is never found in a real ecosystem. Caterpillars are not the only animals that eat oak leaves, small birds are not the only animals that eat caterpillars, and so on. Food chains are actually interconnected in a complex pattern called a **food web**. In a food web, energy is passed between many different organisms. (See Figure 7-2.)

However, no matter how complex a food web is, energy always moves in one direction—from a lower to a higher trophic level. It does not get recycled. As energy moves through each trophic level, some of it is used for life processes and some of it is lost as heat. The most available energy is present at the lowest trophic level (producers); the least is present at the highest trophic level (upper level consumers). For this reason, additional energy must constantly enter an ecosystem, mainly in the form of sunlight.

Figure 7-2 Food chains actually interconnect in complex patterns to form a food web, in which the energy passes between many different organisms.

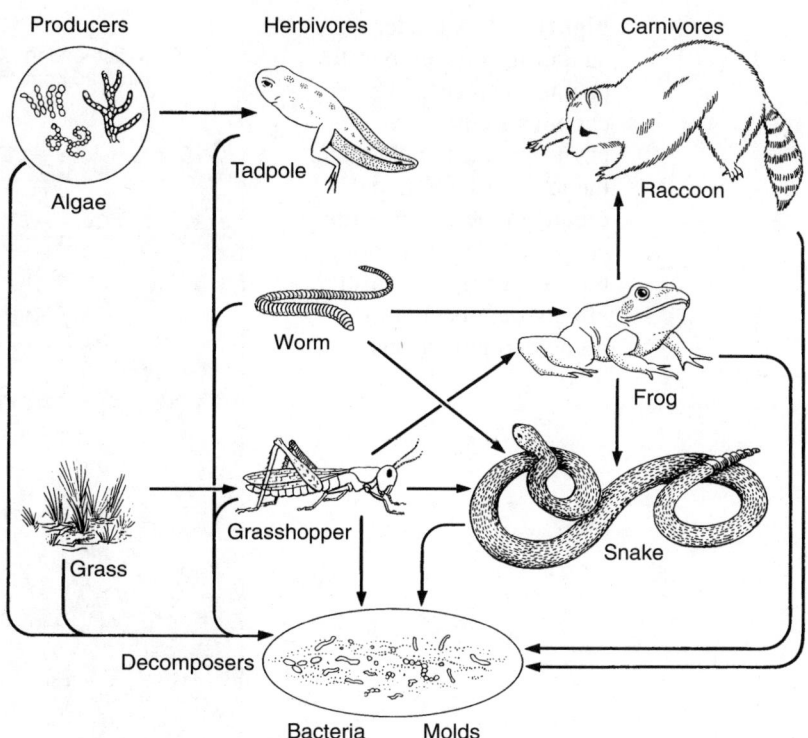

Unfortunately, there is a hidden danger in many food chains. If a long-lasting chemical such as DDT enters the environment, it may get passed on from one trophic level to the next. The level of the chemical in each organism increases as it is moves along the food chain. For example, little fish may contain some DDT, the larger fish even more, and finally the fish-eating birds the most. This effect is called **bioaccumulation**, or *biomagnification*. While harmless at very low levels, a chemical may have serious effects at higher levels. In fact, that is why DDT almost destroyed populations of fish-eating eagles and ospreys, before its use was banned in the United States. (See Figure 7-3.)

The Energy Pyramid

Ecologists use an **energy pyramid** to describe the flow of energy through an ecosystem. The wide base of the pyramid represents the amount of usable energy in the producers, that is, the energy from the sun that is stored in all of the plants. The next step up in the pyramid shows the energy that the first-level consumers get from the producers. This layer is smaller than the energy level for the producers because only about 10 percent of all energy gets passed from one trophic level up to the next. This is true as we move up the pyramid, from each level of consumers to the next and then to the top level. (See Figure 7-4 on page 90.)

An energy pyramid can provide an important lesson in how to feed the ever-increasing human population. Throughout the world, much more food energy is present at the producer level (crops) than at the consumer level (livestock). People may have to make choices based on such questions as: Which type of food is more abundant and available for everyone? Which type of food makes a more efficient use of resources?

Chapter 7 / Energy and Matter in Ecosystems

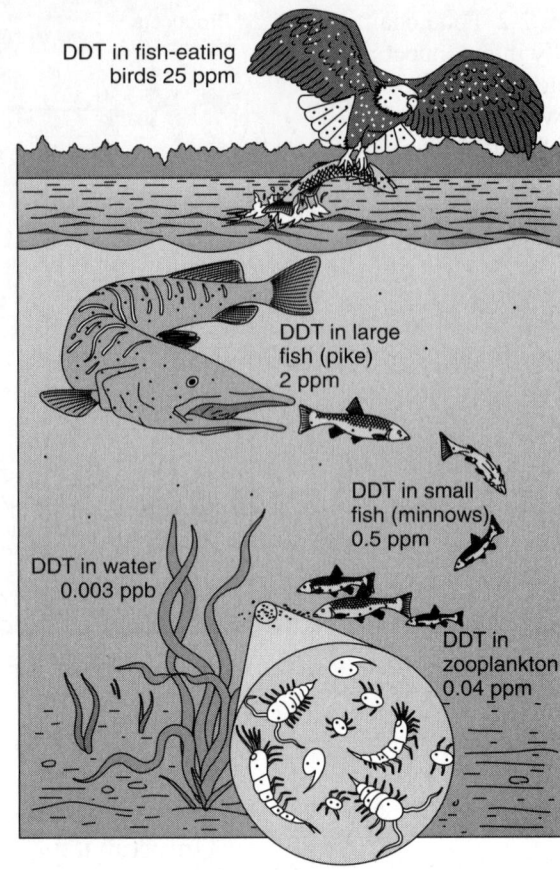

Figure 7-3 A hidden danger may develop in food chains when certain chemicals enter the environment. For example, the level of DDT in each organism increases as the chemical moves up along the food chain. Harmful effects can occur at the highest trophic levels.

THE RECYCLING OF MATERIALS IN ECOSYSTEMS

In many parts of the United States, people are now required to recycle consumer wastes such as glass and plastic bottles, newspapers, and metal. Recycling, although a new idea for people, is not a new idea in nature. Natural ecosystems have recycled materials since life began on Earth. In fact, life could not continue without this recycling of matter.

All substances are made up of chemical elements. Of the dozens of elements that occur naturally, only a few are found in significant amounts in organisms. These include **carbon, hydrogen, oxygen, nitrogen**, phosphorus, and sulfur. The amount of these elements on Earth today is about the same

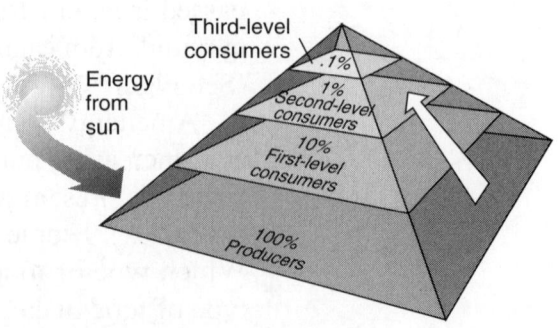

Figure 7-4 This pyramid of energy shows that the amount of available energy decreases at each higher trophic level because only 10 percent of all energy gets passed from one level to the next.

as when the planet formed. Because they are needed by living things and their supply does not increase, these elements have to be recycled again and again.

The Carbon and Oxygen Cycles

How do these elements get recycled in nature? Let us first look at carbon, since all organisms are made of molecules that contain this element. The carbon in organic molecules is obtained from CO_2 in the air. Producers such as trees and grasses take in CO_2 from the air during photosynthesis. They use the carbon from the CO_2 gas to build their carbohydrates (sugars and starches). Consumers obtain carbon from producers and from other consumers that serve as food. The plants and animals then return carbon to the atmosphere when they release CO_2 through respiration, thus completing carbon's recycling.

Recycling of carbon also occurs after a plant or an animal dies. This important part of the recycling process occurs through the actions of **decomposers**, mainly bacteria and fungi. Decomposers are *heterotrophs*, organisms that are unable to make their own food. They get their nutrients by feeding on the **residue**, or remains, of dead organisms. As they carry out their life processes, decomposers also release CO_2 into the atmosphere. Animals that feed on, or **scavenge**, remains help recycle the carbon, too. (See Figure 7-5.)

Oxygen is also recycled between living things. Animals and other organisms need oxygen for respiration—the process that releases the chemical energy stored in food. Land animals obtain oxygen for respiration from the air they breathe. Aquatic animals like fish use the oxygen that is dissolved in the water they live in.

Almost all the oxygen in Earth's atmosphere originally came from the metabolic activities of plants. During photosynthesis, plants and algae give off oxygen as a product. Animals breathe in the oxygen given off by plants, just as plants take in the CO_2 released by animals and decomposer organisms. This is natural recycling.

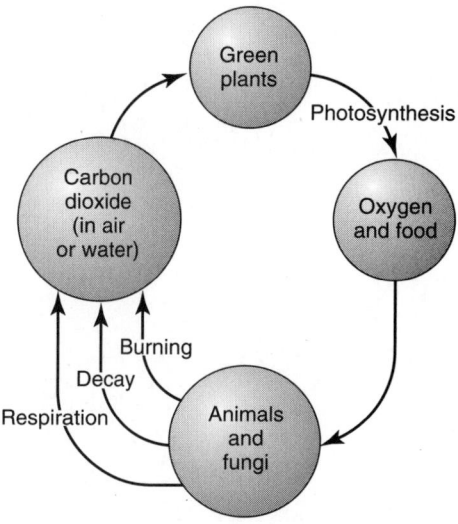

Figure 7-5 Carbon and oxygen are recycled between living things (and the environment) through such processes as photosynthesis, respiration, and decay. Producers, consumers, and decomposers are all part of this natural recycling process.

Nitrogen, a gas that is common in the atmosphere, is another essential element for all living things. It is combined and recombined through a complex cycle between biotic and abiotic parts of the environment. Nitrogen-fixing bacteria, present in the roots of such plants as peanuts, are necessary for transforming the atmospheric nitrogen gas into a form that plants can use. Phosphorus is another important element that is recycled by natural processes; it enters the soil, where it can be taken in for use by plants.

CHANGE AND STABILITY: THE IMPORTANCE OF BIODIVERSITY

The tendency of an ecosystem to resist change and remain the same is known as **stability**. Ecologists have important questions to ask about the stability of ecosystems. Do entire communities in an ecosystem stay the same? What causes a particular community to change? Is the number of species that make up a community critical to its stability? The amount of variety in a community is called *species diversity,* or **biodiversity**. A community with only a few species of plants and animals has low biodiversity. A community with many species has great biodiversity. A tropical rain-forest community, for example, may have the greatest biodiversity of any land community on Earth. Ecosystems that are high in biodiversity are thought to be more stable, since there are more species—of producers, consumers, and decomposers—fulfilling a variety of roles that add to the overall stability. However, the removal of any one species from the environment may still have a very negative effect on the rest of that ecosystem. (See Figure 7-6.)

Figure 7-6 Tropical rain forests, such as this one in Central America, may have the greatest biodiversity of any ecosystem on Earth, containing millions of species of plants and animals.

Biodiversity is a major concern of ecologists today. Many species that were present on Earth just a few decades ago are now extinct. For the most part, human actions, such as overhunting and habitat destruction, have caused these extinctions. As species disappear, biodiversity decreases. Scientists are concerned about the effects of decreased biodiversity on the health of ecosystems.

Does an ecosystem need a certain number of species interacting with each other to remain healthy? How many species can a community lose without being harmed? For example, if several insect species in a forest were to die off, would the forest survive? Would the plants that the insects used to eat grow too quickly? Would bird populations suffer, with fewer insects to eat? Finally, how much loss of biodiversity can occur before Earth's ecosystems stop functioning properly? Greater biodiversity increases the chances that, in case some environmental catastrophe occurs, at least some organisms would survive to reestablish a healthy community. This is a serious concern for people and for all species on Earth. Studies that investigate biodiversity and stability in specific communities are being conducted to try to answer important questions such as these.

HABITAT DESTRUCTION

There is one main reason why biodiversity is decreasing. Many species are disappearing because of *habitat loss*. Humans are using, and changing, many places where wild organisms formerly lived. For example, in the Midwest, many fields contained low-lying areas that remained filled with water all

Figure 7-7 Habitat destruction threatens the biodiversity and stability of ecosystems. Part of this rain-forest habitat has been destroyed to make room for a banana plantation.

year. Many birds, such as ducks and geese, were able to rest and find food in these ponds during their migrations each spring and fall. However, the farmers could not grow wheat and corn in these wet places. So the farmers filled in the ponds. This was a critical loss of habitat for the migrating water birds. The populations of the ducks and geese decreased; and biodiversity was reduced.

This is only one example of the loss of a habitat affecting biodiversity. Habitat and species loss have occurred in many other places, such as on a river where a dam is built. Fish that survive only in moving water die in the still water of a lake that is formed behind a dam. Today, the greatest habitat destruction is occurring in the world's tropical rain forests. It is estimated that most of Earth's biodiversity will be lost if the rain forests are destroyed. Sadly, this is happening while scientists are trying to identify and classify the many organisms still being discovered in these forests. In addition, researchers fear that many tropical species, many of which may contain substances that could prove to be valuable medicines, are being lost forever before even being discovered (See Figure 7-7.)

Chapter 7 Review

Multiple Choice

1. A basic trait of ecosystems is that there must be a continuous input of additional
    A. energy   C. oxygen
    B. carbon   D. matter

2. The source of energy for most ecosystems is
    A. rain   C. flowing water
    B. wind   D. the sun

3. In an ecosystem, which component is *not* recycled?
    A. water   C. energy
    B. oxygen   D. carbon

4. The first trophic level consists of organisms that
    A. use energy to make their own food
    B. eat first-level consumers only
    C. eat producers and consumers
    D. add matter to the ecosystem

5. Organisms that eat plants are called both consumers and
    A. producers   C. carnivores
    B. herbivores   D. scavengers

6. What is always transferred in a food chain?
    A. toxins   C. water
    B. energy   D. oxygen

7. Which sequence indicates a correct flow of energy?
    A. herbivore → sun → carnivore
    B. sun → producer → herbivore
    C. producer → sun → carnivore
    D. carnivore → herbivore → sun

8. Which energy transfer is *least* likely to be found in nature?
    A. consumer to consumer
    B. host to parasite
    C. producer to consumer
    D. predator to prey

9. In most habitats, the removal of carnivores will have the most immediate effect on a population of
    A. producers   C. decomposers
    B. herbivores   D. microbes

10. Which group of terms are *all* directly associated with the larger fish in the following illustration?

    A. herbivore, prey, autotroph, host
    B. carnivore, predator, heterotroph, multicellular
    C. predator, scavenger, decomposer, consumer
    D. producer, parasite, fungus, fish

11. In a food web, energy always moves
    A. in a continuous cycle among trophic levels
    B. back and forth between two trophic levels
    C. from a lower to a higher trophic level only
    D. from a higher to a lower trophic level only

12. A student could best demonstrate knowledge of how energy flows throughout an ecosystem by
    A. labeling a diagram that illustrates ecological succession
    B. drawing a food web using specific organisms living in a pond
    C. conducting an experiment that demonstrates photosynthesis
    D. making a chart to show the role of bacteria in the environment

Base your answers to questions 13 and 14 on the diagram below and on your knowledge of biology.

13. Which organism carries out autotrophic nutrition?
    A. frog   C. grasses
    B. snake   D. grasshopper

14. The base of an energy pyramid for this ecosystem would be the
    A. frog
    B. snake
    C. grasses
    D. grasshopper

15. A food web is more stable than a food chain because a food web
    A. transfers all of the producer energy to herbivores
    B. includes alternative pathways for energy flow
    C. reduces the number of niches in the ecosystem
    D. includes more consumers than producers

16. Which trophic level contains the most available food energy?
    A. producers
    B. first-level consumers
    C. second-level consumers
    D. third-level consumers

17. The hidden danger in many food chains is that
    A. some prey items taste better and thus are eaten too often
    B. harmful chemicals can be passed from one level to another
    C. some foods become poisonous after being eaten too often
    D. the food chains interconnect to form enormous food webs

18. The diagram below represents a pyramid of energy in an ecosystem. Which level in the pyramid would most likely contain members of the plant kingdom?

    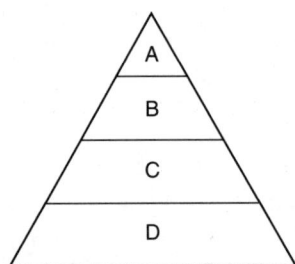

    A. level *A*
    B. level *B*
    C. level *C*
    D. level *D*

19. In an ecosystem, nutrients can be recycled if they are transferred from herbivores to carnivores to
    A. hosts
    B. decomposers
    C. prey
    D. autotrophs

20. Carbon is recycled in nature when
    A. consumers take in carbon dioxide and then release oxygen
    B. producers take in carbon dioxide and consumers release it
    C. decomposers take in carbon dioxide and release oxygen
    D. scavengers and decomposers take in carbon dioxide

21. Which statement best illustrates the recycling of materials in a self-sustaining ecosystem?
    A. In summer, growing plants remove magnesium from the soil to make chlorophyll. In autumn, these plants release magnesium when they die and decompose. In spring, new plants will grow in this same area.
    B. DDT is sprayed on a forest ecosystem to control the mosquito population. After a year, the level of DDT is found to be much higher in the tissues taken from a hawk than in the tissues taken from a mouse in this ecosystem.
    C. Trees do not live in a desert ecosystem where there is not enough water present in the sandy soil to support their growth. Trees can live in a desert oasis.
    D. Plants trap the sun's energy in the chemical bonds of organic molecules. This energy is then used for the plants' metabolic activities.

22. The organisms that help recycle elements by breaking down organic matter include
    A. grass and algae
    B. bacteria and algae
    C. bacteria and fungi
    D. plants and fungi

23. Vultures, which are classified as scavengers, are an important part of an ecosystem because they
    A. hunt herbivores, thus limiting their population size in an ecosystem
    B. cause decay of dead organisms, which releases usable energy to living organisms
    C. feed on dead animals, which aids in the recycling of environmental materials
    D. are the first level in food webs, making energy available to all other organisms

24. Oxygen is needed for respiration, the process that
    A. releases the chemical energy stored in food
    B. uses carbon dioxide to produce sugars
    C. breaks down the bodies of dead organisms
    D. releases oxygen as a waste into the air

25. The oxygen that humans breathe is actually
    A. a waste product of respiration
    B. a waste product of photosynthesis
    C. given off by decomposers
    D. produced within the sun's core

26. The tendency of an ecosystem to stay the same is called
    A. diversity
    B. resistance
    C. sterility
    D. stability

27. Which change in conditions would cause an ecosystem to become unstable?
    A. Only heterotrophic organisms remain after a change in the region.
    B. A variety of nonliving factors are used by the living organisms.
    C. A slight increase occurs in the number of heterotrophs and autotrophs.
    D. The biotic factors and abiotic resources interact more frequently.

28. The ecosystem that would have a better chance of surviving after environmental conditions change would be one with
    A. a great deal of genetic diversity
    B. very little or no genetic diversity
    C. plants and animals but no bacteria
    D. animals and bacteria but no plants

29. Unlike a desert, a tropical rain forest typically has
    A. low biodiversity
    B. great biodiversity
    C. a small variety of organisms
    D. a small number of organisms

30. The loss of biodiversity is often related to
    A. the search for medical cures
    B. too much rain in rain forests
    C. a loss of natural habitat
    D. evolution not occurring

Analysis and Open Ended

31. Briefly describe the two main components of any ecosystem.

32. How do producers differ from consumers?

33. Why is a food web a more accurate description than a food chain of interactions in a community?

34. Explain why the amount of available energy in trophic levels can be shown as a pyramid.

35. The following diagram represents a model of a food pyramid. Which statement best describes what happens in this food pyramid?

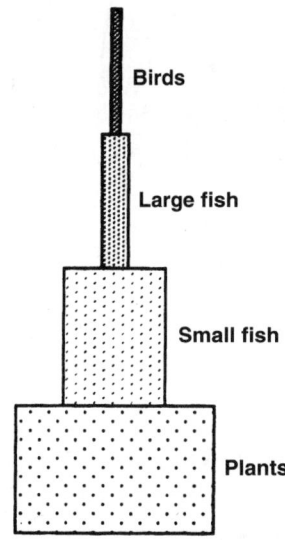

   A. More organisms die at higher levels than at lower levels, decreasing the mass at the top.
   B. When organisms die at higher levels, their remains sink, increasing the mass at lower levels.
   C. Energy is lost to the environment at each level, so less mass can be supported at each higher level.
   D. Organisms decay at each level, so less mass can be supported at each higher level.

36. According to the food chain shown in Figure 7-1 on page 88, a third-level consumer is one that
    A. is capable of photosynthesis
    B. feeds directly on producers
    C. feeds on first-level consumers only
    D. feeds on second-level consumers

Refer to the diagram below to answer questions 37 to 40.

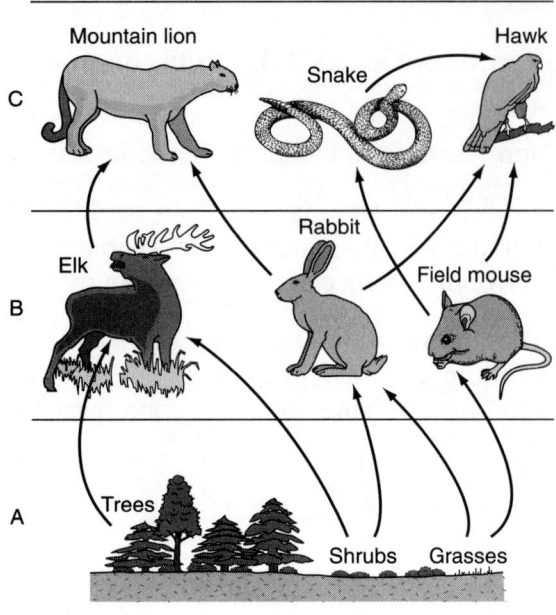

37. Which pattern does the diagram illustrate?
    A. a food chain
    B. a food web
    C. a food cube
    D. a food pyramid

38. The organisms shown in level *B* are classified as both
    A. producers and prey
    B. consumers and prey
    C. scavengers and predators
    D. decomposers and prey

39. Which populations would contain the greatest amount of available energy?
    A. rabbits and field mice
    B. hawks and rabbits
    C. trees and shrubs
    D. hawks and snakes

40. All of these organisms living together make up a natural
    A. population
    B. species
    C. community
    D. consumer

41. Which statement about the producers in the following marine food web is correct?

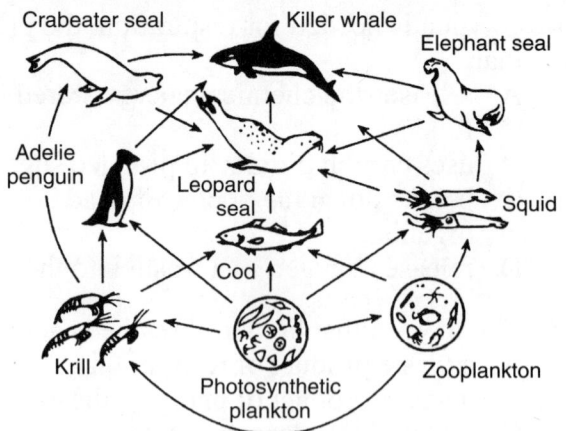

   A. An increase in the producers would most likely decrease available energy for the squid.
   B. If all the producers were destroyed, the number of heterotrophs would increase and the ecosystem would become stable.
   C. The three different species of seals are the most important producers in this ecosystem.
   D. There is only one group of producers, so they must be numerous enough to supply the energy needed to support the food web.

42. What problem can occur in a food chain when a chemical pollutant enters the ecosystem?

43. According to Figure 7-4 on page 90, as energy moves up each trophic level in an ecosystem, the amount of it that is available becomes
    A. 10 percent more than it was before
    B. 10 percent of what it was before
    C. 50 percent more than it was before
    D. 50 percent of what it was before

Base your answers to questions 44 to 46 on the passage below and on your knowledge of biology.

In nature, energy flows in only one direction. Transfer of energy must occur in an ecosystem because all life needs energy to live, and only certain organisms can change solar energy into chemical energy.

Producers are eaten by consumers, which are, in turn, eaten by other consumers. Stable ecosystems must contain a variety of predators to help control the populations of consumers. Without the population control provided by predators, some organisms would soon overpopulate.

44. Draw an energy pyramid that illustrates the sentence, "Producers are eaten by consumers, which are, in turn, eaten by other consumers." Include *three* different, specific organisms in your energy pyramid.

45. Explain the phrase "only certain organisms can change solar energy into chemical energy," which appears in the first paragraph. In your answer be sure to identify: (a) the type of organisms being described in the statement; (b) the type of nutrition carried out by these organisms; and (c) the process used to carry out this type of nutrition.

46. Explain why an ecosystem with a variety of predator species might be more stable over time than an ecosystem with only one predator species.

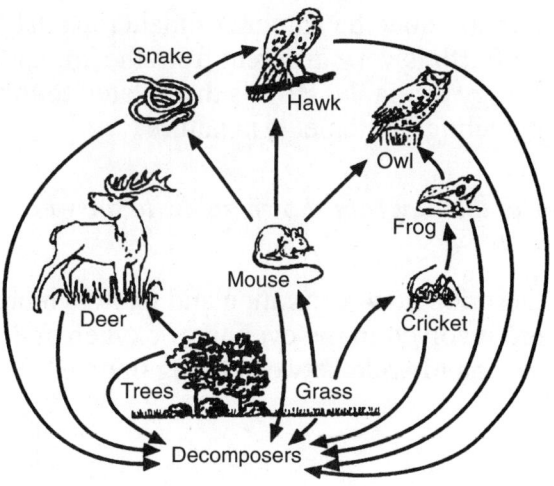

Refer to the diagram of a food pyramid below to answer questions 47 and 48.

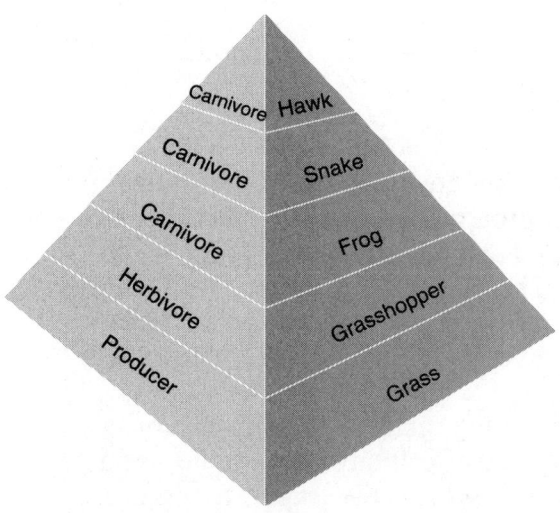

47. The consumer level that would have the largest amount of stored energy is that of the
    A. hawk   C. frog
    B. snake   D. grasshopper

48. The trophic level that has the smallest amount of stored energy would be that of the
    A. top carnivore   C. herbivore
    B. middle carnivore D. producer

Base your answers to questions 49 to 51 on the following diagram of a food web and on your knowledge of biology.

49. If the population of mice were reduced by disease, which change will most likely occur in the food web?
    A. The cricket population will increase.
    B. The snake population will decrease.
    C. The trees and grasses will decrease.
    D. The deer population will decrease.

50. What is the original source of energy for this food web?
    A. enzymatic reactions
    B. chemicals in sugar molecules
    C. energy from sunlight
    D. chemical reactions of bacteria

51. State *one* example of a predator-prey relationship shown in the food web. Indicate which organism is the predator and which is the prey.

52. Use the following terms to complete the boxes in the flowchart below: *decomposers; producers; carnivores; herbivores.*

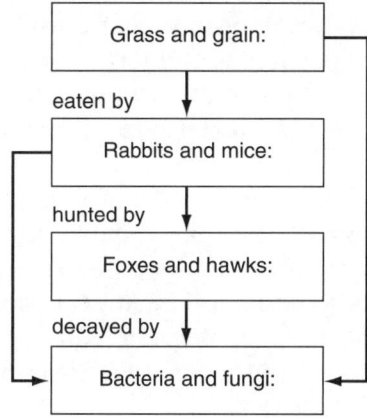

Chapter 7/Energy and Matter in Ecosystems **99**

53. Explain how the amount of matter available on Earth is very different from the amount of energy available. How is this related to the recycling of elements in nature?

Refer to the diagram shown below to answer question 54.

54. Describe how respiration and photosynthesis are involved in the cycling of oxygen and carbon dioxide between living things.

55. Why are ecologists concerned about the number of species in ecosystems? Why would most agree with the statement: "A forest ecosystem is more stable than a cornfield."

56. In what ways are humans responsible for the current decrease in biodiversity?

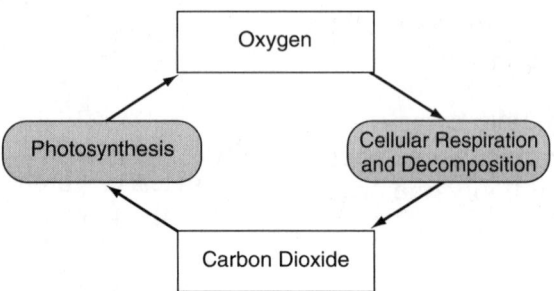

Reading Comprehension

Base your answers to questions 57 to 60 on the information below and on your knowledge of biology. Use one or more complete sentences to answer each question.

The Everglades is a vast, wide freshwater marsh that covers much of the southern part of Florida. The Everglades begins at the northern edge of Lake Okeechobee, with the overflow of rainwater out of the lake, and extends all the way to the southern tip of the state just before the Florida Keys. The vast majority of the Everglades is covered by a dense growth of saw grass. A slow, steady flow of water moves through the saw grass from north to south. The Everglades is, therefore, also called a "River of Grass."

In 1996, the federal government endorsed the Everglades restoration project. The project is one of the largest ecological restoration efforts anywhere in the world. Hundreds of millions of dollars are being spent to protect the fragile Everglades ecosystem. Included in this project is the plan to stop sugarcane production in 100,000 acres of ecologically sensitive farmlands.

Much of the water that flows through the Everglades has become contaminated by pesticides and fertilizers, which are used to increase crop yields on farms in the area. One of the main goals of the restoration project is to let large areas of land act as natural water filters to remove some of the waterborne contaminants.

Another important part of the Everglades project will restore the natural north-south flow of water. The natural pattern of water flow through the Everglades was disrupted by the canals, pumping stations, and water-control structures that were built to create flood-control and water-supply systems for southern Florida. In fact, these unnatural attempts to control the Everglades' water flow have been harmful to the entire ecosystem. Today, planning is under way to find alternatives that can meet flood-control and water-supply needs while ensuring the long-term health of the Everglades.

57. Why is the Everglades also known as a "River of Grass?"

58. In two or more complete sentences, discuss the types of human activities that have had harmful effects on the Everglades.

59. Describe some steps that are being taken to undo the damage to the Everglades and restore it to its original condition.

60. Based on the information in this essay, what change has occurred in peoples' attitudes toward this vast marsh?

Chapter 8

Organisms and Disease

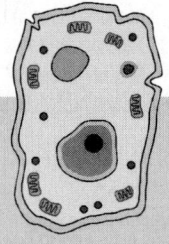

Standard 5.5.12 A4 Relate disease in humans and other organisms to infections or intrinsic failures of system.

DISEASE: A LACK OF HOMEOSTASIS

To stay healthy, organisms need to maintain homeostasis; that is, they must have a carefully controlled set of internal conditions. Maintaining these conditions—including pH, temperature, water and salt balance, and levels of carbon dioxide and oxygen—allows an organism's cells to function normally. Organisms can tolerate changes that occur within very definite limits. But changes outside normal limits disrupt homeostasis, producing illness, **disease**, and even death.

There are many reasons why the body may be pushed beyond its normal limits. These reasons, or *factors,* are often the causes of disease. An inherited defect in a genetic trait might be the cause of a disease. Thus, the disruption of homeostasis in such a disease would be caused by a factor inside the body. Many other diseases result from some influence outside the body, that is, from the environment.

Factors That Cause Disease

Diseases may be caused by one of the following factors, or by a combination of several of these:

- *Inheritance.* Defective genetic traits can be passed from parents to offspring. Often, neither parent has or expresses the disease, but both may carry a single form (recessive allele) of the gene for the same disease. It is the combination of these two recessive genes in the child that gives him or her the disease. A well-known example of an inherited disease is sickle-cell anemia, in which the protein that carries oxygen in red blood cells is flawed. (See Figure 8-1.)

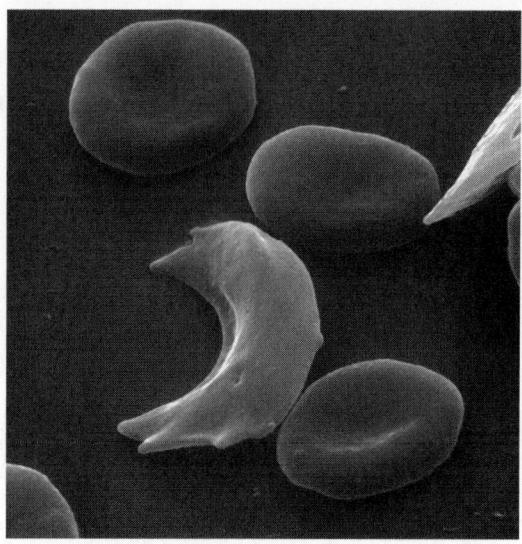

Figure 8-1 Sickle-cell anemia is an inherited disease that disrupts homeostasis because the protein that carries oxygen in red blood cells is flawed.

- *Microorganisms.* Microscopic organisms that cause diseases are called **pathogens**; they include certain fungi, bacteria, viruses, and protozoa. Some diseases caused by pathogenic microorganisms may be passed in a variety of ways from one person to another. These are called **infectious diseases**. Microorganisms most often enter the body through respiratory pathways, the digestive system, or pathways of the excretory system. Infections may also occur through breaks in the skin. Tuberculosis is an infectious disease caused by certain bacteria. (See Figure 8-2.)

- *Pollutants and poisons.* Chemical agents present in the environment may upset the body's normal functioning and produce disease. These pollutants and poisons include coal dust, asbestos, lead, phosphorus, mercury, and many other substances or elements. For example, when asbestos fibers enter the respiratory system, they cause asbestosis, a disease of the lungs; years later, this may result in cancer in the lungs and chest.

- *Organ malfunction.* A disease may develop when one or more of the body's organs **malfunction**. When an organ such as the liver, lung, heart, stomach, or kidney does not function properly, vital life processes cannot be carried out well and serious effects on the body result.

- *Harmful lifestyles.* The way one lives can also be an important factor in causing disease. Specifically, tobacco, alcohol, and drugs in the body can

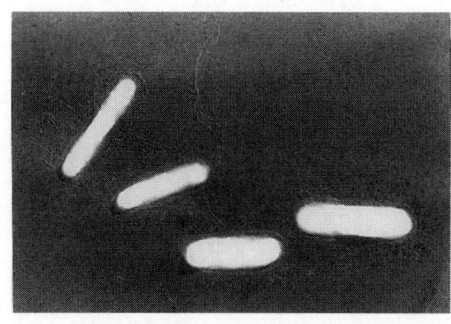

Figure 8-2 These bacteria are an example of microorganisms that can cause infectious diseases, such as tuberculosis.

Chapter 8 / Organisms and Disease

disrupt homeostasis, producing illness. In addition, overeating, poor nutrition, not exercising, having unsafe sexual practices, and living with stress can lead to certain diseases. Hypertension, or high blood pressure, is one such disease.

DISEASE-CAUSING MICROORGANISMS

Viruses

A **virus** is a microscopic nonliving particle that consists of an outer protein coat, called a *capsid,* and a molecule of nucleic acid, either DNA or RNA. Scientists do not actually consider a virus a living organism because it cannot reproduce on its own. Viruses can reproduce only when they are inside another living cell, which is called the **host** cell.

Viruses can be classified based on the type of cell they infect. Animal viruses infect animal cells; plant viruses infect plant cells; and **bacteriophages** (meaning "bacteria eaters") are viruses that infect bacterial cells. The specific proteins of the protein coat on a virus determine which cells the viruses can infect. The host cell must have a specific receptor site on its membrane for the protein of the virus to bind to it. For example, a virus that infects an eye cell cannot infect a muscle cell, even of the same organism. (See Figure 8-3.)

A **virulent virus** is a virus that has the ability to cause disease. The life cycle of a virus depends on the type of virus it is. There are two types, or stages, of viruses: lytic viruses and lysogenic viruses. A *lytic virus* infects the host cell and uses the host cell's proteins and genetic material to make new viral particles. When the new virus particles reach a certain number, the host cell lyses (breaks open) and the new viruses are released. A *lysogenic virus* infects the host cell and combines its own DNA with the host cell's DNA. Each time the DNA of the host cell is replicated, the viral DNA is replicated too. Thus the replicated DNA consists of DNA from two different organisms: the host cell and the virus. A lysogenic virus can become lytic under certain conditions that cause stress to the host cell.

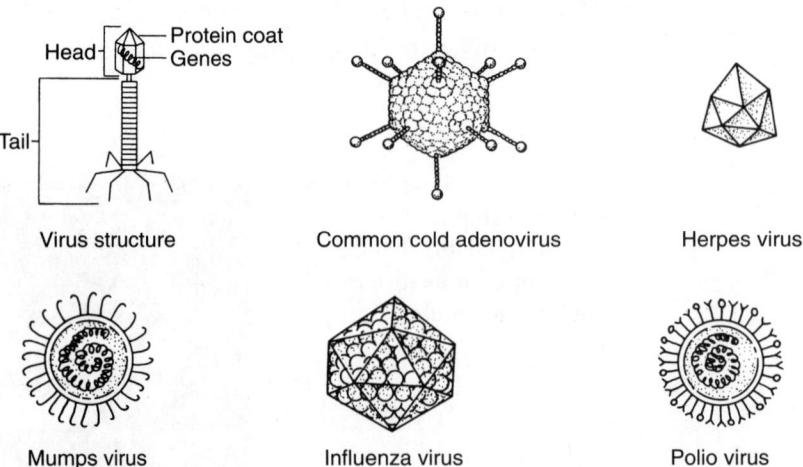

Figure 8-3 Typical structure of a virus, and some different types of viruses.

Viral infections are difficult to treat with drugs. Viruses can spread from a person or animal to another person. Direct contact with an infected person can easily spread some viruses to an uninfected person. Animals can also transmit viruses, such as rabies, to a human if the animal bites or scratches the person. Some viruses can be transmitted by indirect contact, too. For example, if you touch a doorknob that was touched by a person with a viral disease, you may pick up the virus and become infected if you then touch your mouth or nose.

Viruses are easily spread through the air. A person with the common cold can spread the virus when he or she sneezes or coughs. Droplets are released into the air, and an individual nearby may pick up the virus if it makes contact with that person's mouth or nose. Some viruses can travel through the air in particles, and other people can pick up the virus when they breathe in the contaminated air. For example, chicken pox and SARS are examples of viral diseases that can spread through particles in the air.

Certain viruses, such as hepatitis A, can be spread through food and water. If food is handled by someone infected with that virus, the food can become contaminated. Insects can transmit viruses from an animal to a person. For example, some mosquitoes carry the West Nile virus. If a mosquito picks up that virus when it bites an infected bird, and then bites another animal or human, the virus in the mosquito's salivary glands is transmitted to the new host.

Treatment for viral infections is limited, because antibiotics are not effective against viral agents. The best medicine is lots of rest, pain medicine, and fluids. Prevention of many viral diseases is possible if proper sanitation and hygiene are practiced. Food- and water-testing can detect some dangerous viruses, such as polio and rotavirus. Measures to reduce certain insect populations can decrease the spread of viruses. For example, viruses that cause yellow fever and dengue fever can be controlled if exposure to mosquitoes is reduced. In addition, lab technicians and scientists who work with dangerous viruses should wear protective clothing.

Vaccination is a method used to prevent many viral diseases. A *vaccine* is made from a killed or weakened microorganism. When a vaccine is given to a person, the body's white blood cells produce *antibodies* that act against the specific microorganism. People receive vaccines to protect against viral diseases, such as polio and influenza (the flu), and against some bacterial diseases, such as tetanus, too.

Bacteria

Bacteria are microorganisms that have a cell membrane and a cell wall. Surrounding the cell wall of some bacterial species is a protective layer known as a *capsule*. Bacteria are identified by their shapes, which exist in four different forms: *bacilli* are rod-shaped; *cocci* are round, or sphere- shaped; *spirilla* are spiral-shaped; and *vibrios* are curved rod-shaped. (See Figure 8-4.)

Bacteria have many beneficial uses. In the human intestines, bacteria produce essential vitamins and also help break down food molecules. However, many species of bacteria cause infection and disease in other organisms. *Pneumococcus,* a spherical bacterium, causes bacterial pneumonia.

Figure 8-4 Three of the characteristic shapes of bacteria.

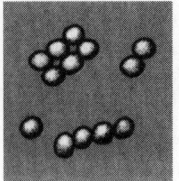 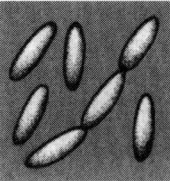 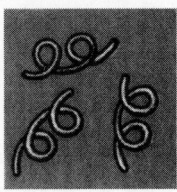

Ball-shaped (cocci) **Rod-shaped** (bacilli) **Corkscrew-shaped** (spirilla)

Streptococcus, another spherical bacterium, can cause strep throat. *Mycobacterium tuberculosis* is the rod-shaped bacterium that causes tuberculosis. Bacteria in the genus *Neisseria* cause gonorrhea. *Treponema pallidum* is a spiral bacterium that causes syphilis.

Antibiotics are used to kill bacteria and prevent bacterial infections from developing in people and animals. Some antibiotics work by interfering with the protein synthesis ability of the bacterial cells. Bacteria need water, nutrients, and a proper temperature to grow and reproduce. To control bacterial growth in food, water can be removed, salt and other chemical preservatives can be added, and food can be stored at colder temperatures that stop, or *inhibit,* bacterial growth. Dried salami, for example, can be stored for months without any spoilage. Food can be boiled to kill most bacterial species, before being stored in sealed cans. Chemical preservatives are usually added to breads, cakes, juices, and sweets to prevent them from spoiling due to bacterial growth. Acids are added to pickled foods to prevent spoilage. Ultraviolet radiation kills most bacteria; and the addition of oxygen can actually kill some species of bacteria. *Sterilization* is the process of subjecting materials to temperatures above 212°F to kill any bacteria present.

Protozoa

Protozoans are unicellular organisms found in many different habitats. The protozoa are divided into four phyla, based mainly on the way the organisms move; these include the Zooflagellates, Sarcodina, Sporozoa, and Ciliophora. Within these phyla are many species that live as **parasites** on or within the cells of other organisms, causing a variety of illnesses. (See Figure 8-5.)

The *Zooflagellates* are protozoa that move by means of one or more flagella (long whiplike tails). Some are parasitic; species in the zooflagellate genus *Trypanosoma* cause African sleeping sickness, known as *trypanosomiasis.* The trypanosome is transmitted to humans by the bite of an infected tsetse fly. The *Sarcodina* are protozoa that move and obtain food by means of pseudopodia (extensions of the cytoplasm). Within the sarcodine genus *Amoeba,* the species *Entamoeba histolytica* causes the intestinal disorder *amebic dysentery* in humans. The *Sporozoa* are parasitic protozoa that reproduce by the formation of spores; some have a complex life cycle that depends on different hosts. For example, the species *Plasmodium vivaxi,* which causes the disease malaria in humans, requires two hosts in its life cycle. This organism is transmitted to humans by the *Anopheles* mosquito. When a mosquito bites an infected human, the parasites enter the mosquito's stomach and reproduce; when the mosquito bites another human, the parasites enter that person's bloodstream. The *Ciliophora* are protozoa that move and

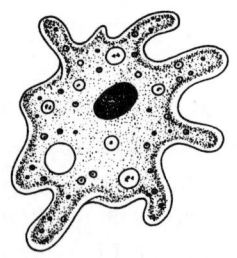

Ameba

Figure 8-5 An example of a disease-causing protozoan, the ameba.

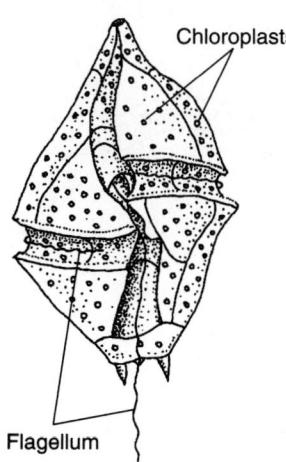

Figure 8-6 A dinoflagellate; these algae can release a toxic chemical.

obtain food by means of cilia (short hairlike structures). This phylum also includes some parasitic species.

Algae: The Dinoflagellates

The *dinoflagellates* are unicellular algae in the phylum *Pyrrophyta*. They have thick walls made of cellulose and two flagella, which they use to spin through the water. Most species reproduce asexually. The explosive reproduction of some species of red-pigmented dinoflagellates causes a phenomenon known as *red tide*. These algae also release a toxic chemical, which becomes concentrated in shellfish, poisoning the shellfish and other marine animals. People who eat shellfish contaminated by the red tide become seriously ill and can even die. (See Figure 8-6.)

Fungi

Some fungal species cause disease in plants and animals. Parasitic plant fungi, such as rusts and smuts, destroy large areas of crop. Ergot, a disease of rye, is caused by a fungus in the phylum *Ascomycota*. Humans who eat rye infected with this fungus experience severe spasms and convulsions. Fungi in the phylum *Deuteromycota* cause the diseases ringworm and athlete's foot in humans.

OTHER DISEASE-CAUSING ORGANISMS

Parasitic Worms

There are three phyla of worms: Platyhelminthes, Nematoda, and Annelida. Within the phylum *Platyhelminthe,* or "flatworms," there are three classes, two of which include disease-causing species. The parasitic flatworms called **flukes** are in class *Trematoda*. Most flukes have two suckers that enable them to attach to their host cells; some flukes have a *cuticle* that protects them from the host organism. Flukes have a pharynx that they use to suck in fluids from the host. Many have several intermediate hosts in their life cycles. For example, the blood fluke, *Schistosoma,* has two hosts, snails and humans. The blood fluke enters a human through the skin and produces many eggs, which are released from the body with its waste products. If the eggs are released into water, they develop into larvae that can enter snails. The larvae reproduce asexually in the snails and are released back into the water where other humans can pick them up through their skin. Many people suffer from intestinal, urinary, and blood infections caused by these flukes. The parasitic flatworms called **tapeworms** are in class *Cestoda*. Tapeworms have suckers and hooks on their head region, or *scolex,* which enable them to attach to the host cells. They lack a mouth and digestive system; instead they absorb nutrients from the host directly through their body. The beef tapeworm is an example of a tapeworm that has two hosts, cattle and humans. People become infected, usually in the intestine, when they eat infected beef that has been undercooked.

Figure 8-7 Examples of different parasitic worms: the tapeworm and liver fluke are flatworms; the hookworm is a type of roundworm.

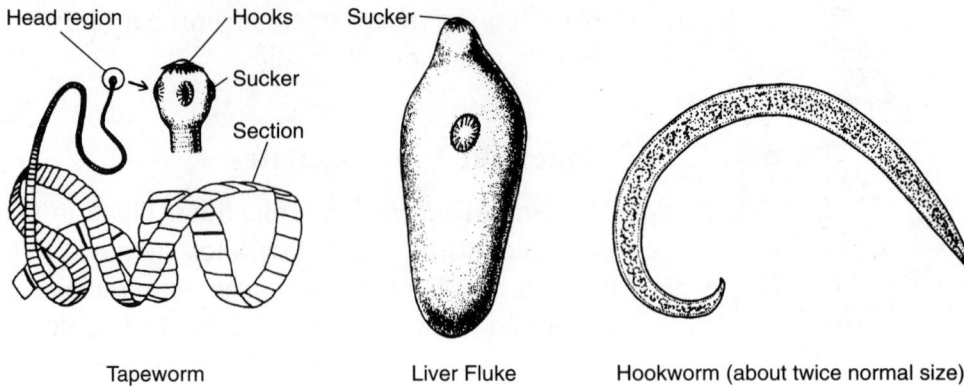

The phylum *Nematoda* includes the *roundworms,* which live in soil and water. Some species of roundworms are parasitic and cause the diseases trichinosis and hookworm. The roundworm that causes *trichinosis* lives in the intestines and muscles of humans and pigs. The worms form cysts in the pig's muscles. When a person eats infected pork without properly cooking it, the cysts develop into worms that reproduce and infect the person's muscles, causing great pain. *Ascaris* roundworms can enter a human as eggs in contaminated food or water. Later, the adult worms infect the person's intestine. The *hookworms* have well-developed hooks in their mouth. When in its larval form, the hookworm penetrates human skin, travels through the blood to the lungs, travels back out of the lungs to the throat, passes down into the intestines, and then attaches to the intestinal tissue. The hookworm sucks blood from its human host and, as a result, can cause anemia. Its fertilized eggs are released in human wastes, and the larvae can develop and start a new cycle. (See Figure 8-7.)

The phylum *Annelida,* or "segmented" worms, includes the leeches, which are parasitic. Leeches usually live among plants in ponds and streams, where they feed on the blood of other animals. The leech has two suckers that it uses to attach to the host, and three sharp teeth for biting. The chemicals in the saliva of leeches—which help the host's blood flow to the leech during feeding—are now being used by doctors to destroy blood clots.

THE BODY'S DEFENSES AGAINST DISEASE

Our bodies are surrounded by microorganisms, or **microbes**, trying to get into us. Some of them succeed, through the nose, through cuts in our skin, or along with the food we eat. Many of these microbes can cause serious problems if they survive and reproduce inside us without challenge. Controlling these microscopic invaders is as important to homeostasis as is regulating body temperature and chemistry.

The first line of defense against infection consists of *physical barriers* that block the entry of microorganisms. The skin is the main physical barrier in our body. A second line of defense, called **inflammation**, is present when

Figure 8-8 Inflammation is the body's second line of defense against infectious bacteria that get through the skin's first (physical) defense.

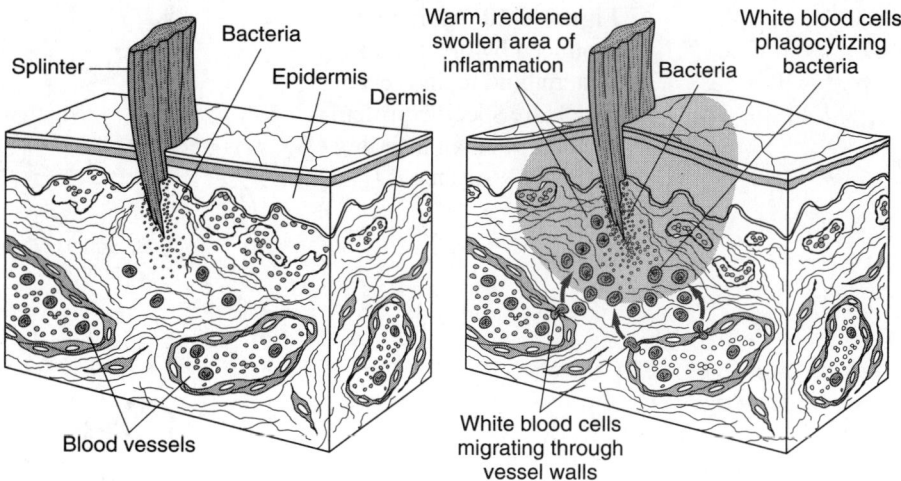

microorganisms get through our physical barriers. For example, when we get a cut or scrape on the skin, the injured area may become warm, reddened, and perhaps swollen with pus. (See Figure 8-8.) Chemicals released by the damaged tissues act like an alarm, causing an increase in blood flow to the site of the injury. Special **white blood cells** that arrive engulf the microbes, destroying them by ingesting them. All of this activity helps prevent a more serious infection from developing. (See Figure 8-9.)

Vertebrates have evolved a very important system that attacks specific invaders. This is the **immune system**. The immune system recognizes who the "bad guys" are and goes after these invaders to try to keep them from disrupting normal body functions.

THE HUMAN IMMUNE SYSTEM

The immune system defends our bodies against very specific microscopic invaders. Each invader—usually a bacterium or virus—has specific protein molecules attached to its surface. These protein molecules are called **antigens**. It is these molecules that are detected by the body's immune system.

When the immune system detects an antigen, it produces **antibodies**, molecules that bind to that particular antigen. Once the antibodies bind to the antigen, it can be destroyed by the body. (See Figure 8-10.) As mentioned previously, *vaccinations* use weakened microorganisms, or parts of them, to

Figure 8-9 During inflammation, the white blood cells engulf and destroy the invading bacteria, thereby preventing a more serious infection.

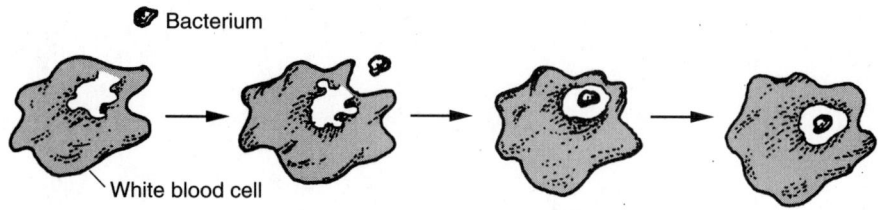

Chapter 8/Organisms and Disease

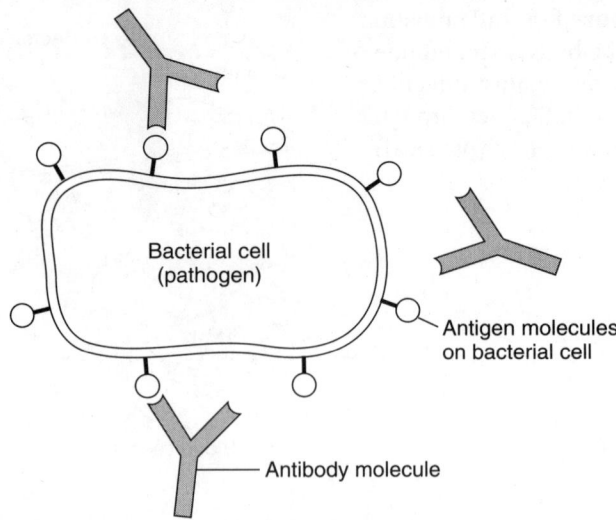

Figure 8-10 Antibodies produced by the immune system bind to antigens, which are specific protein molecules on an invading pathogen's surface.

stimulate the immune system to recognize harmful antigens and produce antibodies against them. This reaction, or immune response, provides the body with **immunity**—the ability to resist an infection—by preparing it to fight subsequent invasions by the same harmful microorganisms.

B Cells and T Cells

The immune system also includes **B cells** and **T cells**, special kinds of white blood cells that are produced in bone marrow, the thymus gland, the spleen,

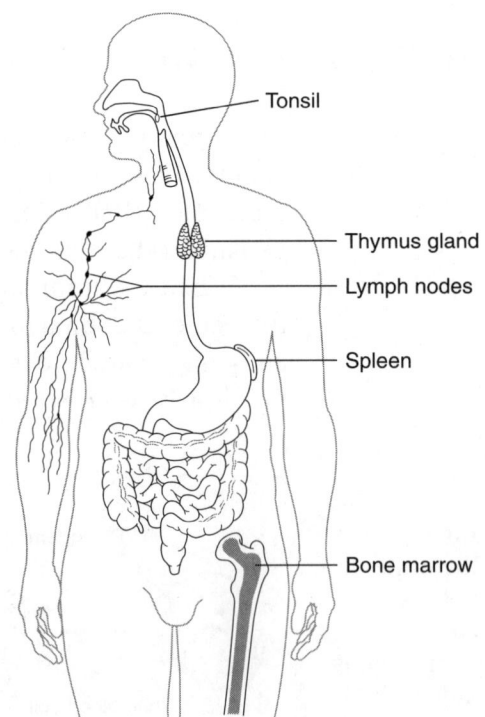

Figure 8-11 Special white blood cells, called B cells and T cells, are produced in the tonsils, thymus gland, lymph nodes, spleen, and bone marrow.

Figure 8-12 Over time, a person's body comes to have many different B cells. The memory B cells, which remain in the body after their first exposure to an antigen, can instantly make antibodies when they encounter the same antigen again.

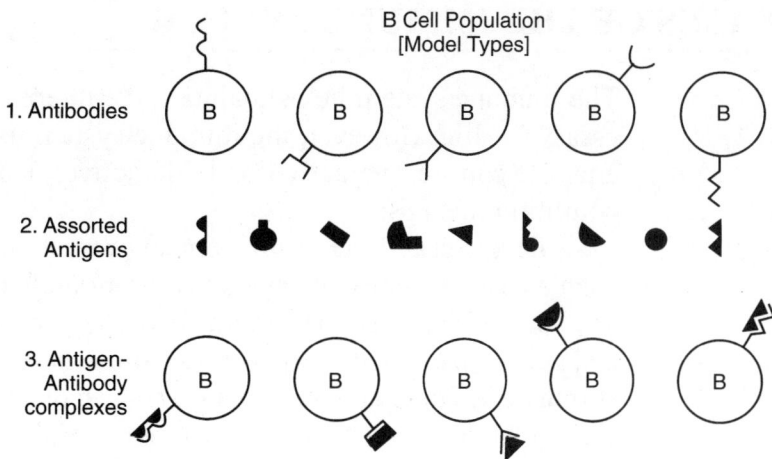

the lymph nodes, and the tonsils. (See Figure 8-11.) B cells are the ones that respond to specific antigens by beginning to produce antibody proteins that will bind only with that antigen.

As time goes on, the body comes to have many different types of B cells, each producing antibodies for one specific antigen. After having been invaded once by an antigen, some special B cells that recognize that antigen remain in the body for the rest of one's life. These are called *memory B cells*. Because they are already present in the body, you instantly start making antibodies the moment you encounter the same invading microorganisms again. That is why individuals usually do not get measles or chicken pox a second time. The immune system remembers the first exposure to the disease and is ready to defend the body. (See Figure 8-12.)

One type of T cell is called *killer T cells*. Through protein receptors on their surface, they can recognize cells in the body that have been infected with invading microorganisms. The killer T cells punch holes in the membranes of the infected cells, sometimes injecting poison into them. (See Figure 8-13.) Another important type of T cell, called *helper T cells,* assists both B cells and killer T cells. Without helper T cells, the other members of the immune system cannot do their job. Just how important helper T cells are is shown by the fact that they are the cells that are destroyed by the human immunodeficiency virus (HIV), which results in the disease called *AIDS*.

Figure 8-13 Killer T cells can recognize cells in the body that have been infected by invading microorganisms. Here, some killer T cells are shown attacking and destroying an infected body cell.

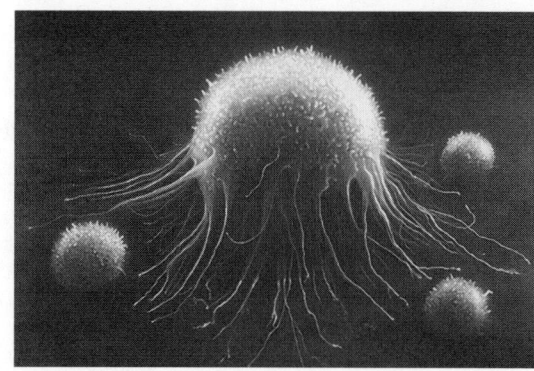

Chapter 8 / Organisms and Disease

DISEASES OF THE IMMUNE SYSTEM

The immune system helps maintain the internal dynamic equilibrium necessary for life. However, the immune system itself can become out of balance. It can be overactive or underactive, and in either case the body's equilibrium is upset.

Allergic reactions result from an overactivity of the immune system. In such a case, the body responds inappropriately to common substances such as dust, mold, pollen, or certain foods. The immune system begins making a special type of antibody to these substances, which normally would not stimulate it. These antibodies cause cells in the body to release chemicals, including **histamines**, which then cause many allergic symptoms, such as extra fluid in the nasal pathways, difficulty breathing, or hives. The allergies are often treated with **antihistamines**, drugs that stop the release of histamines.

Sometimes an overactive immune system begins to attack its own normal body tissues. These are called **autoimmune diseases** and they are very serious. They include rheumatoid arthritis and lupus erythematosus. In all autoimmune diseases, the body is literally rejecting its own tissues. The immune system may also attack transplanted organs. Medications are then taken to try to prevent organ rejection.

The condition known as *AIDS* is an **immunodeficiency disease**, which means that the body's immune system is underactive because it is weakened; in this case, helper T cells are destroyed by HIV. (*Note: AIDS* stands for *a*cquired *i*mmuno*d*eficiency *s*yndrome.) As a result, the body cannot protect itself from other diseases, such as pneumonia, tuberculosis, and cancer, that may attack it—a condition that is usually fatal. (See Figure 8-14.)

Finally, **inflammation**, which protects us from infection when we are young, may actually contribute to crippling diseases when we get older. For example, researchers now suspect that many heart attacks occur when a rupture develops in the wall of an artery, brought on by overactive immune system cells causing inflammation.

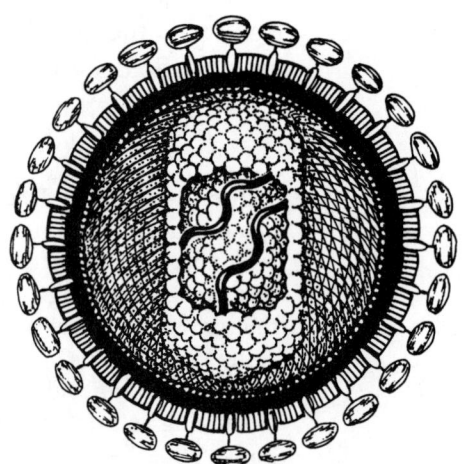

Figure 8-14 Structure of the human immunodeficiency virus, or HIV, which causes the disease called AIDS.

Chapter 8 Review

Multiple Choice

1. A disruption of homeostasis can result in all of the following *except*
   A. illness
   B. death
   C. disease
   D. stability

2. Infectious diseases result from
   A. genetic defects
   B. microorganisms
   C. pollutants
   D. organ malfunctions

3. The inhalation of particles such as asbestos fibers and coal dust can result in respiratory diseases. In such a case, the main cause of the disease would be
   A. microorganisms
   B. pollutants
   C. genetic defects
   D. organ malfunction

4. Scientists do not consider viruses as true living organisms because they
   A. contain both DNA and RNA
   B. have an outer protein coat, or capsid
   C. can reproduce only on dry surfaces
   D. cannot reproduce outside a living cell

5. An example of a disease caused by a virus is
   A. strep throat
   B. cholera
   C. chicken pox
   D. tuberculosis

6. A vaccine is usually made up of
   A. a live, deadly virus
   B. a weakened antibody
   C. white blood cells
   D. killed or weak microbes

7. Bacteriophages are viruses that infect
   A. fungal cells
   B. bacterial cells
   C. animal cells
   D. plant cells

8. Lysogenic viruses are viruses that do *not*
   A. have the ability to cause disease
   B. ever go through the lytic cycle
   C. infect a host cell with their DNA
   D. usually break open the host cell

9. Which of the following diseases is caused by bacteria?
   A. AIDS
   B. tuberculosis
   C. measles
   D. polio

10. Antibiotics are effective against all of the following microorganisms *except*
    A. bacilli bacteria
    B. bacteriophages
    C. spirilla bacteria
    D. cocci bacteria

11. *Plasmodium* is a sporozoan that causes the disease called
    A. dysentery
    B. red tide
    C. potato blight
    D. malaria

12. Parasites that transmit diseases to humans through insect bites are species of
    A. fungi
    B. bacteria
    C. protozoa
    D. algae

13. Dinoflagellates that release a toxic chemical can cause illness in humans who eat contaminated
    A. mushrooms
    B. shellfish
    C. algae products
    D. pork and beef

14. Ringworm is a skin disease that is caused by a
    A. flatworm
    B. fungus
    C. bacterium
    D. viral particle

15. Blood flukes are flatworms that travel between two hosts in their life cycle, a human and a
    A. cow
    B. pig
    C. snail
    D. mosquito

16. The parasitic worm that lacks a mouth and instead absorbs nutrients through its body is the
    A. tapeworm
    B. earthworm
    C. ringworm
    D. roundworm

17. The hookworm is a type of roundworm that sucks blood from its human host, resulting in
    A. malaria
    B. anemia
    C. ringworm
    D. allergies

18. The body's main physical barrier against infection is
    A. the skin
    B. white blood cells
    C. red blood cells
    D. inflammation

19. Produced in several parts of the body, B cells and T cells are special kinds of
    A. white blood cells
    B. red blood cells
    C. antigens
    D. hormones

20. Allergic reactions are most closely associated with
    A. the action of circulating hormones
    B. an overreaction of the immune system
    C. a low blood sugar level
    D. the shape of red blood cells

21. White blood cells can prevent a serious infection by
    A. filling the damaged tissues with pus
    B. repairing the skin after it has been cut
    C. ingesting the harmful microorganisms
    D. constructing barriers against microorganisms

22. Certain microbes, foreign tissues, and some cancerous cells can cause immune responses in the human body because all three contain
    A. antigens
    B. lipids
    C. enzymes
    D. cytoplasm

23. When the immune system detects an antigen, it
    A. pushes it out of the body immediately
    B. produces antibodies that bind to the antigen
    C. produces antigens that cancel its bad effects
    D. destroys the antigen by cutting it in half

24. Which activity would stimulate the human immune system to provide protection against a particular microbe?
    A. receiving antibiotic injections after surgery
    B. being vaccinated against chicken pox
    C. following a well-balanced diet and exercising
    D. receiving hormones that come from cow's milk

25. The following diagram represents one possible immune response that can occur in the human body. The structures that are part of the immune system are represented by the parts labeled

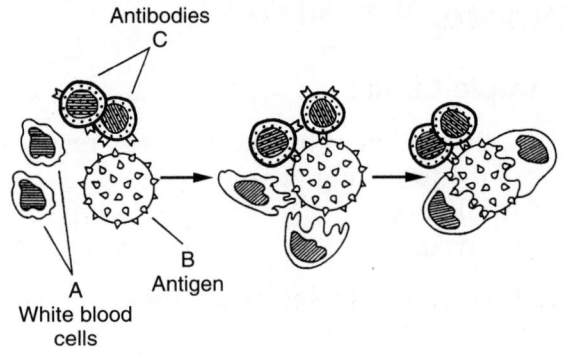

    A. *A* only
    B. *A* and *C*
    C. *B* and *C*
    D. *A* and *B*

26. Which of the following is *not* a characteristic of white blood cells?
    A. They destroy some microbes by engulfing them.
    B. They carry oxygen atoms throughout the body.
    C. They make antibodies that bind with antigens.
    D. They punch holes in membranes of infected cells.

27. The killer T cells function to
    A. produce antibodies that kill invading microorganisms
    B. destroy body cells that are infected by microorganisms
    C. bind with the infected cells and repair their membranes
    D. destroy invading microorganisms before they infect cells

28. Which condition would most likely result in a human body being unable to defend itself against pathogens and cancerous cells?
    A. a genetic tendency toward a disorder such as diabetes
    B. the production of antibodies in response to an infection in the body
    C. a parasitic infestation of ringworm on the body
    D. the presence in the body of the virus that causes AIDS

29. The human immunodeficiency virus (HIV) is particularly devastating to the immune system because it destroys
    A. all the white blood cells in the body
    B. all the red blood cells in the body
    C. the memory B cells and killer T cells
    D. the helper T cells, which assist B cells and killer T cells

30. When an overactive immune system starts to attack its own body tissues, it causes serious conditions known as
    A. antihistamine diseases
    B. allergic reaction diseases
    C. autoimmune diseases
    D. immunodeficiency diseases

Analysis and Open Ended

31. How can a child inherit a disease if neither parent appears to have it?

32. Explain what a pathogen is. Your answer should include the following information:
    - what the *four* types of pathogens are that can cause diseases
    - the *term* for such diseases when they are passed from one person to another
    - the *three* common ways that pathogens can enter the body

33. What is the difference between an inherited disease and an infectious disease?

34. Briefly describe the relationship between organ malfunction and disease.

35. Explain why a harmful lifestyle can lead to disease. Give *one* example. In what way are the factors that cause such a disease preventable or controllable?

36. Describe the first line of defense against infection in the body. Include the following:
    - what kind of defense it consists of
    - what organ carries out this function
    - what it is defending against and how

37. Refer to the diagram below to answer this question. When bacteria enter a cut, this process occurs as part of the body's second line of defense, known as

    A. infection   C. invasion
    B. inflammation   D. immunity

38. A part of the hepatitis B virus can be synthesized in the laboratory. This viral particle can be identified by the immune system as a foreign substance, but it is not capable of causing the disease. Immediately after this viral particle is injected into a human, it
    A. stimulates the production of enzymes that are able to digest the hepatitis B virus
    B. synthesizes specific hormones that provide immunity against the hepatitis B virus
    C. triggers the formation of specific antibodies that protect against the hepatitis B virus
    D. breaks down key receptor molecules so that the hepatitis B virus can enter body cells

39. A researcher needs information about antigen–antibody reactions. He or she could best find information on this topic by searching for which phrase on the Internet?
    A. Protein Synthesis
    B. Energy Sources in Nature
    C. White Blood Cell Activity
    D. DNA Replication

40. Use the following terms to fill in the missing words in the flowchart below, which describes the immune system's reaction to microscopic invaders: *body develops immunity; microbes destroyed by body; antibodies; antigens.*

    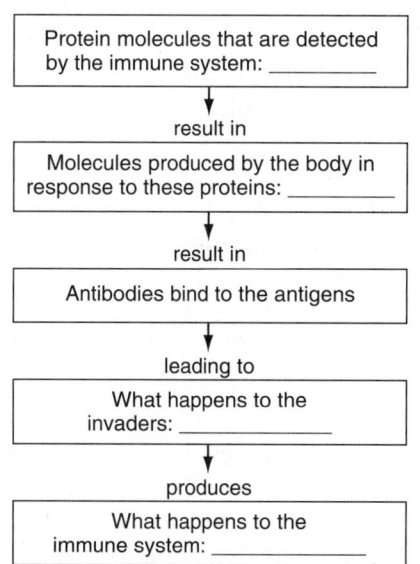

41. How does the immune system help to maintain homeostasis?

42. Briefly define the term *immunity.* How do vaccines provide people with immunity?

Base your answers to questions 43 and 44 on the table and on your knowledge of biology.

Volunteer	Injected with Dead Chicken Pox Virus	Injected with Dead Mumps Virus	Injected with Distilled Water
A	✔		
B		✔	
C			✔
D	✔	✔	

43. None of these volunteers ever had chicken pox. After the injection, there most likely would be antibodies to chicken pox in the bloodstream of
    A. volunteers A and D only
    B. volunteers A, B, and D
    C. volunteer C only
    D. volunteer D only

44. Volunteers A, B, and D underwent a medical procedure known as
    A. cloning   C. electrophoresis
    B. vaccination   D. chromatography

Refer to the following list, which describes three ways of controlling viral diseases in humans, to answer question 45.

- administer a vaccine containing dead or weakened viruses, which stimulates the body to form antibodies against the virus
- use chemotherapy (chemical agents) to kill viruses, similar to the way in which antibiotics act against bacteria
- rely on the action of interferon, which is produced by cells to protect the body from pathogenic viruses

45. Based on this information, which activity would provide the greatest protection against viruses?
    A. producing a vaccine that is effective against interferon
    B. using interferon to treat a number of diseases caused by bacteria
    C. developing a method to stimulate the production of interferon in cells
    D. synthesizing a drug that prevents the destruction of bacteria by viruses

46. Compare and contrast the functions of B cells and killer T cells.

47. Explain how HIV affects the immune system. Your answer should include the following: (a) what "HIV" stands for and what the term *immunodeficiency* means; (b) which cells in the immune system are affected and in what way; and (c) how this affects the rest of the immune system's functioning.

48. In what way are allergic reactions and autoimmune diseases similar to one another (in terms of the immune system)? In what important way are they different?

Reading Comprehension

Base your answers to questions 49 to 51 on the information below and on your knowledge of biology. Use one or more complete sentences to answer each question.

> Vaccination protects the body from disease. The United States government is committed to the goal of vaccinating all children against common childhood diseases. In fact, children must be vaccinated against certain diseases before they can even enter school. However, some parents feel that vaccines may be dangerous and do not want their children to receive them.
>
> Recently, many parents have been choosing not to vaccinate their children against childhood diseases such as diphtheria, whooping cough, and polio. For example, the parents of a newborn baby may be concerned about having their child receive a potent three-in-one shot such as the DPT vaccine, which protects a child against

> diphtheria, whooping cough, and tetanus. Since bacteria, not viruses, cause these particular diseases, the parents may think that antibiotic therapy is a safe and effective alternative to vaccination.

49. Explain to these concerned parents what is in a vaccine and what it does in the body.
50. Describe how the process of vaccination promotes immunity to various diseases.
51. Explain the difference between antibiotics and vaccines in preventing bacterial diseases.

Unit III
Diversity and Biological Evolution

STANDARD 5.5.12 B
Characteristics of Life

All students will gain an understanding of the structure, characteristics, and the basic needs of organisms and will investigate the diversity of life.

Enduring Understanding II Evolution provides the central scientific understanding of the history of the modern living world. Evolutionary processes allow some species to survive through long-term Earth changes, while leading to the extinction of others. Organisms that inherit characteristics advantageous for survival in their physical environment reproduce and increase the proportion of individuals with similar traits.

Chapter 9

Classification of Organisms

Standard 5.5.12 B1 Explain that through Earth's history, present species developed from earlier distinctly different species.

CLASSIFICATION OF LIVING THINGS

Earth is estimated to be about 4.5 billion years old. Life on the planet is thought to have begun as simple, unicellular (single-celled) organisms, about 3.5 billion years ago. Then, approximately 1 billion years ago, through the process of evolution, increasingly complex multicellular (many-celled) organisms began to appear. (Scientists base this knowledge on the fossil record.)

During the constant process of change in living things, some characteristics (that do not aid survival) may be are lost while other characteristics (that do aid survival) are gained. As a result, Earth is populated by many different kinds of organisms. So far, scientists have classified approximately two million *species,* and they continue to identify new ones each year. To study so many life-forms, biologists have organized them into numerous groups based on their similarities. Hence, **classification** is the grouping and naming of organisms according to their evolutionary relationships and shared characteristics.

The branch of biology that deals with classification of life-forms is known as **taxonomy**. Scientists called *taxonomists* use the tools of classification to identify and find relationships among organisms. Today, taxonomists classify organisms based on their biochemical and genetic information (nucleotide and amino acid sequences), embryological development, fossil record, and evolutionary relationships, in addition to their *morphology,* or body structure. For example, organisms that have similar protein structures are usually closely related because they share a recent common ancestor.

EARLY CLASSIFICATION AND NAMING SYSTEMS

The first classification system was developed by the Greek philosopher Aristotle. He classified organisms into two major groups: plants and animals. Plants were classified based on their different kinds of stems: small and soft (herbs); medium and woody (shrubs); or large and woody (trees). Animals were classified into three groups based on where they lived—on land, in the water, or in the air.

In the eighteenth century, the Swedish botanist Carolus Linnaeus developed a new system for classifying organisms. Like Aristotle, Linnaeus classified all organisms into either the plant or the animal kingdom. However, he classified animals based on their similarities in morphology, rather than by location. For example, bats were now grouped with mammals (and not with birds, which are other "air animals"), because they have hair, give live birth (after internal development), and nurse their offspring with milk. (See Figure 9-1.)

Bat

Figure 9-1 Although bats can fly, they are classified as mammals, not birds, because they share common traits with other mammals; they have hair, give live birth, and nurse their offspring with milk.

In addition, Linnaeus gave each organism a two-word, Latin scientific name. This system is called **binomial nomenclature**, and it is still used today. *Binomial* means "consisting of two terms," and *nomenclature* means "a system of names." The first word in binomial nomenclature is the name of the *genus* to which an organism belongs. A **genus** is a group that has one or more different species classified within it; these are closely related species that evolved from a common ancestor. The second word is the name of the unique *species*. A **species** is a group of similar organisms that are capable of producing fertile offspring with each other. For example, *Panthera leo* is the scientific name for the species you know as the lion. *Panthera tigris* is the scientific name for the tiger. Both species are classified in the same genus *Panthera*, along with several other big cats. Both the genus and species names are italicized. The first letter of the genus name is always capitalized; the first letter of the species name is lowercase. Humans are classified as *Homo sapiens*.

TAXONOMIC GROUPINGS

Taxonomists classify organisms into seven major groups, or *taxa* (singular, *taxon*). The seven groups are: kingdom, phylum, class, order, family, genus, and species. This classification system works in the following way: Each **kingdom**—the largest taxonomic group—is divided into related *phyla* (plural for *phylum*); each *phylum* into related classes; each *class* into related orders; each *order* into related families; each *family* into related genera; and each *genus* (plural, *genera*) into related *species*. An easy way to remember the order of taxa, from most general to most specific, is the following sentence: **K**ing **P**hillip **c**ried **o**ut **f**or **g**ood **s**oup (representing **k**ingdom, **p**hylum, **c**lass, **o**rder, **f**amily, **g**enus, and **s**pecies).

Figure 9-2 shows how two different animals (a tiger and a wolf) are classified within the seven taxonomic groups. The taxa are presented in ranked

Figure 9-2 Taxonomic classification of two mammals, from kingdom (the largest group) to species (the smallest group).

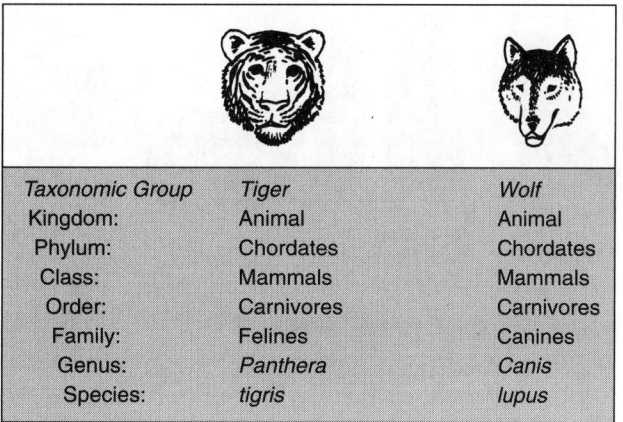

(hierarchical) order, going from the largest, most general group (kingdom) to the smallest, most specific group (species). Note, for example, that organisms belonging to the same family (all felines *or* all canines) are more closely related than are organisms belonging to the same order (all carnivores), which is the next largest taxon.

Phylogenetic Taxonomy

Modern taxonomists can use two different approaches to place an organism into the correct taxonomic category. The first approach is known as **systematics**. In this approach, a *phylogenetic tree,* or family tree, is used to show the evolutionary relationships between different groups of organisms. (*Note:* The term *phylo* means "tribe" or "group.") Systematics stresses common ancestry and relies on the amount of differences within a group to construct a phylogenetic tree. The tree is based on evidence from the fossil record, morphology, embryological development, biochemistry, and genetic studies. (See Figure 9-3 on page 124.)

The second approach is known as **cladistics**. In this approach, a scientist constructs a diagram based on specific characteristics of a group of organisms. This diagram, called a *cladogram,* is used to determine the evolutionary relationships among the different groups based on traits called *shared derived characteristics.* (See Figure 9-4.) A derived characteristic is a trait that evolved only within the specific group under study. For example, if the group being studied is birds, having feathers is the shared derived characteristic, since feathers evolved within the bird group.

MODERN CLASSIFICATION: THE FIVE-KINGDOM SYSTEM

Since the time of Linnaeus, scientists have discovered many new species and uncovered more information about known species. As a result, biologists now realize that there are more than just two kingdoms of living

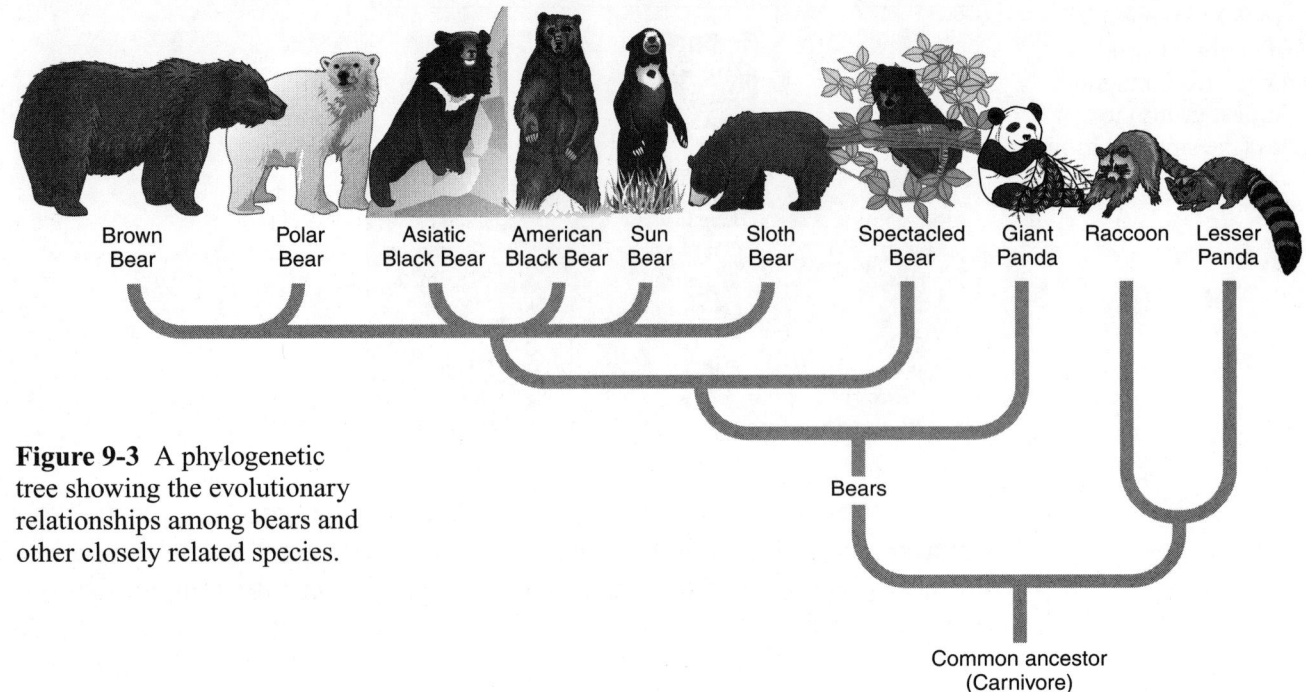

Figure 9-3 A phylogenetic tree showing the evolutionary relationships among bears and other closely related species.

things. In one commonly used modern system of classification, all living things are placed within one of the following five kingdoms: Monera, Protista, Fungi, Plantae, and Animalia. In the five-kingdom system, the Monera include all *prokaryotic* organisms, which are the unicellular life-forms that lack true organelles. The four other kingdoms include all *eukaryotic* organisms, the life-forms whose cells do contain various organelles. (See Figure 9-5.)

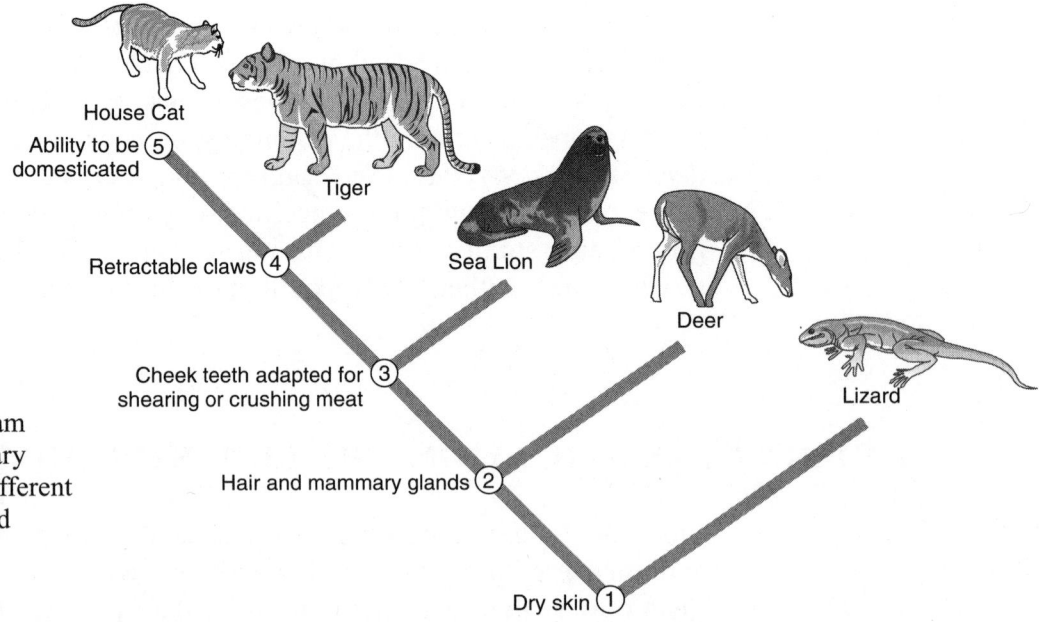

Figure 9-4 A cladogram showing the evolutionary relationships among different species based on shared derived characteristics.

124 Preparing for the New Jersey Biology Competency Test

Figure 9-5 One commonly used classification system groups all living things within five main kingdoms.

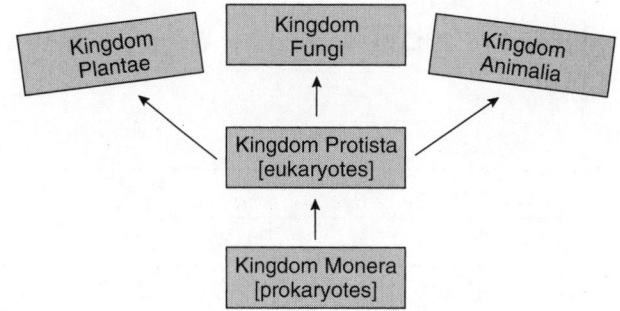

Prokaryotes and Eukaryotes

Cells can be grouped into either of two major types: *prokaryotes* and *eukaryotes*. The distinction is based on the presence or absence of a nucleus (with a nuclear membrane), which separates the DNA from the rest of the cell, and the presence or absence of other membrane-bound organelles.

The cells of **prokaryotes** lack a nucleus (no nuclear membrane) and other membrane-bound organelles, such as lysosomes and mitochondria. As such, prokaryotic cells are less complex than eukaryotic cells. Many functions of a prokaryotic organism are carried out by its cell membrane, which all cells have. *Bacteria* are examples of prokaryotic cells. (See Figure 9-6.)

The cells of **eukaryotes** contain a nucleus (surrounded by the nuclear membrane) and other membrane-bound organelles, which are the specialized structures that carry out important cell functions. (See Table 9-1.) The presence of membrane-bound organelles allows many chemical reactions to occur simultaneously in the same cell without interference. The cells of protists, fungi, plants, and animals are all eukaryotic cells. (Refer back to Figures 4-1 and 4-2 on page 52).

THE SIX-KINGDOM SYSTEM OF CLASSIFICATION

Recently, taxonomists have begun to classify organisms within six kingdoms of life: Eubacteria, Archaebacteria, Protista, Fungi, Plantae, and Animalia. In the six-kingdom system, the Monera are split to form two separate kingdoms of prokaryotic organisms, based on important differences between them. (See Figure 9-7 on page 126.)

Figure 9-6 A bacterial cell is an example of a moneran.

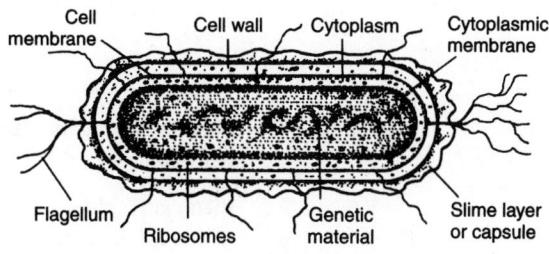

Chapter 9/Classification of Organisms **125**

Table 9-1 A Comparison of Organelles in Eukaryotic and Prokaryotic Cells

Organelles	Prokaryotic Cell	Eukaryotic Cell
Chromosomes	yes (1 strand only)	yes (more than 1 strand)
Nuclear membrane	no	yes
Nucleus	no	yes
Mitochondria	no	yes
Ribosomes	yes	yes
Lysosomes	no	yes
Golgi apparatus	no	yes
Endoplasmic reticulum	no	yes
Chloroplasts (with chlorophyll)	no (but some have chlorophyll)	yes (in plant and alga cells)
Cell wall	yes (some)	yes (some)
Centrioles	no	yes

Kingdom Archaebacteria

The prefix *archae* means "ancient." *Archaebacteria* (formerly within Monera) are unicellular prokaryotic organisms that have special cell membranes. Unlike Eubacteria (the other prokaryotic kingdom), Archaebacteria do not contain peptidoglycan in their cell walls; and their biochemistry is different from that of the Eubacteria. Most Archaebacteria live in extreme, harsh environments such as hot sulfur springs, deep hot vents in the ocean, and the Dead Sea in Israel.

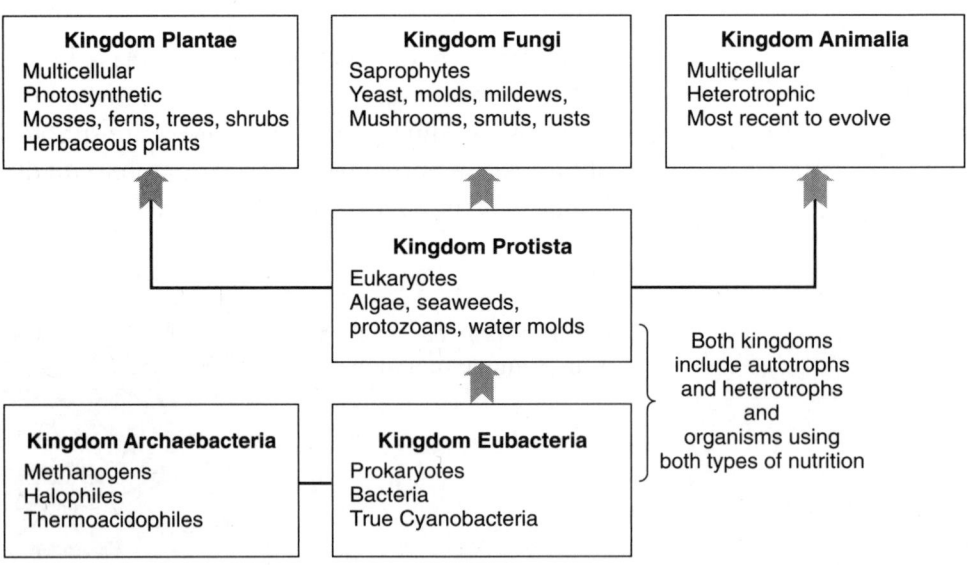

Figure 9-7 Many scientists now group organisms within a six-kingdom classification system.

Figure 9-8 The three main shapes of bacteria.

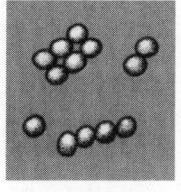
Ball-shaped
(cocci)

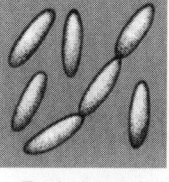
Rod-shaped
(bacilli)

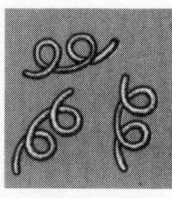
Corkscrew-shaped
(spirilla)

Kingdom Eubacteria

The prefix *eu* means "true." *Eubacteria* (formerly within Monera) are unicellular prokaryotic organisms. These true **bacteria** include the disease-causing varieties as well as many harmless types. Most Eubacteria have cell walls that contain a rigid substance called *peptidoglycan;* and the amino acid that is used in protein synthesis is different from that of the Archaebacteria. The RNA sequences and DNA replication and transcription system of the Eubacteria are also very different from those of the Archaebacteria. (See Figure 9-8.)

Kingdom Protista

The *Protista* include both eukaryotic unicellular and eukaryotic multicellular organisms. **Protists** are found in aquatic or damp environments. Their cell types are the most diverse compared with the cells of other eukaryotic organisms. Protists include the following life-forms: plantlike organisms, such as *algae,* which are autotrophs; animal-like organisms called *protozoa,* such as the ameba and paramecium, which are heterotrophs; and funguslike organisms, which are the slime molds. The *euglena* is an unusual plantlike protist that can be either autotrophic or heterotrophic, depending on the environmental conditions in which it lives. Protists reproduce both sexually and asexually. The protozoans (*proto* means "first"; *zoa* means "animal") can be either free-living or parasitic; some have flagella or cilia for locomotion. Biologists think that protists most likely gave rise to all other eukaryotic organisms. (See Figure 9-9.)

Kingdom Fungi

Members of the *Fungi* are all eukaryotic multicellular organisms, except for yeasts, which are unicellular. Most **fungi** have cell walls composed of *chitin,*

Figure 9-9 Some different species of protists.

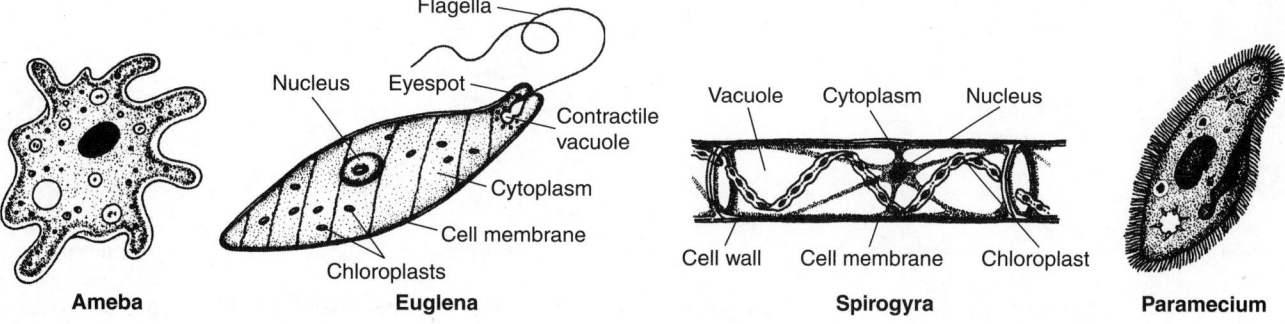

Ameba | Euglena | Spirogyra | Paramecium

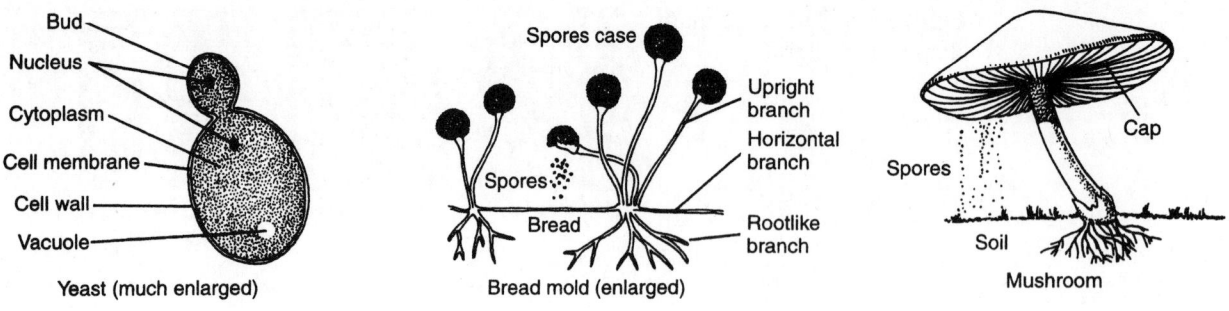

Figure 9-10 Examples of different types of fungi.

a nitrogen-containing polysaccharide. All fungi are heterotrophs. Unlike plants, they do not contain chlorophyll. Instead, fungi absorb the nutrients they need from organic matter in their environment. Many fungi are **saprophytes**, organisms that obtain nutrients from dead or decaying plants and animals. Fungi reproduce both sexually and asexually. Examples of fungi include mushrooms, yeasts, rusts, smuts, and molds. (See Figure 9-10.)

Kingdom Plantae

The *Plantae* are all eukaryotic, multicellular autotrophic organisms. The cell walls of plants consist mainly of *cellulose,* a polysaccharide. Plant cells contain chlorophyll within chloroplasts, which are needed for the production of food (glucose) through photosynthesis. Plants have a life cycle that includes a sexually reproducing generation and an asexually reproducing generation; this is known as **alternation of generations**. Examples of plants include mosses, liverworts, ferns, gymnosperms (pinecone-bearing plants), and angiosperms (flower-bearing plants). (See Figures 9-11a and 9-11b.)

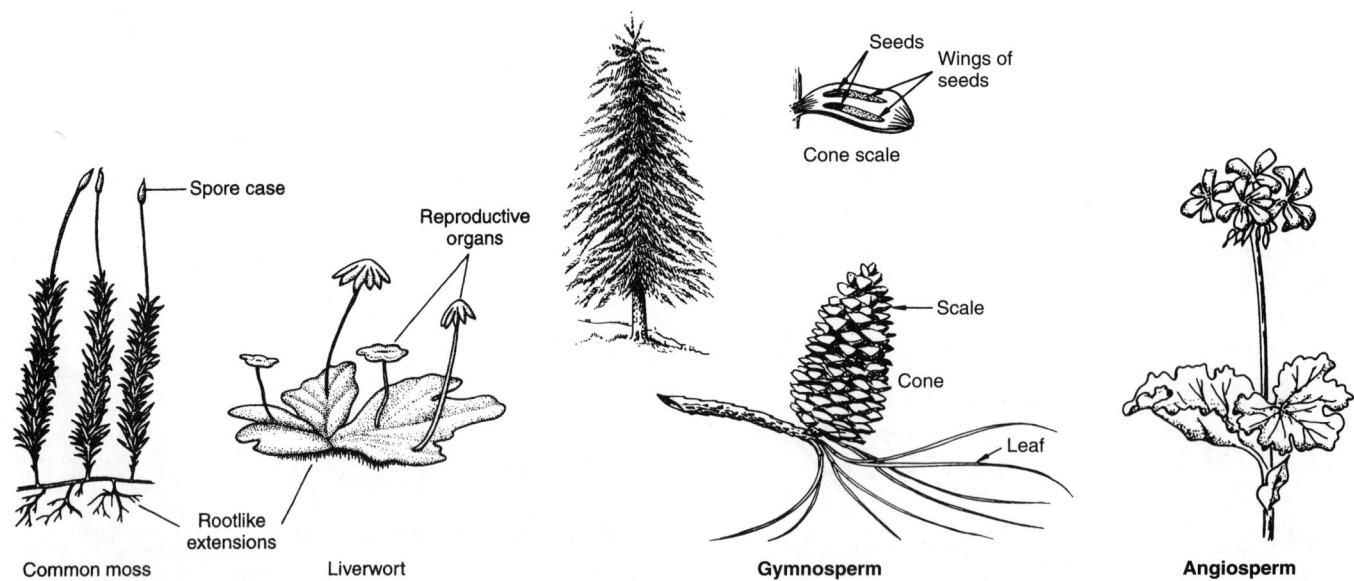

Figure 9-11a Examples of two different types of plants: a moss and a liverwort.

Figure 9-11b Examples of a gymnosperm (cone-bearing plant) and an angiosperm (flower-bearing plant).

Figure 9-12 Examples of different members of the animal kingdom.

Kingdom Animalia

Members of the *Animalia* are eukaryotic, multicellular heterotrophic organisms. Animal cells do not possess cell walls. Animals have a digestive cavity that functions in the digestion and absorption of food. Most animals reproduce only sexually, although some animals do reproduce asexually. For example, cnidarians (such as the jellyfish) have a life cycle that includes a *polyp* generation, which reproduces asexually. In sexual reproduction, the nuclei of the male and female sex cells fuse in the process of *fertilization,* and a zygote is formed. The **zygote** is the fertilized egg cell that contains the diploid number of chromosomes, and that develops into an embryo. Examples of animals include worms, insects, sponges, corals, lobsters, birds, fish, frogs, lizards, whales, cats, and humans. (See Figure 9-12.)

THE DICHOTOMOUS KEY

Biologists have developed a precise method to help them classify and identify unknown organisms. This classification tool is called a **dichotomous key**, and it uses a logical approach to classify an organism. Each dichotomous key is composed of a list of observable, alternative characteristics that leads, step-by-step, to the correct identification of the organism. The term *dichotomous* means "dividing in two" and refers to the fact that there are always two choices to pick from at every step of the key. At each step, one of the two descriptions is eliminated, which narrows the possibilities. For example, at some step in a key for identifying plants, there would be a choice between the types of structures used for reproduction, such as spores versus seeds, and then (for seeds) pinecones versus flowers. The key would also have choices based on traits such as the type of leaf structure, whether the stem is soft or woody, and so on. In this way, an unknown plant can be identified. (See Figure 9-13.)

Figure 9-13 Important differences in the structures of dicot and monocot plants; key features such as these would be used to identify a plant by means of a dichotomous key.

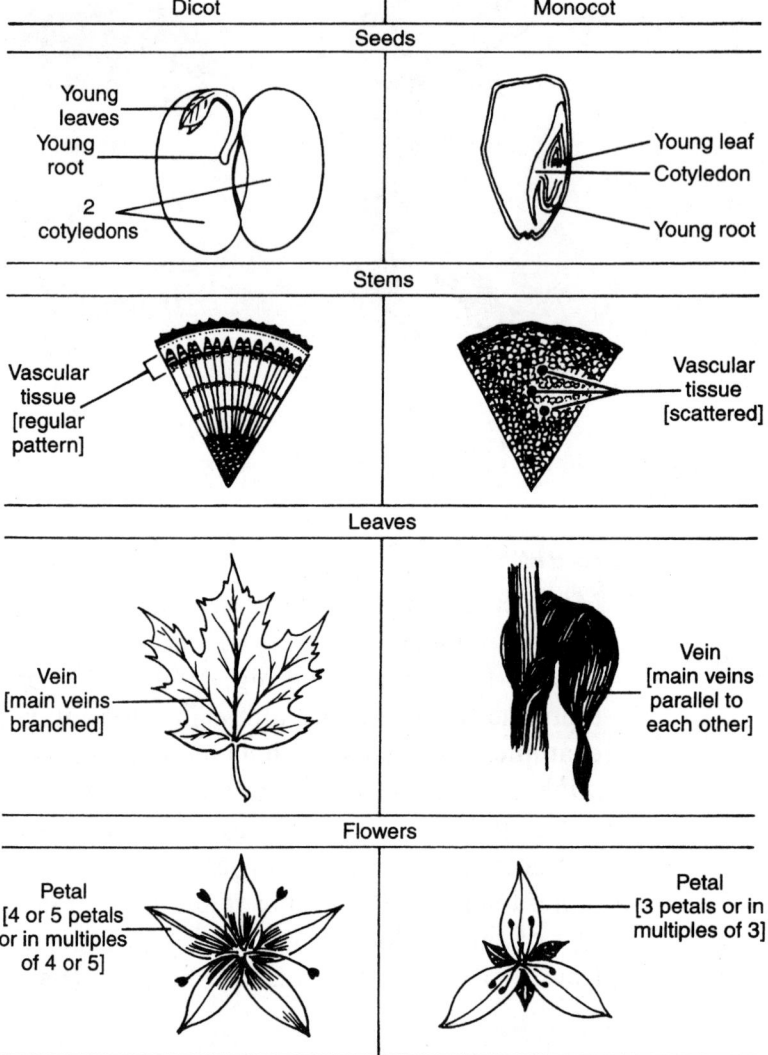

Chapter 9 Review

Multiple Choice

1. The largest and least specific taxon is the
   A. species C. class
   B. kingdom D. genus

2. Norway maples, sugar maples, and red maples are probably classified within
   A. the same species and same genus
   B. different species and different genera
   C. the same species but different genera
   D. the same genus but different species

3. The scientific name for the red maple tree is *Acer rubrum*. This name includes its
   A. class and phylum
   B. family and species
   C. genus and species
   D. genus and order

4. Which is the correct order of taxa, going from the smallest groups to the largest groups?
   A. phylum, kingdom, genus, class
   B. genus, species, phylum, class
   C. species, genus, phylum, kingdom
   D. phylum, class, genus, species

5. Organisms that are alike and capable of producing fertile offspring with each other are placed in the same
   A. genus   C. species
   B. family   D. kingdom

6. Which kingdom includes organisms that have both plantlike and animal-like characteristics?
   A. Plantae C. Animalia
   B. Fungi   D. Protista

7. Except for yeasts, the members of Kingdom Fungi are all
   A. prokaryotic and unicellular
   B. prokaryotic and multicellular
   C. eukaryotic and unicellular
   D. eukaryotic and multicellular

8. Which kingdom contains autotrophic organisms whose life cycles include alternation of generations?
   A. Eubacteria C. Animalia
   B. Plantae   D. Protista

9. Which of the following groups does *not* include any autotrophic organisms?
   A. algae   C. fungi
   B. protists D. plants

10. The branch of biology that deals with the classification of life-forms is called
    A. embryology
    B. taxonomy
    C. morphology
    D. biochemistry

11. Organisms in kingdoms Archaebacteria and Eubacteria differ from each other in
    A. their biochemistry and genetics
    B. the materials in their cell walls
    C. the type of environments they inhabit
    D. all of the above characteristics

12. A specific trait, such as fur or feathers, that is used to determine an organism's classification group is called a
    A. homologous characteristic
    B. compared characteristic
    C. shared derived characteristic
    D. primitive characteristic

13. A group that has one or more different species is a
    A. genus   C. class
    B. family D. phylum

Analysis and Open Ended

Base your answer to question 14 on the table below and on your knowledge of evolution.

Row	Organism X	Organism Y
(1)	Simple multicellular	Simple unicellular
(2)	Complex multicellular	Simple multicellular
(3)	Simple unicellular	Simple multicellular
(4)	Complex multicellular	Complex unicellular

14. Organism *X* appeared on Earth much earlier than organism *Y* did. Many scientists think that organism *X* appeared between 3 and 4 billion years ago, and that organism *Y* appeared about 1 billion years ago. Which row in the chart above most likely describes both organisms *X* and *Y* correctly?
    A. Row 1   C. Row 3
    B. Row 2   D. Row 4

Base your answers to questions 15 and 16 on the chart below and on your knowledge of taxonomic groups.

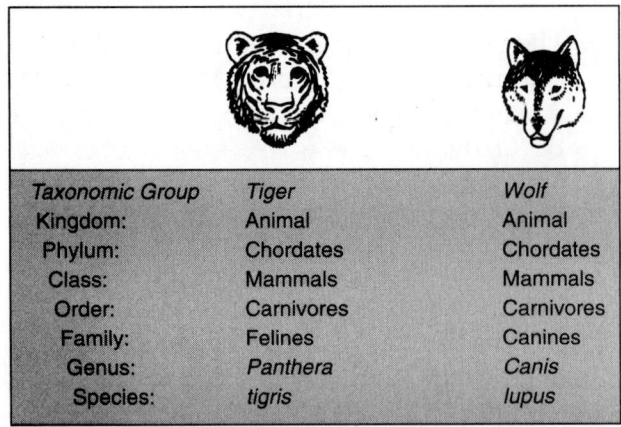

15. The tiger and the wolf are classified within the
    A. same species but different orders
    B. same genus but different species
    C. same order but different families
    D. same family but different orders

16. According to the chart, the largest (and least specific) division used in this classification system is the
    A. species   C. class
    B. family   D. kingdom

17. Refer to the following phylogenetic tree. Which letter represents the most recent common ancestor of organisms 2 and 4?

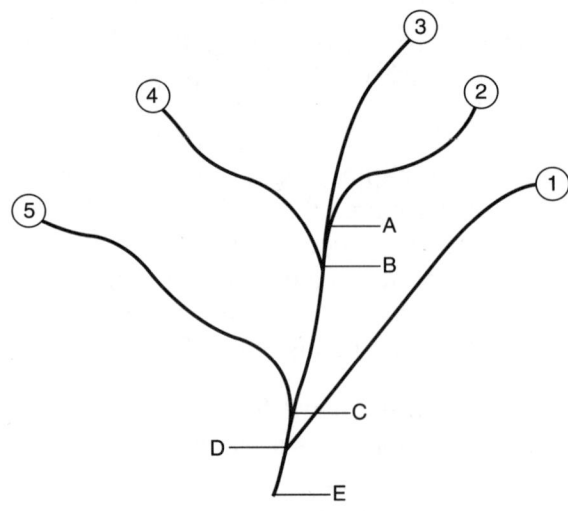

   A. letter *A*   C. letter *C*
   B. letter *B*   D. letter *D*

18. What are the two main differences between prokaryotic and eukaryotic cells? Explain how these differences determine how the cells carry out their life functions.

Base your answers to questions 19 and 20 on the following diagram and on your knowledge of evolution and classification.

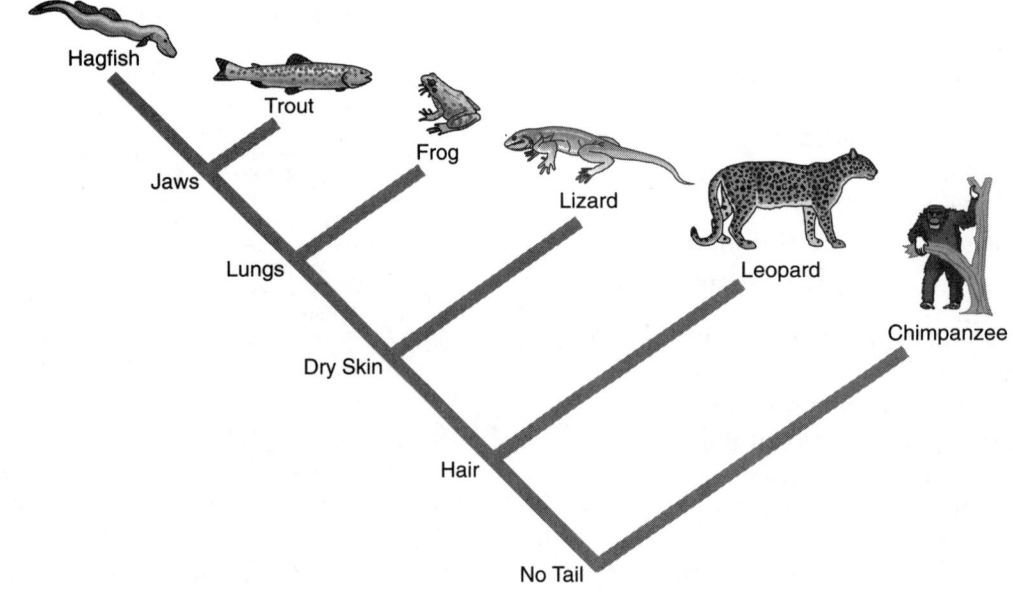

19. The diagram, which is used to show evolutionary relationships, is known as a
    A. phylogenetic tree
    B. family tree
    C. cladogram
    D. dichotomous key

20. Pick one of the animals shown in the diagram. Explain why it best illustrates the trait that it is associated with, rather than the trait before or after it; describe how that trait distinguishes it (in terms of evolution and classification) from the other animals in the diagram.

21. Briefly explain how a dichotomous key is used to identify an unknown organism. Choose any organism and give an example of two valid choices that may be given at one step in such a key.

Reading Comprehension

Base your answers to questions 22 to 24 on the information below and on your knowledge of biology. Use one or more complete sentences to answer each question.

> It seems impossible to imagine all organisms of a single species reproducing at the same time. But this is exactly what happens when all trees of a species simultaneously release pollen into the wind to fertilize the female flowers of that species. Still, in animal species, simultaneous reproduction of many individuals is a rare occurrence. However, one of the most spectacular underwater events involves the mass spawning of the millions of small organisms that make up a coral reef. A coral reef is a stony structure made of minerals removed from the water, over a long period of time, by tiny coral animals that live on the outer edges of the reef. Reefs are found only in the clear, warm, shallow waters of the tropics. During their lifetime, coral organisms remain in one place. They catch and remove food particles from the water that surrounds them. One good place to observe the mass spawning of coral animals is the Flower Gardens National Marine Sanctuary in the Gulf of Mexico.
>
> Scientists think that mass spawning increases the chances of successful fertilization in three ways. First, with so many eggs and sperm in the water at the same time, fertilization is more likely to occur. Second, with gametes from different colonies of one species being released at the same time, cross-fertilization between different colonies is more likely. This increases the genetic variation among the offspring. Finally, with so many fertilized gametes in the water at once, the amount lost to predation is limited.

22. Compare the process of simultaneous reproduction in plants to that in animals such as coral.

23. Why is cross-fertilization between different coral colonies beneficial?

24. What advantage does mass spawning provide for the fertilized eggs?

Chapter 10

The Theory of Evolution

Standard 5.5.12 B1 Explain that through Earth's history, present species developed from earlier distinctly different species.

Standard 5.5.12 B2 Explain how the theory of natural selection accounts for extinction as well as an increase in the proportion of individuals with advantageous characteristics within a species.

THE THEORY OF EVOLUTION

Figure 10-1 Charles Darwin proposed the theory of evolution to explain how organisms change over time. Here he is shown in his later years.

Discoveries in modern science have shown that, over many thousands of years, populations of living things change. In 1859, the English naturalist Charles Darwin proposed a scientific theory to explain how organisms change over time. It is called the *theory of evolution*. (See Figure 10-1.)

The **theory of evolution** explains how the immense variety of living things on Earth has developed from ancestral forms during the past three billion years. This theory is considered to be the most important unifying idea in biology. It offers an explanation, based on fossil and other scientific evidence, of how more complex life-forms could have evolved from simpler ones over time; how Earth came to be populated by the millions of different kinds, or **species**, of organisms alive today; and how these species are related to one another. It is interesting to note that, as a result of the changes that occur in living things and in the environment over time, most of the species that once lived on Earth are now **extinct**; that is, they are no longer alive today.

A Struggle for Existence

During its lifetime, a female elephant may produce six offspring. Darwin calculated that, over hundreds of years, millions of elephants could descend from one original pair of elephants. Similarly, if a plant produced only two seeds a year, in 20 years there could be a million new plants descended from the original parent plants. However, this does not happen. In fact, only a relatively small number of offspring of any species survive to produce their own offspring. Darwin realized that Earth cannot support huge increases in populations; thus, they do not increase that dramatically. He concluded that there is a "struggle for existence" in which only a few offspring of any type survive to maturity and reproduce.

In this struggle for existence, there is **competition** among organisms for various resources. Lack of food and space are two factors that can limit an organism's chances to survive and reproduce. Plants must have minerals, sunlight, and space in order to make their own food and grow. Animals also need resources such as food, water, space, and shelter in order to survive. Competition for available resources exists between individuals of the same species living in the same area and between members of different species living in the same area.

Genetic Variations

Through the process of *reproduction,* characteristics are passed on from parents to offspring. The resulting offspring resemble their parents. However, all the offspring produced by one pair of parents are not identical. The offspring inherit different combinations of their parents' characteristics, or *hereditary traits*. These differences among the offspring are called *genetic variations*.

As you have learned, there are two types of reproduction. In *asexually* reproducing organisms, a single parent organism splits in two to produce two identical new organisms. In *sexually* reproducing organisms, a male and a female organism mate to produce offspring. That is, sexual reproduction involves the combining of genetic material from two individuals. The traits in the offspring are the result of a new assortment, or **recombination**, of traits inherited from both parents. Thus, sexual reproduction produces greater genetic variability among offspring than asexual reproduction does.

It is very important to recognize the difference between hereditary characteristics and changes that occur to an individual during its lifetime. For example, eye color is inherited. It is a hereditary trait. By contrast, becoming extremely muscular due to weight lifting is an *acquired* trait. It is not inherited. A characteristic that is hereditary can be passed along to offspring. An acquired characteristic cannot be passed on to offspring. Thus, only changes that are in the sex cells, or **gametes**, of the parents can become the basis for evolutionary change.

In addition to the recombination of genetic information that occurs during sexual reproduction, genetic variation can arise from mutations. A **mutation** is a sudden change that occurs in the genetic material of an organism. These changes occur randomly and spontaneously, and may be caused by **radiation** or chemicals. A mutation in the gametes may produce a small change in the resulting offspring, a major effect in the offspring, or no noticeable effect in the offspring at all.

DEVELOPMENT OF DARWIN'S THEORY

Darwin recognized that there is variation among individuals produced by the same parents. He wondered how the differences *within* a group of organisms could somehow lead to differences *between* groups of organisms. Darwin

Figure 10-2 The first finches to arrive on the Galápagos Islands came from the South American mainland, hundreds of kilometers away.

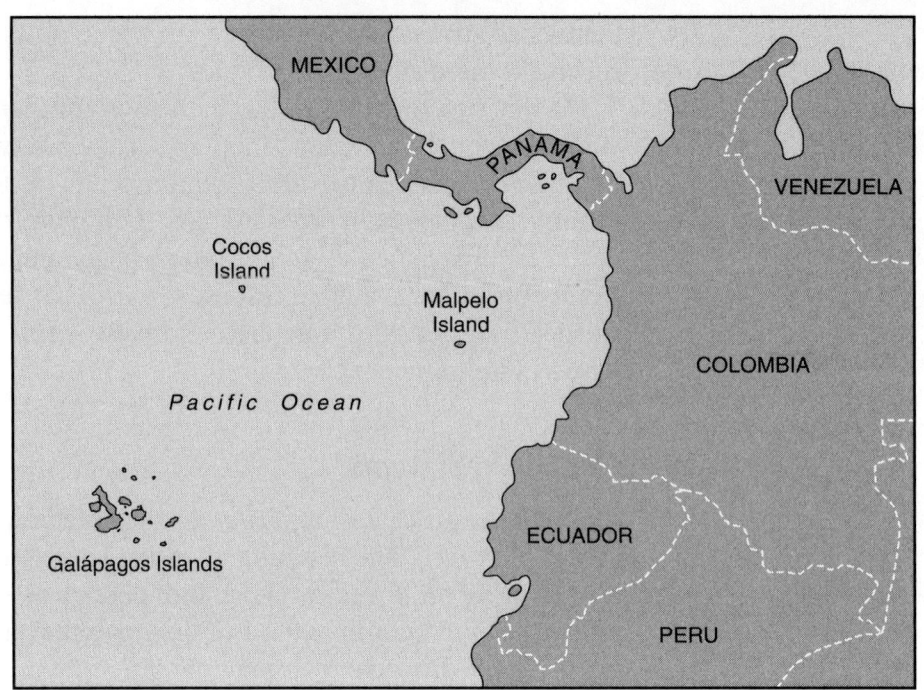

Ground finch eats seeds

Catches insects with beak

Tool-using finch digs insects from under bark with thorn

Figure 10-3 The beaks of these Galápagos finches show adaptations to different environments and a variety of food sources.

asked this question about many species he had observed during his travels, such as a group of small birds called *finches,* which he had collected on the Galápagos Islands. He realized that there probably were small differences among the first finches that arrived on these islands from the South American mainland. (See Figure 10-2.) How did these minor variations within the ancestral finch species lead to the significant differences that now exist between the groups of finches? In other words, how did they develop into separate *species?* (See Figure 10-3.) Darwin recognized two important facts that play a role in the development of new species:

- There is a struggle for existence, which limits the number of offspring that survive.
- There are differences among offspring due to their individual, inherited variations.

Perhaps more importantly, Darwin posed this question: What determines which individuals survive to reproduce and thus become the parents of the next generation of offspring? His answer to this question formed the basis for his theory of evolution. It also revolutionized our understanding of how various forms of life could have come to be.

Evolution by Natural Selection

The special characteristics that make an organism well suited to a particular environment are called **adaptations**. How do organisms evolve the adaptations that enable them to survive so well in a particular environment?

Darwin attempted to answer this question. He developed an answer by combining what he knew about the inheritance of traits with what he observed about an organism's struggle for existence. He concluded that whatever slight variations an organism had that gave it an advantage over other individuals in that environment, would make it more likely to survive. That is what is meant by "survival of the fittest." An organism that was more likely to survive also would be more likely to reproduce and pass on its genetic variations to future offspring. Those individuals that did not have such useful adaptations would be less likely to reproduce and pass on their characteristics. Darwin used the term **natural selection** to describe the way that environmental conditions determine which organisms survive to reproduce. Over time, the proportion of those more-fit individuals would increase in a population. This occurs because there is an increase in the gene frequency for the traits that give the surviving individuals an advantage. In this way, new species evolve as populations undergo significant changes in their characteristics.

ANTIBIOTIC RESISTANCE IN BACTERIA: A TYPE OF NATURAL SELECTION

A drug called *penicillin* can kill some bacteria that cause diseases in people, while leaving human cells unaffected. Penicillin was the first **antibiotic** discovered. Later, many more antibiotics with the ability to kill different bacteria were discovered. (*Note:* The prefix *anti* means "against" and *biotic* means "life"; the word refers to a drug's ability to kill bacterial life-forms.)

Over time, scientists noticed that some strains of bacteria that were once killed by antibiotics were no longer affected. The bacteria had developed a **resistance** to penicillin and some of the other antibiotics. How did the bacteria develop this resistance? Were there genetic variations that made some bacteria naturally resistant to the antibiotics, without having had any previous exposure to them? (See Figure 10-4.)

If some bacteria were resistant to antibiotics from the start, they would have a survival advantage when such chemicals were added to their environment. In fact, this is what happened. By killing off nonresistant bacteria,

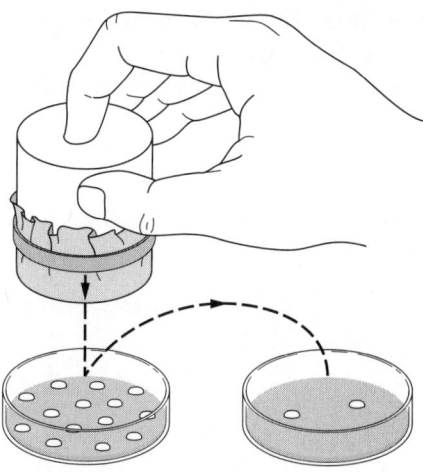

Figure 10-4 In this experiment, colonies of bacterial cells were transferred to a dish that contained an antibiotic. The only bacteria that survived were those that already had a natural resistance to antibiotics.

the antibiotics had decreased the competition for food that existed in the original population. The environment would "naturally select" for those resistant bacteria that could survive exposure to the chemical. Many more resistant bacterial cells could survive, grow, and reproduce. The result would be an entire strain of bacteria that has resistance to antibiotics. Similar results have been noted in some insects exposed to **pesticides**; those that are resistant can survive to reproduce. These are examples of natural selection at work—where external factors affect the survival of individuals within a population.

ARTIFICIAL SELECTION

People who raise dogs to perform certain tasks intentionally select, train, and breed those pups that have the characteristics best suited to their intended function. As noted in Chapter 11, this selection by people of organisms with specific characteristics is known as *selective breeding,* or **artificial selection**. It is a process that is similar to natural selection. However, humans—not the natural environment—select the organisms that have certain desirable traits and decide which ones will breed and pass on those traits to their offspring. Plant and animal breeders have practiced selective breeding for centuries, resulting in a variety of domestic animal breeds and crops that are quite different from their wild ancestors. (See Figure 10-5.)

Figure 10-5 The selective breeding, or artificial selection, by people for specific characteristics produces different breeds of animals, such as these dogs.

Chapter 10 Review

Multiple Choice

1. How did Darwin explain the fact that only a small number of offspring of any species survive to reproduce?
   A. Each species acts to limit the size of its own population.
   B. Every species is limited to a certain number of offspring.
   C. The members of a species allow only specific offspring to reproduce.
   D. Offspring must compete for available resources in order to survive.

2. Which statement best describes competition? It exists between individuals
   A. of the same species living in the same area only
   B. of different species living in the same area only
   C. of different species living in different areas only
   D. of the same species and different species living in the same area

3. Suppose two animals live in the same location and eat the same kind of food. What adaptation would decrease the competition between them?
   A. Both animals eat at the same time.
   B. Both animals breed at the same time.
   C. One animal has hair and the other has feathers.
   D. One eats during the day and the other eats at night.

4. Heredity is best described as
   A. a behavioral difference among offspring
   B. the struggle for existence among living things
   C. traits that are passed from one generation to the next
   D. the gradual change in organisms over many years

5. A couple had two children, one with blue eyes and the other with brown eyes. This difference is an example of
   A. natural selection
   B. artificial selection
   C. genetic variation
   D. acquired characteristics

6. Which description relates to an acquired characteristic?
   A. Jamal is tall and thin.
   B. Olivia has curly, blond hair.
   C. Brittney has a widow's peak like her father.
   D. Jose has large muscles from doing exercises.

7. What happens during asexual reproduction?
   A. Two organisms join together to become one new organism.
   B. A single parent organism splits to produce two organisms.
   C. Two organisms mate to produce a new single offspring.
   D. A single organism forms from the halves of two organisms.

8. In sexually reproducing organisms, the offspring inherit a combination of genetic traits from the
   A. mother only
   B. father only
   C. mother and father
   D. grandparents only

9. When compared to asexual reproduction, sexual reproduction produces
   A. less genetic variation among offspring
   B. greater genetic variation among offspring
   C. offspring that are identical to their parents
   D. offspring that are identical to one another

10. A mutation usually results from
    A. artificial selection carried out by humans
    B. the fact that only the fittest organisms survive
    C. a sudden change in the genetic material of an organism
    D. competition for resources such as food and water

11. When mutations occur in body cells, they can be passed along to
    A. sex cells only
    B. other body cells only
    C. offspring only
    D. gametes only

Chapter 10/The Theory of Evolution **139**

12. Which statement best describes the current understanding of natural selection?
    A. Natural selection influences the frequency of adaptive traits in a population.
    B. Changes in gene frequencies due to natural selection have little effect on evolution.
    C. Natural selection has been dismissed as an important concept in evolution.
    D. New combinations and mutations of genetic material are due to natural selection.

13. The Florida panther, a member of the cat family, has a population of fewer than 100 individuals and has limited genetic variation. Based on this information, a valid inference would be that the panthers
    A. will probably begin to evolve very rapidly
    B. can easily adapt to drastic changes in their environment
    C. are less likely to survive any changes in their environment
    D. will evolve to become more resistant to diseases

14. Which statement represents the major concept of the biological theory of evolution?
    A. A new species moves into a habitat whenever another species becomes extinct.
    B. Present-day organisms on Earth developed from earlier, different organisms.
    C. Every period of time in Earth's history had its own group of organisms.
    D. Every location on Earth's surface has its own unique group of organisms.

15. Which concept is *not* a part of the theory of evolution?
    A. Present-day species developed from earlier, different species.
    B. Complex organisms have developed from simpler organisms.
    C. Some species die out when environmental conditions change.
    D. Change occurs based on the needs of an individual organism.

16. Which statement best describes a rapid biological adaptation that has actually occurred?
    A. Pesticide-resistant insects have developed in certain environments.
    B. Paving large areas of land has decreased habitats for certain organisms.
    C. Scientific evidence indicates that large dinosaurs once lived on land.
    D. Characteristics of sharks have remained unchanged for a very long time.

17. When a breeder allows only the strongest and fastest horses to reproduce, she is practicing
    A. artificial selection
    B. natural selection
    C. artificial mutation
    D. asexual reproduction

18. Unlike in natural selection, in artificial selection
    A. genetic information is passed down from one generation to the next
    B. humans, not the natural environment, decide which organisms will reproduce
    C. the natural environment, not humans, decides which organisms will reproduce
    D. mating is random and all organisms may pass their traits on to their offspring

19. People can develop new varieties of cultivated plants by carrying out
    A. random breeding for all traits
    B. selective breeding for all traits
    C. random breeding for specific traits
    D. selective breeding for specific traits

20. The situation that would most likely result in the highest rate of natural selection in a population would be the reproduction of organisms
    A. by an asexual method in an unchanging environment
    B. in an unchanging environment that has few predators
    C. that have a low mutation rate in a changing environment
    D. that show genetic differences in a changing environment

21. Selective breeding for particular traits can be used to
    A. develop cultivated plants only
    B. develop domesticated animals only
    C. develop cultivated plants and domesticated animals
    D. breed rare, wild animal species only

22. Behaviors such as nest-building and caring for offspring are genetically determined in most species of birds. The existence of these behaviors is probably due to the fact that
    A. most birds do not have the ability to learn new behaviors
    B. these behaviors helped so many birds survive in the past
    C. individual birds need to learn to survive and reproduce
    D. within their lifetimes, birds developed these behaviors

23. According to the theory of evolution by natural selection, some organisms are more likely than others to survive and reproduce because they
    A. can pass on to offspring new characteristics they acquired during their lifetimes
    B. do not pass on to offspring any new characteristics they have acquired
    C. are better adapted to conditions in the environment than other organisms are
    D. tend to produce fewer offspring than others do within the same environment

24. According to modern evolutionary theory, genes for new traits that help members of a species survive in a particular environment will usually
    A. not change in frequency over time
    B. decrease rapidly in frequency
    C. decrease gradually in frequency
    D. increase in frequency over time

Analysis and Open Ended

25. How has the process of evolution led to the great diversity of species alive today?

26. Explain why the concept of a "struggle for existence" is important to the study of evolution.

27. In terms of evolution, why are the variations among individuals within a population more important than the similarities between them?

28. Give two possible causes of genetic mutations. What cells would they have to affect in order to be passed along to offspring? Explain.

29. Organisms compete for access to various resources. Identify at least three resources for which animals in the same population would compete. How is this "struggle for existence" affected by natural selection?

30. Which concept is best illustrated by the diagram, which shows (in simplified form) some changes in the body size and foreleg structure of horse species over time?

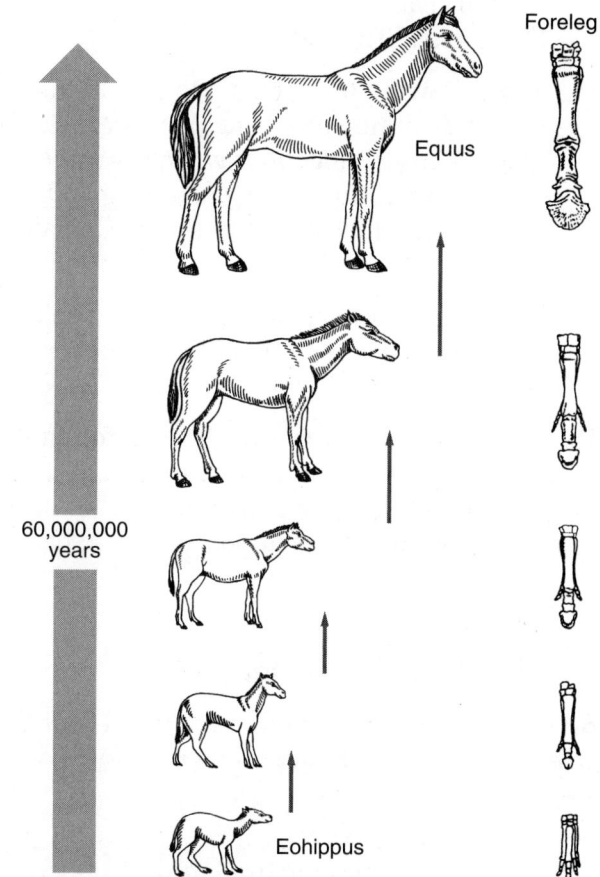

    A. acquired characteristics
    B. artificial selection
    C. genetic recombination
    D. evolution by natural selection

31. Why does sexual reproduction produce greater variation among offspring than asexual reproduction? How is this important for the process of evolution?

32. Briefly explain why mutations are important to evolutionary change in a population.

33. The best title for the chart below would be

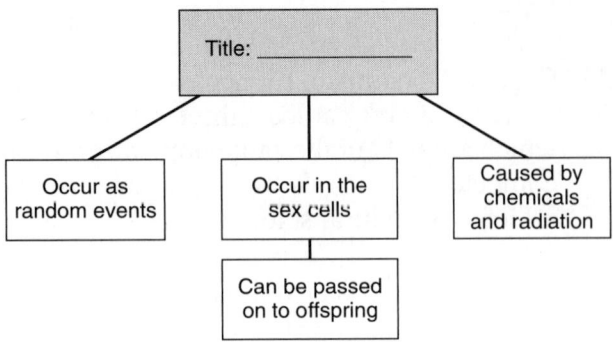

   A. Types of Natural Selection
   B. Characteristics of Mutations
   C. Survival of the Fittest
   D. Asexual Reproduction

34. The following terms relate to important factors in evolutionary change: *struggle for existence; natural selection; environmental change; variation among offspring.* Which concept includes the other three? Explain why.

35. Describe how the process of natural selection can lead to the evolution of new species over time.

36. A patient was given an antibiotic for an infection. The doctor told the patient to take it for 10 days; but the patient took it for only three days and then stopped because he felt better. After several days, the patient became sick with the same infection again. Why?

37. Suppose there are two types of fur color, brown and white, in a species of rabbit that lives in an area with very little snow all year. Most of the rabbits have brown fur. Then the environment changes so that there is snow most of the year. Based on your knowledge of natural selection, you might predict that the proportion of white fur to brown fur in the new climate would change so that
   A. equal numbers of rabbits would have brown fur and white fur
   B. more rabbits would have white fur than brown fur
   C. more rabbits would have brown fur with white patches
   D. more rabbits would have white fur with brown patches

38. Briefly define, and give an example of, an adaptation, and explain how it helps an organism survive.

39. Many pesticides have been used to kill insects that destroy crops or that spread diseases such as malaria. Unfortunately, some of these pesticides are no longer as effective as they once were for getting rid of insect pests. Explain why.

Reading Comprehension

Base your answers to questions 40 to 42 on the information below and on your knowledge of biology. Use one or more complete sentences to answer each question.

Antibiotics are used to treat infections in people and animals. Due to the enormous success of antibiotics, their use is very common worldwide. When we are ill, we have come to expect quick, effective treatment with antibiotics. Physicians often prescribe antibiotics at the earliest sign of an infection.

One result of the widespread use of these medicines is a growing number of antibiotic-resistant strains of bacteria. Some scientists have warned about the alarming possibility of infections that will not be treatable by the antibiotics we have. Already, one disease, tuberculosis—which was largely under control—has reappeared in a strain that is much more difficult to treat with antibiotics.

Recently, scientists became alarmed to find bacteria, in the food that is given to chickens, that are resistant to the most powerful antibiotics. Even though those particular bacteria were harmless, the finding raised the disturbing possibility that these bacteria could pass on their antibiotic resistance to disease-causing bacteria in

> chickens and, ultimately, in humans. One reason it is thought that such drug-resistant bacteria are being found more frequently is the heavy, routine use of antibiotics in farm animals.
>
> This is an issue for everyone to be aware of and concerned about. Science has provided us with a group of wonder drugs to treat diseases that once killed many people. However, we must be thoughtful and wise in our use of antibiotics. The laws of nature—in this case, the process of natural selection that produces resistance to antibiotics—can never be ignored.

40. How is the use of antibiotics a matter of both good news and bad news?
41. Why should people be concerned about the use of antibiotics in farm animals?
42. How is knowledge of the process of natural selection necessary in order to understand the problem of overuse of antibiotics?

Chapter 11

Evidence for Evolution

Standard 5.5.12 B1 Explain that through Earth's history, present species developed from earlier distinctly different species.

Standard 5.5.12 B2 Explain how the theory of natural selection accounts for extinction as well as an increase in the proportion of individuals with advantageous characteristics within a species.

TYPES OF EVIDENCE FOR EVOLUTION

Simple diagrams can be used to represent the evolutionary, or phylogenetic, relationships among different species. See, for example, the diagram in Figure 11-1. Suppose that letters *A, B, C,* and *D* represent four living species. The letters *E, F,* and *G* represent ancestral forms of the species that are most likely extinct. In this case, organisms *B* and *C* are more closely related because they evolved from their common ancestor *E* most recently. *B* and *C* are both equally related to *A,* with the more distant common ancestor *F.* Organism *D* is the least closely related to the others because it evolved from their common ancestor *G* the longest time ago.

No single idea explains the enormous diversity and complexity of life on Earth more powerfully than the theory of *evolution by natural selection* as proposed by Darwin. The types of evidence that support this theory include fossils, the shapes and structures of living organisms, similar features among embryos, the chemicals found in all living things, and the distribution of species on Earth today.

Evidence from Fossils

Fossils are the traces or remains of dead organisms that have been preserved by natural processes. Usually only the hard parts of organisms—that is, the bones, shells, or teeth—become fossilized. A common way fossils are formed is through the gradual replacement of an organism's remains by other substances. This process of fossil formation usually occurs when the organism is buried in sediments. After burial, the organism's hard tissues are slowly replaced by minerals dissolved in underground water. Over time, these minerals harden to form an exact copy of the original organism. In undisturbed layers, or *strata,* of sedimentary rock, each lower layer would be older than

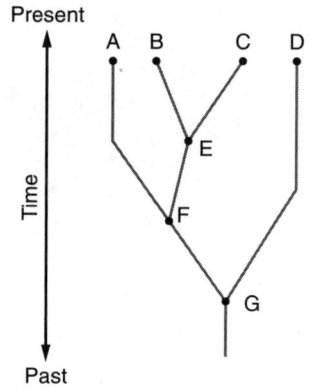

Figure 11-1 A diagram can be used to represent the evolutionary relationships among different living and ancestral species.

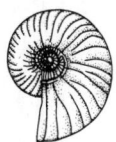
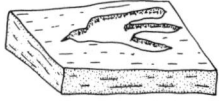

Figure 11-2 Fossils can form when minerals slowly replace hard body parts, such as shells or bones, or when the body of an organism creates an impression in soft mud or clay, which later hardens to form an imprint or mold.

the strata deposited above it. Therefore, scientists can infer that fossils found in lower strata are also older than fossils found in the strata above. Structural similarities between older and more recent fossils provide additional evidence of the evolutionary relationships among ancestral species.

Fossils can also be formed if the body of a plant or animal creates an impression in soft mud or clay. The material then hardens, forming an imprint or mold of the body. This process shows only the original external shape of the organism and not the internal structure, as do some other types of fossil formation. By studying the fossil record, scientists can see that species have changed over time and that most ancient life-forms no longer exist. (See Figure 11-2.)

Evidence from Comparative Anatomy

One way to determine the evolutionary relationships among different organisms is to find some similar parts, called *homologous structures,* in their anatomy that they inherited from a common ancestor. Homologous structures have similar forms, but different functions. For example, similarities exist in the forelimb bones of some very different animals. The wing of a bat, flipper of a whale, front leg of a cat, and arm of a human—although they appear to be quite different—are all made up of the same types of bones. These bones are attached to each other and to other bones in similar ways. The forelimbs indicate that, long ago, these four mammals all evolved from a common ancestral animal. (See Figure 11-3a on page 146.)

Sometimes a structure has little or no function in one organism, but is clearly related to a more developed structure that does function in another organism. This is called a *vestigial* (meaning "lost" or "trace") *structure.* (See Figure 11-3b on page 146.) For example, the appendix in humans is a small sac attached to the place where the small and large intestines meet. In appearance, the appendix is a smaller copy of the cecum, which is a large pouch found in plant-eating mammals such as rabbits. The cecum contains microorganisms that help digest plant materials the rabbit ingests. The fact that

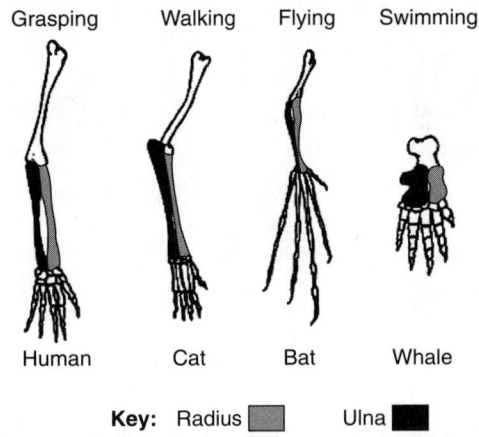

Figure 11-3a The similarities in the bones of each forelimb indicate that these four mammals shared a common early ancestor.

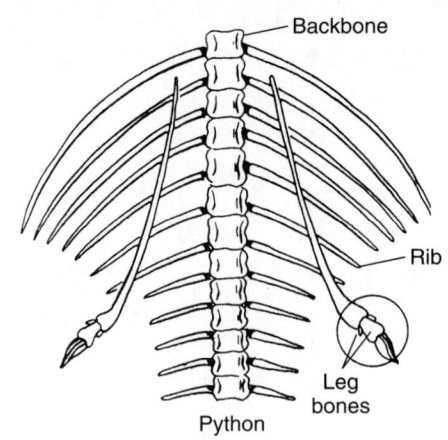

Figure 11-3b The skeletons of some snake species have tiny leg bones that serve no function. These vestigial structures show that snakes evolved from animals that had legs.

a similar organ is still useful in another species is evidence that humans probably evolved from an organism that also had this larger, functional structure.

Evidence from Comparative Embryology

Figure 11-4 illustrates five animals at very early stages in their embryonic development. Although all of these *embryos* resemble one another, they are actually the embryos of different species: a salamander, a chicken, a pig, a monkey, and a human. The similarity of the embryos shown in these diagrams provides evidence that all vertebrates follow a common plan in their early stages of development. This is due to the fact that these animals have similar sets of genes; and this similarity comes from their having had common ancestors.

Evidence from Comparative Biochemistry

The similar *chemistry* of living things, or **biochemistry**, provides some of the strongest evidence that organisms evolved from common ancestors long ago. All organisms store their genetic information, which is passed from one generation to the next in DNA molecules, in almost exactly the same manner. This genetic code shows that all organisms are related in fundamental ways.

Figure 11-4 The very young embryos of these five animals have many similarities, such as gill slits, which shows that they shared a common early ancestor.

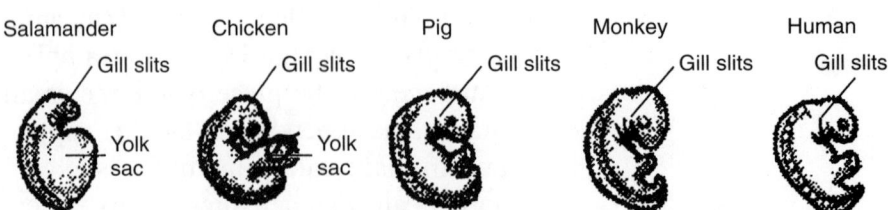

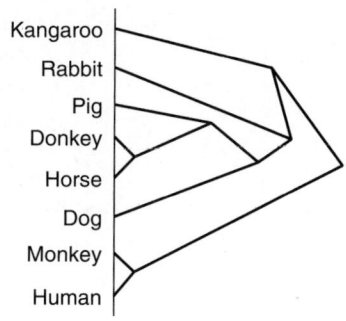

Figure 11-5 The evolutionary relationship between two organisms can be determined by comparing their DNA; the more similar their base sequences, the more recently they evolved from a common ancestor. For example, the donkey and the horse are more closely related than are the pig and the horse.

Proteins, a type of organic molecule in all living things, are made up of smaller units called *amino acids*. The same protein in two different species may be made up of similar but not identical amino acids. Biologists now know that a small number of amino acid differences in the same protein means that the two species are closely related in evolutionary terms. On the other hand, a large number of amino acid differences means that the two species are more distantly related.

By matching a DNA base sequence from one organism to a DNA sequence from another organism, scientists can determine if the sequences belong to organisms of the same, closely related, or distantly related species. Again, the greater the similarity is, the closer the species. The same is true for all other organisms on Earth. The evolutionary relationship between two organisms can be learned by comparing their DNA. The more similar their DNA sequences, the more recently the two organisms evolved from a common ancestor. (See Figure 11-5.)

EVOLUTION WITHIN POPULATIONS

Evolution occurs within a population as frequencies of inheritable traits change. This process occurs due to natural selection from one generation to the next. The peppered moth, carefully studied in England for more than a century, provides one of the best-known examples. When the peppered moth was first studied, most of its population was light colored. The moths were well camouflaged when they rested on trees and rocks covered with light-colored lichens.

In 1845, a dark-colored peppered moth was observed for the first time. At that time, soot and smoke produced by coal-burning factories had begun to pollute the air. The trees and rocks became dark with soot and the lichens began to die. As a result, the light-colored moths were easily seen against the darker backgrounds and became the easy prey of insect-eating birds. By 1900, most of the peppered moth population was dark colored. How did this happen? The darker moths were better camouflaged when resting on the tree trunks and rocks blackened by soot. It is important to understand that the light-colored moths did not change color; they were replaced by increasing numbers of dark-colored moths through natural selection. The dark-colored moths were less likely to be preyed on, and so were more likely to survive to

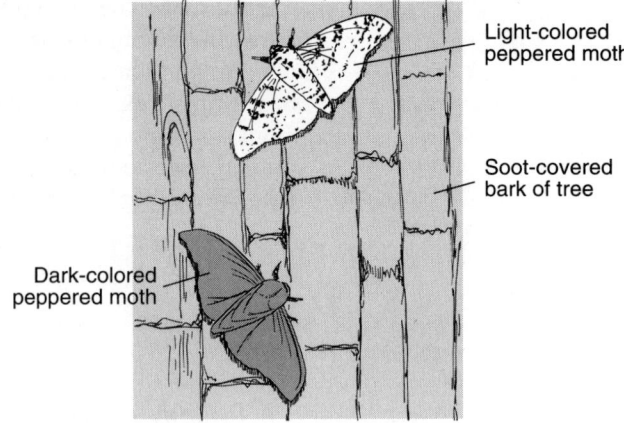

Figure 11-6 The changing frequencies of the light trait and dark trait in the peppered moth population are due to natural selection.

Figure 11-7 Sometimes evolution leads to the development of entirely new groups of species, such as the mammals, which arose during the time of the dinosaurs.

pass on their genetic traits. Interestingly, as air pollution decreased, more light-colored moths were once again seen in the study area. (See Figure 11-6.)

Sometimes evolution leads to the development of new adaptive features, new species, and new groups of species. Examples of new adaptive features include the legs of an amphibian (such as a salamander), the shell-encased eggs of a reptile (such as a turtle), and the large brain of a primate (such as an ape). Examples of entirely new groups of species that arose are the flowering plants and the mammals. (See Figure 11-7.) Another example of large evolutionary developments is both the appearance and the extinction of all dinosaur species.

Chapter 11 Review

Multiple Choice

1. All of the following can be used as evidence to support Darwin's theory of evolution *except* the
   A. similarity of chemicals in all living things
   B. distribution of species on the planet today
   C. shapes and structures of living organisms
   D. distribution of mountain ranges on Earth's surface

2. Which statement is best supported by evidence from the fossil record?
   A. Most of the organisms that lived on Earth in the past are now extinct.
   B. The struggle for existence between organisms results in genetic changes.
   C. Species occupying the same habitat have identical environmental needs.
   D. Structures such as leg bones and wing bones come from the same embryonic tissue.

3. Over time, fossils can be formed when an organism is
   A. buried in sediment; then its hard tissues are replaced by dissolved minerals
   B. buried in sediment; and then an impression of its soft tissues is formed
   C. preserved in soft mud or clay, with its bones and soft tissues all intact
   D. buried in mud; then its bones dissolve and its internal organs remain intact

Base your answer to question 4 on the diagrams below, which show the forelimb bones of three different mammals.

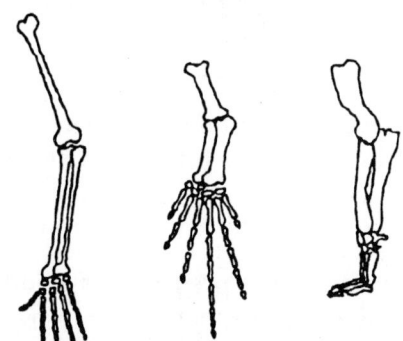

4. For these mammals, the number, position, and shape of the bones most likely indicates that they
   A. developed in the same environment
   B. have an identical genetic makeup
   C. developed from a common earlier species
   D. have identical methods of obtaining food

5. Suppose a scientist suggests that humans are distantly related to rabbits because the human appendix resembles the cecum of a rabbit. The scientist is probably using evidence from
   A. fossil remains
   B. embryology
   C. comparative anatomy
   D. comparative biochemistry

6. Two species that have only a small number of amino acid differences in the same protein probably
   A. are identical in their appearance
   B. share the same parent organisms
   C. are closely related in evolutionary terms
   D. are distantly related in evolutionary terms

7. A scientist using biochemistry to compare the evolutionary relationship between two organisms could analyze the similarities in their
   A. DNA base sequences
   B. embryo development
   C. homologous structures
   D. fossilized imprints

8. After the Industrial Revolution in England, many trees became covered with dark soot. It was noticed that the number of light-colored peppered moths decreased, while the number of dark-colored peppered moths increased. How can this be explained in terms of natural selection?
   A. The dark-colored moths chased the light-colored moths away from the soot-covered trees.
   B. The light-colored moths changed their colors in order to blend in with the soot-covered trees.
   C. Light-colored moths had a genetic variation that gave them an advantage over dark-colored moths.
   D. Dark-colored moths had a genetic variation that gave them an advantage over light-colored moths.

9. Which is an example of an evolutionary change at the population level?
   A. the development of legs on amphibians
   B. the evolution of large brains in primates

C. the replacement of light-colored moths by dark-colored moths
D. the appearance of the flowering plants group

Analysis and Open Ended

Base your answer to question 10 on the diagram below, which shows the evolutionary relationships of several living and extinct mammals.

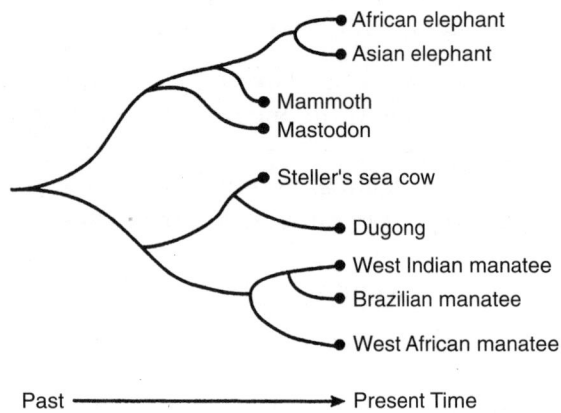

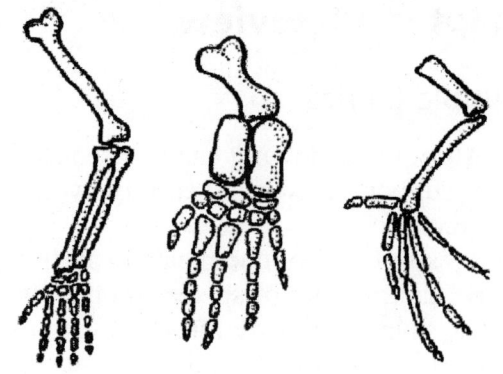

10. According to the diagram, which statement about the African elephant is correct?
    A. It is more closely related to the mammoth than it is to the manatees.
    B. It is not even remotely related to the Brazilian manatee or the mammoth.
    C. It is more closely related to the West Indian manatee than it is to the mastodon.
    D. It is the common ancestor of the dugong and the West African manatee.

11. Explain how a family tree can be used to show evolutionary relationships among organisms.

12. Describe two ways that fossils can form, and tell which parts of an organism are usually fossilized. Your answer should explain the following: (a) how fossils are formed through replacement by minerals; and (b) how fossil imprints or molds are formed.

13. Explain how studies of similarities in the biochemistry of proteins can be useful in determining evolutionary relationships among organisms.

Use the following diagrams, which illustrate the forelimb bones of three different mammals, to answer question 14.

14. Differences in the bone arrangements support the hypothesis that these animals
    A. are probably members of the same species
    B. have adaptations for different environments
    C. most likely have no ancestors in common
    D. all contain the same genetic information

15. Why would an "evolutionary bush" be a more accurate way to illustrate relationships among species than just a linear, or ladder-like, diagram?

16. The diagrams below show the bones in the forelimbs of a cat and a human. The similarities between these appendages suggest that humans and cats

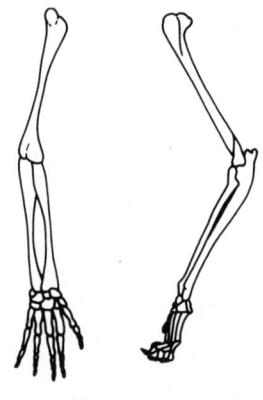

Human   Cat

   A. have identical genetic material
   B. have the same direct ancestor
   C. once shared a common ancestor
   D. evolved in the same environment

17. Explain why the presence of a body structure with no current function can provide evidence of an evolutionary relationship. Give an example.

18. The following diagram represents a series of undisturbed sedimentary rock layers in a

given area. Several layers show representative fossils of different organisms. Relative to those in the other layers, the oldest fossil would be found in the

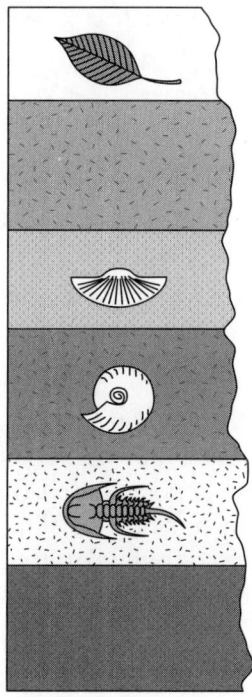

A. first layer, on the top
B. third layer from the top
C. fourth layer from the top
D. fifth layer from the top

19. Refer to Figure 11-6, which shows the light-colored and dark-colored peppered moths. Over time, depending on changes in the environment, the percentage of each color type in their population has varied. Are these changes in frequency due to natural selection, artificial selection, or acquired traits? Explain.

20. The diagrams below represent the embryos of three different vertebrate species. It is thought that they provide evidence of evolution based on their similar

A. sizes
B. fossils
C. structures
D. molecules

21. As stated in the text, "sometimes evolution leads to the development of new adaptive features" within populations. Describe how such a change may eventually result in the development of a completely new species. Give either a real or an imagined example.

22. The three species shown below have similar enzymes, hormones, and proteins; this supports the idea that they share a common ancestor, based on their similar

A. external structures
B. biochemistry
C. feeding habits
D. behavioral patterns

23. The best title for the chart below would be

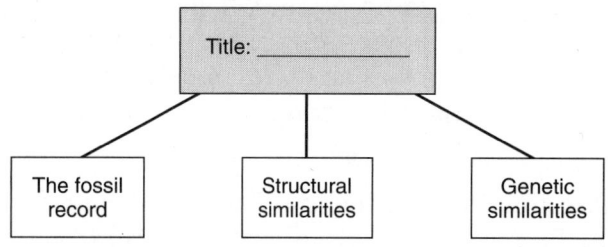

A. Evolutionary Pathways
B. Evidence for Evolution
C. Natural Selection
D. Mutations in Evolution

Chapter 11/Evidence for Evolution **151**

24. Use data from the diagram at right to explain why DNA nucleotide sequencing is important to the study of evolution.

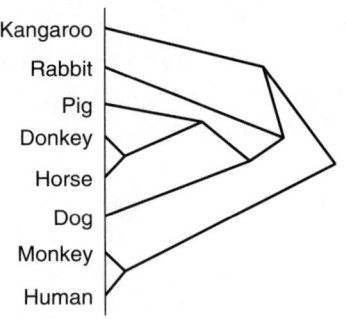

Reading Comprehension

Base your answers to questions 25 to 27 on the information below and on your knowledge of biology. Use one or more complete sentences to answer each question.

> In 1987, a group of scientists thought they could use a better method to study human evolution. Instead of studying bits and pieces of fossil remains, they decided to study the genes contained within the cells of living people. These genes, passed from generation to generation, have stored within them a history of our origins. The molecular biologists decided to examine the DNA that is located in our cells' mitochondria. Unlike ordinary DNA—the genetic material in the nuclei of our cells that we get from both parents—mitochondrial DNA (mtDNA) in our cells comes only from our mother. From one generation to the next, mtDNA never gets mixed with the DNA in the genes we get from our father.
>
> The researchers collected mtDNA from women living in many parts of the world. By studying the similarities and differences in mtDNA among these women, the researchers were able to look back in time to study the origins of human history. Their startling conclusion was that the molecular evidence indicated that all humans alive today are the descendants of a single female who lived in Africa about 200,000 years ago. Some people began to call this person "Mitochondrial Eve."
>
> Since 1987, scientists have disagreed widely on the results. Some scientists claim that the computer program the researchers used for their analysis was not used correctly. Others think that Mitochondrial Eve lived only 150,000 years ago. Still other scientists point to evidence showing that modern humans may have evolved much earlier—and in several parts of the world, not only in Africa.
>
> In 2001, researchers determined that mtDNA from a human fossil found years earlier in Mungo, Australia, showed no linkage to any humans living today. Therefore, "Mungo Man," as the fossil is called, could not have descended from Mitochondrial Eve. "Put the gloves on, Mitochondrial Eve, because Mungo Man has stepped into the ring," began an article on the topic, showing how the debate on our origins continues. This kind of open discussion is what science is all about—questions are asked, answered, and then, when more evidence is found, even more questions arise. For now, the answer to this question about our ancestry remains undecided, although the "Out of Africa" model is still the most popular.

25. How is mitochondrial DNA different from the ordinary DNA found in cell nuclei?

26. Compare how these scientists used mitochondrial DNA with the way other scientists have used fossils.

27. Scientific research often produces some answers and then even more questions. How is this true about the Mungo Man and Mitochondrial Eve research?

Chapter 12

Mechanisms of Evolution

Standard 5.5.12 B2 **Explain how the theory of natural selection accounts for extinction as well as an increase in the proportion of individuals with advantageous characteristics within a species.**

ADAPTATIONS TO THE ENVIRONMENT

Every species lives in a particular place; and every place on Earth has specific conditions, such as average air temperature, monthly rainfall, kinds of minerals in the soil, and wind speeds. Darwin's theory states that those organisms that are best suited to tolerate the conditions of their environment—that is, the ones that have beneficial traits—will be most likely to survive and pass their traits on to their offspring. Since environments on Earth are constantly changing, however slowly, the evolution of living things is an ongoing process.

No individual organism intentionally changes to survive in a particular environment. Different adaptations in living things occur by chance as a result of the genetic variations within a population. Sometimes an adaptation works well for an organism—that is, it helps it survive—and sometimes it does not. The *adaptive value* of a trait is determined by the specific conditions of the environment. For example, at first, the dark coloration of some peppered moths did not aid their survival. In fact, it made those moths more visible to predators whenever they landed on lichen-covered trees and rocks. The dark moths tended to be eliminated by natural selection. Yet that same dark coloration gave some moths an advantage when pollution darkened the surfaces of trees and rocks. In that environment, the trait for darker color was "chosen" by natural selection, and more dark moths lived to pass on their genes. If the main predator of moths did *not* hunt by sight, the darker coloration would have provided neither an advantage nor a disadvantage to the moths. (That is, it would not be selected *for* or *against* by the environment.)

It is through the process of natural selection that species, not individual organisms, evolve. Over time, a species' traits make a remarkable fit with its environment. If they do not, the species will probably not survive in that environment.

Figure 12-1 The different shapes and structures of leaves are physical adaptations of plants to their various environments.

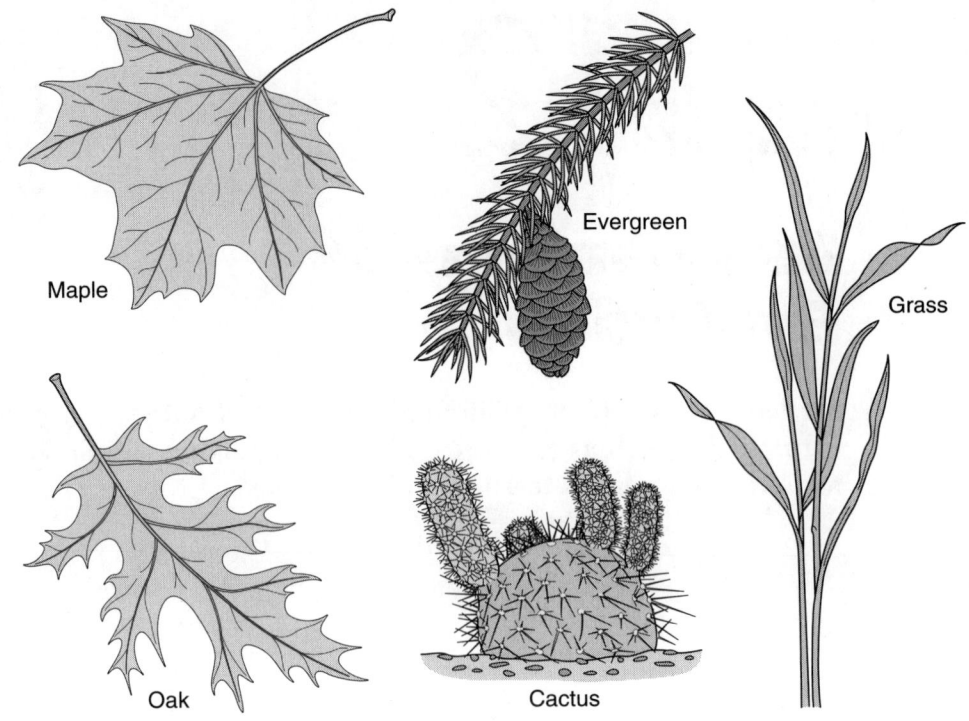

Different Types of Adaptations

Some of the most common types of adaptations are called *physical adaptations;* these involve the size, shape, color, and structure of organisms or the parts of organisms. For example, camels have extremely wide two-toed feet to avoid sinking into the desert sand. Plant leaves are adapted in form and structure to the conditions of their environment, such as temperature, amount of sunlight, and availability of water. These physical adaptations develop over time as plant populations adapt to changes in their environment. (See Figure 12-1.)

An adaptation to conditions in the environment may also involve *behavioral adaptations,* which include the functions and behaviors of an organism. The building of nests by birds is an example of an adaptive behavior. To survive the long, cold winters when food is scarce, some animals, such as the black bear, slow their metabolic functions. This physical adaptation leads to *hibernation,* a type of behavioral adaptation in which the bear retreats to a den to sleep during the winter months.

SPECIATION AND REPRODUCTIVE ISOLATION

In order for a population to change, it must be physically separated from other populations of its kind, usually for a long time. As a result of evolution by natural selection, such a population may change so much that its members are no longer able to reproduce with similar members of any other population. This situation is known as **reproductive isolation**. The population would have undergone *speciation;* that is, it has become a new species.

Figure 12-2 Geographic isolation of the ancestral finches led to the formation of several species of Galápagos finches, each with a different type of beak.

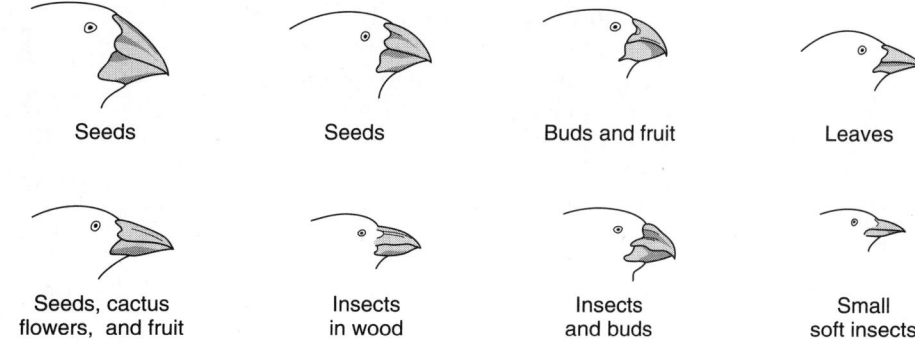

Sometimes a species will evolve to fill a niche in the environment that has become available, possibly due to the extinction of another species. A **niche** includes all the things an organism does to survive, such as how it gets its food, reproduces, finds shelter, and avoids predators. A new species in a particular niche will likely have some physical and/or behavioral adaptations that give it an advantage (over other species) for surviving in that niche.

SPECIATION AND GEOGRAPHIC ISOLATION

The most common type of separation that leads to the formation of new species is **geographic isolation**. An actual physical barrier, such as a river or a mountain, can prevent organisms from moving between related populations. For example, the Galápagos Islands, which were colonized by finches from mainland South America, each have different environmental conditions. Due to natural selection, the birds changed in different ways on each of the islands. Those individuals that had advantageous traits for the particular conditions on their island survived to produce offspring. Over time, the finches evolved into several species, in a process known as *adaptive radiation*. (See Figure 12-2.)

In addition to islands, there are other types of geographically isolated areas in which speciation can occur. Examples include mountaintops, lakes, and forests that are separated by different features of the landscape. (See Figure 12-3.)

Figure 12-3 Three examples of physical barriers that can cause geographic isolation.

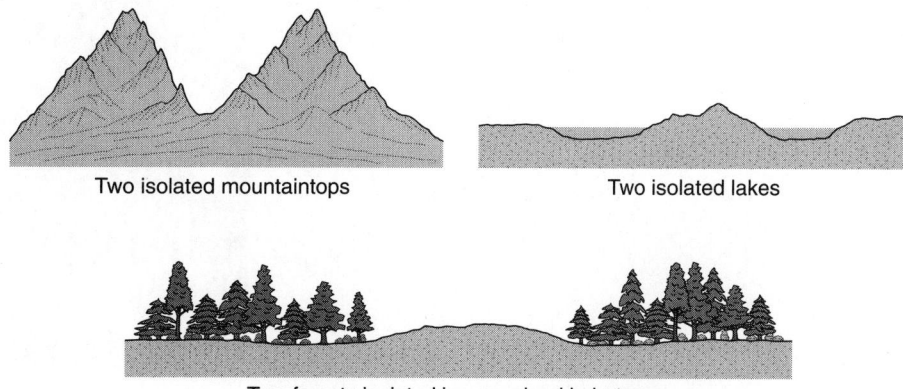

Chapter 12/Mechanisms of Evolution

Figure 12-4 The woolly mammoth went extinct about 10,000 years ago, most likely due to changes in its environment to which it could not adapt.

EXTINCTION OF SPECIES

Another natural process that happens over time is **extinction**, the complete disappearance of a species from Earth. Extinction occurs when the members of a species can no longer reproduce enough offspring to keep that species in existence. This inability to produce sufficient offspring may occur when members of the population cannot adapt to changes in their environment. For example, a significant increase in average temperature or decrease in yearly rainfall may affect an organism's ability to find food, survive, and reproduce. (See Figure 12-4.) The extinction of a species may also arise from problems that occur within a population. Harmful genetic traits that become widespread in a population—particularly a population that already has been reduced due to other events—may cause a species' extinction.

At particular times in Earth's history, *mass extinctions* have occurred. These extinctions are usually due to natural events that drastically change the planet's climate. For example, the impact of an asteroid or large comet is thought by many scientists to have caused the mass extinction of the dinosaurs and numerous other species. (See Figure 12-5.) In a mass extinction, thousands of species disappear forever. Largely due to these mass extinctions, about 99 percent of all plant and animal species that have ever existed on Earth have since become extinct.

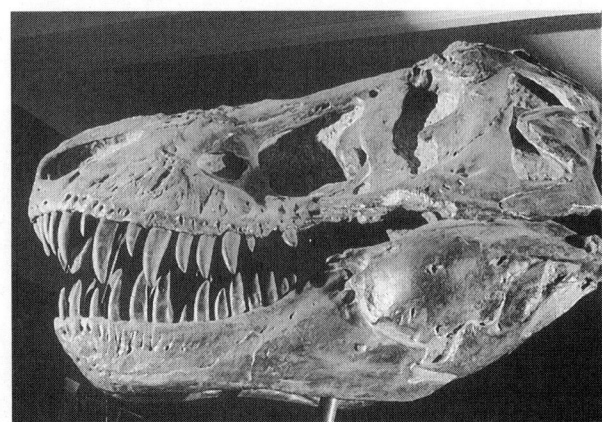

Figure 12-5 Mass extinctions are usually due to natural events. Dinosaurs, such as this *T. rex,* went extinct about 65 million years ago—probably because of an asteroid or comet impact that caused drastic changes to Earth's climate.

Chapter 12 Review

Multiple Choice

1. Which example describes an adaptation that aids survival?
   A. Being tall enables a giraffe to feed on the leaves of trees that other animals cannot reach.
   B. A person's poor eyesight makes it difficult for him to see without wearing glasses.
   C. A white peppered moth is clearly visible against the background of a dark-colored tree.
   D. The broad leaves of a maple tree shrivel up when placed in the hot climate of a desert.

2. The needle-shaped leaves of evergreen trees prevent the loss of valuable moisture. This is a type of
   A. behavioral adaptation
   B. physical adaptation
   C. geographic isolation
   D. reproductive isolation

Base your answer to question 3 on the following diagrams, which illustrate the change that occurred in the frequencies of different fur colors in a rabbit population over a 10-year period.

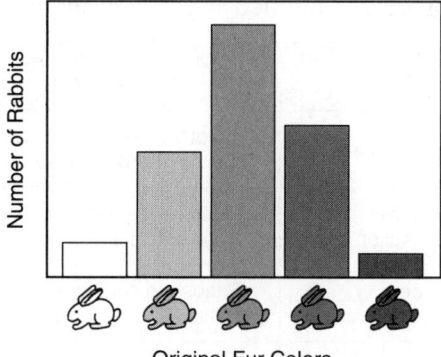
Original Fur Colors

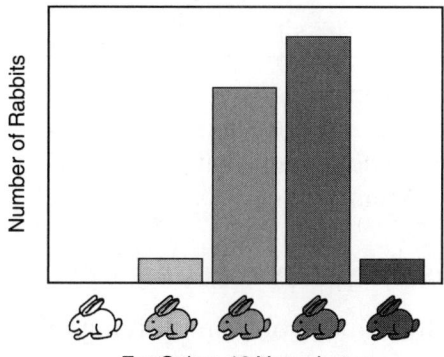
Fur Colors 10 Years Later

3. Which circumstance would best explain this change over time?
   A. a decrease in the mutation rate of the rabbits with black fur
   B. an increase in the advantage of having white fur
   C. a decrease in the advantage of having white fur
   D. an increase in the chromosome number of rabbits with black fur

4. In an area in Africa, temporary pools form where rivers flow during the rainy months. Some fish have developed the ability to use their fins as "feet" to travel on land from one of these pools to another. Other fish in these pools die when the water dries up. What might happen in this area after many years?
   A. The fish using their fins as "feet" will be present in increasing numbers.
   B. The fish using their fins as "feet" will develop real feet like salamanders.
   C. The other types of fish will develop the same fins that function as "feet."
   D. All of the different fish will survive and produce equal numbers of offspring.

5. One explanation for the variety of organisms present on Earth today is that over time
   A. each niche has changed to support a certain variety of organism
   B. new species have evolved that fill available niches in the environment
   C. evolution has caused the appearance of organisms that are all similar
   D. the environment has remained unchanged, causing rapid evolution

6. The original Galápagos finches eventually branched into 13 new species as a result of
   A. the eruption of volcanoes on the islands
   B. their geographic isolation on different islands
   C. the extinction of other species on the islands
   D. artificial selection by researchers on the islands

7. According to modern evolutionary theory, genes responsible for new traits that help a species survive in a particular environment will usually
   A. not change in frequency over time
   B. decrease rapidly in frequency

C. decrease gradually in frequency
D. increase in frequency over time

Base your answer to question 8 on the information in the following paragraph.

As the Colorado River formed the Grand Canyon, a population of squirrels gradually became separated. The conditions on the northern portion of the canyon were different from those on the southern portion. The squirrels on the northern portion evolved into a different species of squirrel.

8. The situation described above is an example of
   A. extinction, in which a species could no longer survive on Earth
   B. migration, in which a species traveled to a new environment
   C. distribution, in which members of the same species are distributed randomly
   D. isolation, in which a new species evolved due to its geographic separation

9. The complete disappearance of a species from Earth is known as
   A. isolation
   B. speciation
   C. extinction
   D. adaptation

10. The woolly mammoth became extinct when the individuals of this species
    A. learned to adapt to changing climate conditions
    B. could not adapt to a changing climate and reproduce
    C. moved to an area with a different environment
    D. survived beyond a specific period of geologic time

Base your answer to question 11 on the table below and on your knowledge of the process of evolution.

Habitat	Number of Toes	Type of Horse Species
Plains	One toe (hoof)	Modern horse (*Equus*)
Forest	Four toes	Ancestral horse (*Eohippus*)

11. You could infer that modern horses have fewer toes than ancestral horse species had because the
    A. changed habitat wore down their side toes as they ran faster over the plains
    B. ancestral horse species mated with a population of mutant one-toed horses
    C. people who first rode them preferred horses that had one large hoof per foot
    D. changed habitat favored survival of faster horses, which had reduced side toes

12. Of all the species that have ever existed on Earth, approximately what percentage is now extinct?
    A. 5 percent   C. 50 percent
    B. 20 percent   D. 99 percent

13. According to the theory of evolution, the rates of evolution for Earth's different species are all
    A. identical, because all of Earth's species live on the same planet
    B. identical, because all species are equally at risk of becoming extinct
    C. different, because each species adapts to its particular environment
    D. different, because each species has access to unlimited resources

14. Which process listed in the table below is correctly matched with its explanation?

Process	Explanation
Extinction	Adaptive characteristics of a species are not adequate
Natural selection	The most complex organisms survive
Gene recombination	Genes are copied as a part of mitosis
Mutation	Overproduction of offspring takes place within a certain population

   A. Extinction
   B. Natural selection
   C. Gene recombination
   D. Mutation

Analysis and Open Ended

15. Why is the evolution of living things considered to be an ongoing process?

Base your answer to question 16 on the diagrams below.

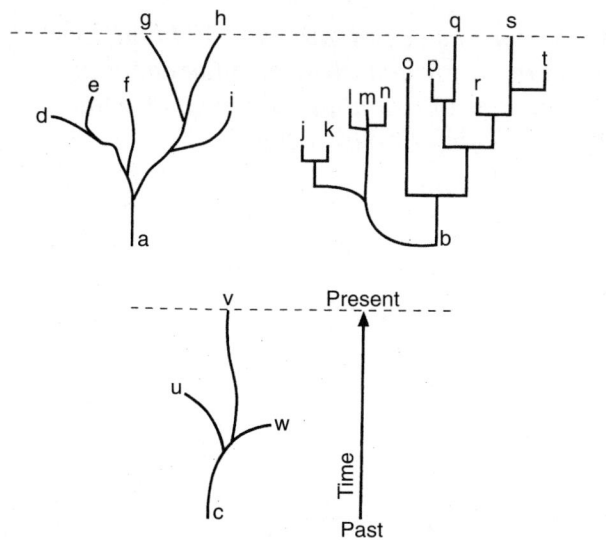

16. According to some scientists, patterns of evolution can be illustrated by diagrams such as those shown above. Which statement best explains the patterns seen in these diagrams?
    A. The organisms at the end of each branch still can be found living in the environment today.
    B. Evolutionary changes give rise to a variety of organisms, some of which continue to change while others die out.
    C. The organisms that are living today have all evolved at the same rate and have all undergone the same kinds of changes.
    D. These patterns can only be used to illustrate the evolution of organisms that are now extinct.

17. What determines the "adaptive value" of a trait within a population? Give an example, either real or imagined.

18. Describe one type of behavioral adaptation of an organism. Why might this behavior change over time?

19. How does natural selection affect the development of behaviors? In what way is this related to the evolutionary process?

Base your answer to question 20 on the data in the following paragraph and map.

Thousands of years ago, a large flock of hawks was driven from its normal migratory route by a storm. The birds scattered and found shelter on two distant islands, shown on the map below. The environment of Island A is very similar to the hawk's original nesting region. But, the environment of Island B is very different from that of Island A. The hawks have survived on these two islands to the present day with no interbreeding between the two populations.

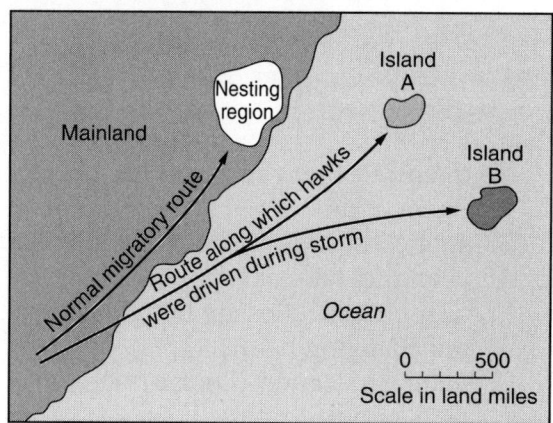

20. Which statement most likely predicts the present-day condition of these two hawk populations?
    A. The hawks that adapted to Island B have changed more than those on Island A.
    B. The hawk populations on Islands A and B have undergone identical mutations.
    C. The hawks that landed on Island A have evolved more than those on Island B.
    D. The hawks on Island A have given rise to many new species of hawks.

21. The following diagrams indicate that the frequency of unspotted beetles is decreasing relative to the frequency of spotted beetles in this population of insects. Possible explanations for the changing frequencies of these traits include all of the following *except* that

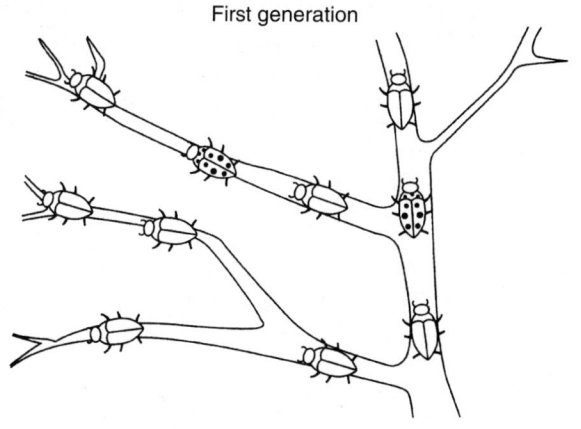

Chapter 12/Mechanisms of Evolution 159

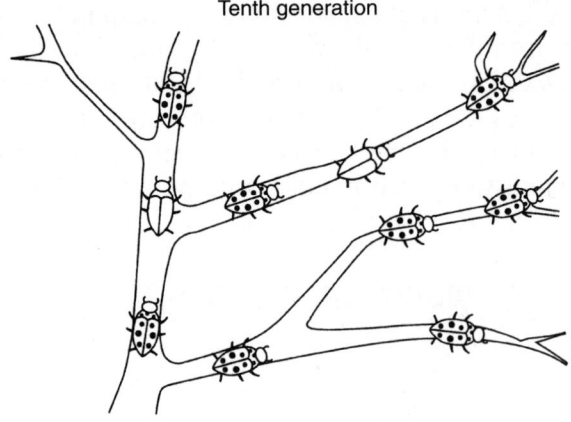

Tenth generation

A. the beetles' environment has been changing over time
B. spotted beetles are better adapted to the changing habitat
C. unspotted beetles are better adapted to the changing habitat
D. natural selection is occurring, which affects survival rates

Answer questions 22 and 23 based on the information in the paragraph below.

The variation of organisms within a population increases the likelihood that at least some members of the species will survive changing environmental conditions. A large population of houseflies was sprayed with a newly developed, fast-acting insecticide. While most of the houseflies were killed off, some houseflies that were resistant to the new insecticide survived.

22. The changing environmental condition in this case was the
    A. original population of houseflies
    B. appearance of resistant houseflies
    C. newly developed fast-acting insecticide
    D. houseflies that were exposed to the spray

23. Which of the following items represents the variation that enabled some flies to survive the changing conditions?
    A. The insecticide was new and fast acting.
    B. Most of the flies were killed by the spray.
    C. Some flies were resistant to the spray.
    D. Only some of the flies were sprayed.

Base your answers to questions 24 and 25 on the graph below, which illustrates the changing percentages of two physical varieties within a species' population over time.

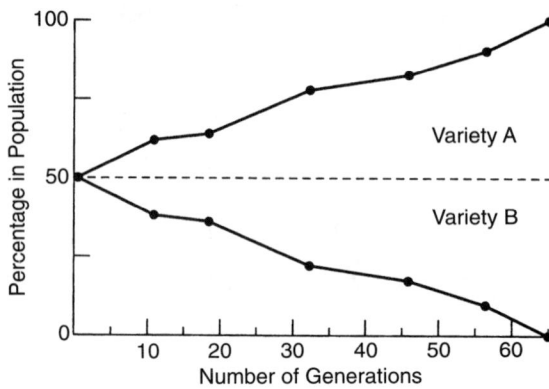

24. Which variety will most likely contribute to this population's physical traits in the future?
    A. variety A only
    B. variety B only
    C. both varieties A and B
    D. neither variety A nor B

25. What is the probable reason that the percentage of variety A is increasing while the percentage of variety B is decreasing?
    A. There is no opportunity for variety A to mate with variety B.
    B. Variety A has some adaptive feature that variety B does not have.
    C. Variety B has some adaptive feature that variety A does not have.
    D. There is no genetic variation between variety A and variety B.

26. Distinguish between a situation in which variations within a population enable some organisms to survive changing conditions and a situation in which a mass extinction occurs. What is the main difference in each situation's cause and effect?

Reading Comprehension

Base your answers to questions 27 to 29 on the information below and on your knowledge of biology. Use one or more complete sentences to answer each question.

> Domesticated ferrets have become popular as pets in recent years. Their clever antics can be a source of amusement. The unusual appearance of the small, long, slender, furry body of a pet ferret at the end of a leash always draws the attention of onlookers. However, there is a much more serious story about a wild species of ferret—a story of life, death, and near-extinction.
>
> The black-footed ferret is the only ferret species native to North America. These animals lived mostly in western parts of Canada and the United States. The black-footed ferret has a black face, a black tip on its tail, and—as its name suggests—black feet. It is a carnivore and survives mainly on a diet of prairie dogs. The only places where it has ever been found are near prairie dog tunnels. Prairie dogs are not well liked by farmers because their tunnels interfere with the planting of crops. Ranchers do not like them either, because they think their cattle can fall and injure themselves when they step in the openings of prairie dog tunnels. Farmers often put poison in the tunnels to kill prairie dogs; and when prairie dogs are eliminated, black-footed ferrets also die. As a result, ferrets became so rare that by 1979 they were thought to be extinct in the wild.
>
> However, to the great delight of wildlife biologists, a small population of black-footed ferrets was discovered living in a field near Meeteetse, Wyoming, in 1981. A species thought to be extinct was, in fact, still here! However, by the end of 1987, scientists counted only 18 survivors, so these animals were captured. In time, a captive-breeding program produced 400 individuals, which were released in seven different areas. Black-footed ferrets have now been reintroduced into the wild in Wyoming, Montana, South Dakota, and Nebraska. By 2002, they were also being released 100 miles south of the United States border, in Mexico, where many healthy prairie dogs live. The hope of wildlife biologists is to reach a goal of 1500 free-living black-footed ferrets by 2010.
>
> The successful reintroduction of black-footed ferrets to the wild—along with the protection of prairie dogs and their grasslands habitat—has brought this interesting animal back from the brink of extinction. Still, we will never know what other species were unintentionally eliminated when the early settlers eradicated 90 percent of the North American prairie dog population.

27. Describe the connections and conflicts between farmers, prairie dogs, and black-footed ferrets.
28. Why did scientists in 1987 capture all the wild black-footed ferrets in Wyoming?
29. How does the protection of prairie dogs provide protection to black-footed ferrets?

Unit IV
Reproduction and Heredity

STANDARD 5.5.12 C
Characteristics of Life

All students will gain an understanding of the structure, characteristics, and the basic needs of organisms and will investigate the diversity of life.

Enduring Understanding III Information passed from parent to offspring is coded in deoxyribonucleic acid (DNA) molecules. The molecular structure of the DNA molecule is consistent in all livings things and similar in members of a species: variance in the sequence of DNA bases in an organism gives it its unique characteristics. The information in DNA provides instructions for assembling protein molecules in cells.

There are predictable patterns of inheritance. Asexual reproduction produces offspring that have the same genetic code as the parent and leads to less variation in a species.

Sexual reproduction produces offspring with a mixture of DNA, increasing the genetic variation of an organism, and therefore the species.

Chapter 13
DNA and Heredity

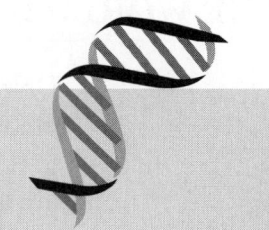

Standard 5.5.12 C1 Describe how information is encoded and transmitted in genetic material.

DNA: THE GENETIC MATERIAL

Heredity is the process by which organisms pass on their genetic information to their offspring. During the 1940s and 1950s, several scientists conducted research to determine if it was the protein or the DNA within a cell's *chromosomes* that contained the genetic material. As a result of careful experimentation and chemical analyses, they discovered that **DNA (deoxyribonucleic acid)** contains the information on which all life depends; that is, DNA *is* the genetic material. This substance, which serves as the genetic material, has the most significant job in the world: to carry on life itself. In order to carry out this job, the genetic material must do the following:

- It must be able to store information that can be passed on from one generation of cells to the next. It must be able to store enough information to make an organism like a tree or like you.
- It must be able to make a copy of itself in order to pass its information on again and again.
- It must be strong and stable so that it does not easily fall apart and perhaps cause harmful changes to its store of information.
- It must be able to mutate, or change, slightly from time to time. These changes allow a species to produce the variations on which natural selection acts, which can lead to the evolution of new species.

We can now look at how the DNA molecule is built and how it functions to do these jobs.

THE WORLD LEARNS OF THE DOUBLE HELIX

DNA is made up of smaller **subunits**. These subunits, or **nucleotides**, include four types of *bases,* which occur in two pairs. A chemist, Erwin Chargaff, had discovered that the amounts of bases adenine (A) and thymine (T) are always the same (A pairs with T); and the amounts of bases guanine (G) and cytosine (C) are always the same (G pairs with C), too. In addition, x-ray pictures of DNA (called *x-ray diffraction patterns*) taken by researchers Rosalind Franklin and Maurice Wilkins indicated that the DNA molecule has a spiral, or *helical,* pattern. In 1953, based largely on these findings, scientists James Watson and Francis Crick constructed a model and described the structure of DNA, for the first time, as a double helix. (See Figure 13-1.)

To understand the double-helix structure of DNA, picture a ladder that has been twisted. The two sides of the ladder are parallel to each other, and the steps of the ladder link the two sides together. The sides of the ladder are the backbone of the DNA molecule (composed of alternating sugar and phosphate molecules). Stretching between the two sides (forming the steps) are the pairs of bases. The Watson-Crick model showed that the only possible way all the parts could fit was for each large adenine base to be matched opposite a smaller thymine base. Similarly, the large guanine base had to be opposite a smaller cytosine base. (See Figure 13-2.)

So, a molecule of DNA consists of two strands, opposite each other, connected by matching base pairs. If we look at one strand, we can describe it in terms of the order, or sequence, of its subunits. Because the subunits are in a long line, the order of the subunits is called a *linear sequence.* This linear sequence of nucleotides builds the DNA molecule, which may be very

Figure 13-1 Scientists James Watson (left) and Francis Crick (right), shown in 1953 with their model of part of a DNA molecule.

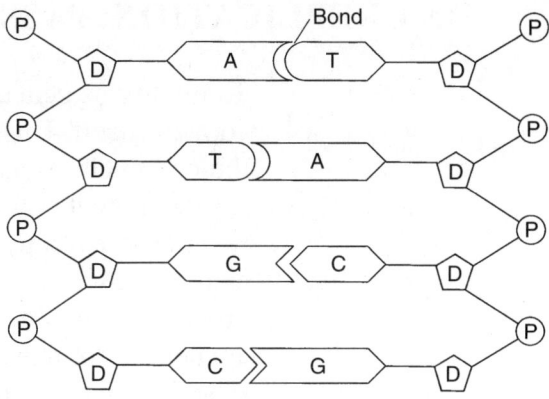

Figure 13-2 The structure of a DNA molecule—the nucleotide subunits include four types of bases (A, T, C, and G).

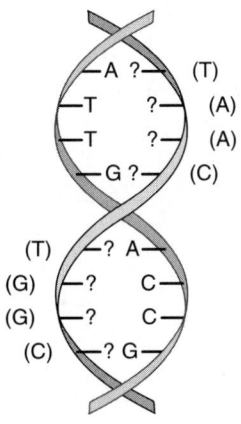

Figure 13-3 From the sequence of bases on one strand of DNA, we can determine the sequence on the opposite strand: A pairs with T, and C pairs with G.

long. (Recall that long molecules such as DNA, which can contain thousands of nucleotides in a sequence, are called *polymers*.)

Imagine walking along a single strand of DNA. The bases in the subunits may occur in any order. The linear sequence on a short molecule of DNA might be A-T-T-G-A-C-C-G. Now imagine walking along the opposite strand, starting at the same place. Opposite the A in the first strand is a T. Because we know the sequence of bases in the first strand we automatically know the sequence of bases in the other strand. In this example, beginning with the T, the sequence must be T-A-A-C-T-G-G-C. This is the key to how the DNA molecule copies itself. The process by which DNA copies itself depends on the matching base pairs in the subunits of each strand. What is so important about the order of the subunits in a strand of DNA? The sequence of bases in the subunits *is* the genetic information that the strand of DNA contains. (See Figure 13-3.)

DNA: A Library of Information

In some ways, the bases in DNA are like the letters of an alphabet, only the DNA "letters" are chemical letters. Because there are only four letters (A, T, G, and C) in the DNA alphabet, scientists thought that DNA was too simple to contain the complex genetic information of life. But what is also significant in DNA is the sequence of the letters, not just the letters themselves. Using these four letters in long sequences, nature can create an almost unlimited variety of genetic messages.

When you realize that human DNA consists of three billion pairs of bases, you can begin to imagine how much information can be stored in the DNA of our cells. All of the information for constructing our bodies, determining all of our characteristics or traits, and keeping our bodies functioning is stored in the linear sequences of bases in our DNA. To make use of the genetic information stored in DNA, organisms must change that information into proteins. Proteins are made up of *amino acids,* subunits that—like nucleotide bases—are joined in a linear sequence. The sequence of DNA subunits is used to direct the synthesis of proteins that have the correct sequence of amino acid subunits. In other words, through a chemical process, the order of the nucleotides determines the order of the amino acids in the proteins that are built.

DNA REPLICATION: PASSING IT ON

To qualify as genetic material, DNA has to be able to **replicate** (or make a copy of) itself. This process of DNA replication occurs during the middle of the cell cycle. What we already know about DNA's structure is enough to explain how it replicates.

To make a copy, you need an original, sometimes called a **template**. Because DNA is a double helix, it has templates built into it. To begin the process, the double helix unwinds. As with all metabolic activities, *enzymes* are needed for this process. Once the double-stranded molecule is untwisted, it begins to unzip, just like a zipper. Through the activity of an enzyme, the bonds between bases begin to break apart. (See Figure 13-4.)

As the bonds break, each strand of the DNA molecule becomes separate. Many free subunits float around in the cell. Specific enzymes match up

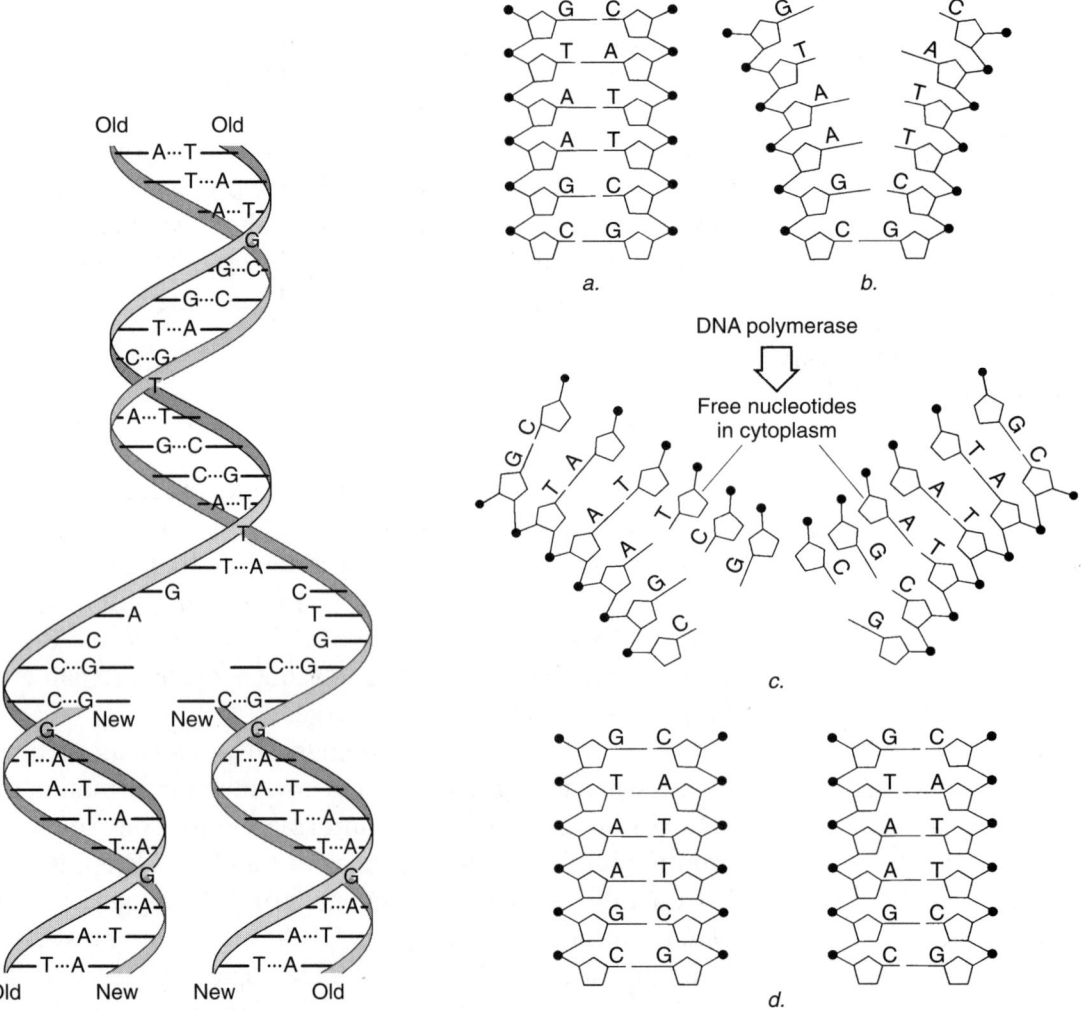

Figure 13-4 During DNA replication, the double helix unwinds, the strands separate, and the new strands form opposite each of the original DNA strands.

Figure 13-5 Through the process of DNA replication, two identical double-stranded DNA molecules are formed.

these free subunits with the existing subunits in each DNA strand. Wherever a T is located on a strand, an A pairs to it; wherever a C is located, a G joins up, and so on. One by one, new subunits are joined together to make a new strand opposite each old strand. The sequence of bases in the old strands determines the linear sequence of subunits in the new strands. When replication is complete, two double-stranded DNA molecules are formed. Each molecule is made up of one old strand joined to a newly synthesized strand. How do the two new DNA molecules compare to the original one? They are identical. DNA replication has occurred. (See Figure 13-5.)

Errors in DNA Replication

In life, nothing is perfect. This is true about DNA replication, too. The enzymes that are responsible for directing the correct pairing of subunits during DNA replication occasionally make mistakes. A nucleotide base may be left out, or the wrong base may be matched up. Sometimes an extra base is added. These mistakes produce errors in the linear sequence in one strand of the DNA molecule. Such an error is called a genetic **mutation**. From what we know about the replication process, once an error occurs in a DNA strand, it may be copied again and again. Thus, a mutation in the genetic material of one cell can easily be passed on to future cells.

A mutation is simply a change. However, many changes in the genetic material are harmful and may make it impossible for future cells, or even the entire organism, to survive. Other mutations cause an unnoticeable change; rather than harming the organism, the mutation seems to produce no effect. Sometimes a mutation gives the organism a sudden advantage that other similar organisms lack. Not only can mutations in DNA be good, but they are actually an important source of the genetic variation that is necessary for natural selection to occur. Much of the evolution of different life-forms on Earth has depended on the chance occurrence of these mutations. (*Remember:* Only mutations within the DNA of gametes can be passed along to offspring; mutations within the DNA of body cells cannot.)

Chapter 13 Review

Multiple Choice

1. One characteristic of genetic material that would *not* be good to have is the ability to
   A. make a copy of itself
   B. fall apart very easily
   C. mutate from time to time
   D. store a lot of information

2. If a set of instructions that determines all of the characteristics of an organism is compared to a book, and a chromosome is compared to a chapter in the book, then what might be compared to a paragraph in the book?
   A. a starch molecule
   B. an amino acid
   C. a protein polymer
   D. a DNA molecule

3. A portion of a molecule is shown in the diagram below. Which statement best describes the main function of this type of molecule?

   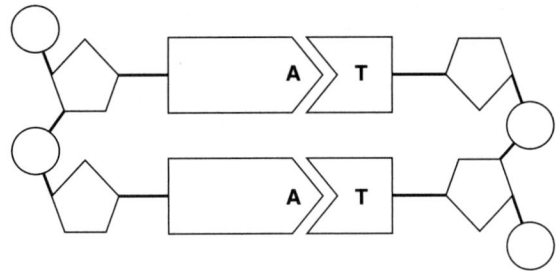

   A. It is a structural part of the cell wall.
   B. It determines what traits may be inherited.
   C. It stores energy for metabolic processes.
   D. It transports materials across the cell membrane.

4. The subunits of proteins are
   A. simple sugars
   B. phosphates
   C. amino acids
   D. enzymes

5. Watson and Crick contributed to the study of DNA by
   A. experimenting with pea plants
   B. recognizing that traits are inherited
   C. discovering the double helix structure of DNA
   D. mapping the entire human genome

6. The genetic code of a DNA molecule is determined by its specific sequence of
   A. ATP molecules
   B. carbohydrates
   C. sugar molecules
   D. nucleotide bases

7. The DNA molecule is formed from subunits arranged in a
   A. sequence with three kinds of bases
   B. circle with four kinds of bases
   C. sequence with four kinds of bases
   D. sequence with four kinds of acids

8. The base pairs in DNA are similar in arrangement to the
   A. sides of a ladder
   B. steps of a ladder
   C. railing of a staircase
   D. surface of a ramp

9. The order of the subunits in a strand of DNA is called a
   A. subunit sequence
   B. linear sequence
   C. strand sequence
   D. nucleotide sequence

10. If one strand of a DNA molecule is G-A-T-C-C-A-T, the sequence of the opposite strand is
    A. G-A-T-C-C-A-T
    B. C-T-A-G-G-T-A
    C. A-T-G-G-A-T-G
    D. T-A-C-C-T-A-G

11. The organization of bases in DNA can best be likened to the
    A. arrangement of letters in a word
    B. kinds of tools in a garage
    C. number of books in a library
    D. colors in a rainbow

12. When DNA separates into two strands, the DNA would most likely be directly involved in
    A. replication
    B. differentiation
    C. fertilization
    D. evolution

13. The sequence of subunits in a protein is most directly dependent upon the
    A. region in the cell where enzymes are produced
    B. type of cell in which starch is found
    C. DNA in the chromosomes in a cell
    D. kinds of materials in the cell membrane

14. In the diagram below, strands I and II represent sections of a DNA molecule. Strand II would normally include (from top to bottom)

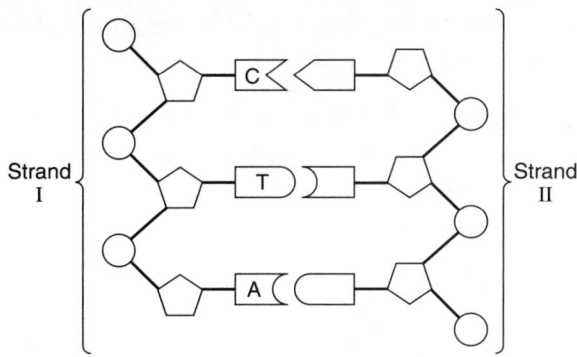

A. AGC
B. TAC
C. TCG
D. GAT

15. During the first step in the replication of DNA, the
    A. double helix unwinds
    B. base template is created
    C. subunits of DNA form pairs
    D. double helix rewinds itself

16. What causes the base pairs of DNA to break apart?
    A. a mutation during replication
    B. the activity of an enzyme
    C. the production of new bases
    D. the introduction of a fifth base

17. After DNA replication, the new DNA molecules are
    A. the reverse of the original
    B. the mirror image of the original
    C. identical to the original
    D. totally different from the original

18. Which statement is true regarding an alteration or change in DNA?
    A. It is always referred to as a mutation.
    B. It is always passed on to the offspring.
    C. It is always advantageous to an individual.
    D. It is always detected by chromatography.

19. A mutation occurs in a cell. Which sequence best represents the order of events for this mutation to affect traits expressed by the cell?
    A. amino acids joining in sequence → a change in the sequence of DNA bases → appearance of characteristic
    B. a change in the sequence of DNA bases → amino acids joining in sequence → appearance of new characteristic
    C. appearance of new characteristic → amino acids joining in sequence → a change in the sequence of DNA bases
    D. a change in the sequence of DNA bases → appearance of new characteristic → amino acids joining in sequence

20. A mutation is considered positive when it
    A. makes it hard for the organism to survive
    B. has absolutely no effect on the organism
    C. changes the organism in an undetectable way
    D. provides a sudden advantage that aids survival

Analysis and Open Ended

21. What four qualities must the genetic material have in order to do its job?

22. List the four bases of the DNA nucleotides and tell which bases pair together.

23. Explain the basic structure of DNA as described by Watson and Crick.

24. Why did scientists once think that DNA was too simple to contain the genetic information of living things? Explain why their reason was not correct.

25. Molecule 1 represents a section of inherited information, and molecule 2 represents part of the substance that is determined by the information in molecule 1. What will most likely happen if there is a change in the first three subunits on the upper strand of molecule 1 shown below?

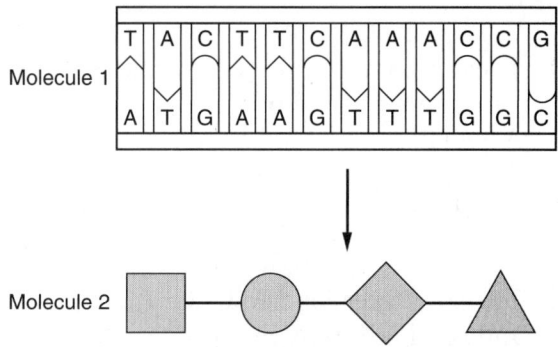

A. The remaining subunits in molecule 1 will also change.

Chapter 13 / DNA and Heredity 171

B. Molecule 1 will split apart, triggering an immune response.
C. A portion of molecule 2 may be formed differently.
D. Molecule 2 may form two strands rather than one.

26. In an experiment, DNA from dead pathogenic bacteria was transferred into living bacteria that were, normally, not pathogenic. These altered bacteria were then injected into healthy mice. The mice died of the same disease caused by the original pathogens. Based on this information, which statement would be a valid conclusion?
   A. DNA is present only in living organisms.
   B. DNA functions only in the original organism from which it comes.
   C. DNA functions only when it is transferred into another organism.
   D. DNA from a dead organism can become active in another organism.

27. Briefly explain how the genetic information is arranged within a DNA molecule.

28. You see a photograph of a famous man and his teenaged son. You notice that they look very much alike, and that they even wear similar eyeglasses. What conclusion can you draw from this observation?
   A. The DNA present in their body cells is identical.
   B. Their percentage of having the same proteins is high.
   C. The base sequences of their genes are all identical.
   D. The mutation rate is the same in their body cells.

Refer to the figure below to answer questions 29 to 31.

29. The diagram at right represents part of a molecule of
   A. ATP
   B. RNA
   C. DNA
   D. FSH

30. The structures labeled G, C, T, and A all represent
   A. nucleic acids
   B. simple sugars
   C. nucleotide bases
   D. phosphates

31. Starting from the top of the diagram, what would be the letters of the missing units on the matching strand?

32. Complete the analogy: Nucleotide bases are to DNA as amino acids are to
   A. sugars
   B. proteins
   C. lipids
   D. nucleic acids

33. How would you explain to someone who has never heard of DNA why it is such an important molecule?

34. How do the nucleotides of the DNA molecule allow it to replicate?

35. Briefly describe the process of DNA replication. Your answer should include the following terms (but not necessarily in this order):
   • template
   • enzymes
   • subunits

Base your answers to questions 36 and 37 on the following chart, which provides information about heredity, and on your knowledge of biology.

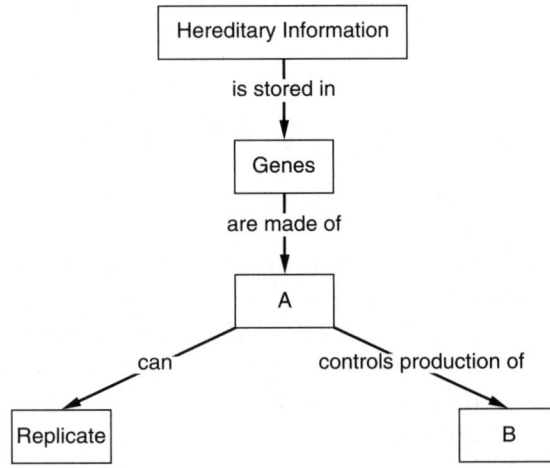

36. The molecule that is represented by box *A* serves as a template. Identify the type of molecule and explain how it is a template.

37. Which types of molecules are represented by box *B*?
   A. bases
   B. proteins
   C. lipids
   D. sugars

38. Mutations can be helpful to organisms; yet people fear the effects of substances that can cause mutations. Explain how mutations can be both helpful and harmful.

Base your answers to questions 39 and 40 on the passage below and on your knowledge of biology.

When making movies about dinosaurs, film producers have sometimes used ordinary lizards and enlarged their images thousands of times. We all know, however, that while they may look like dinosaurs and be related to dinosaurs, modern lizards are not actually dinosaurs.

Recently, some scientists have developed a hypothesis that challenges this view. These scientists suggest that some dinosaurs were actually the same species as some modern lizards that had grown to unbelievable sizes. They think that such growth might be due to a special type of DNA called *repetitive DNA,* often referred to as "junk" DNA because scientists do not understand its functions.

The scientists studied pumpkins that can reach sizes of nearly 1000 pounds and found them to contain large amounts of repetitive DNA. Other pumpkins that grow to only a few pounds in weight have very little of this kind of DNA. In addition, cells that reproduce uncontrollably have almost always been found to contain large amounts of repetitive DNA.

39. State *one* reason why scientists formerly thought of repetitive DNA as "junk."

40. Which fact best supports the hypothesis that large amounts of repetitive DNA are responsible for increased sizes of organisms?
    A. Lizards look very much like little dinosaurs.
    B. Modern lizards may be related to dinosaurs.
    C. Large pumpkins contain a lot of repetitive DNA.
    D. Another term for repetitive DNA is "junk" DNA.

Reading Comprehension

Base your answers to questions 41 to 45 on the information below and on your knowledge of biology. Use one or more complete sentences to answer each question.

> Scientists have learned that the end section of a chromosome—called the telomere—plays an important role in the life of a cell. Normal human cells divide only a limited number of times. Each time a cell divides, it gets a little older. Research has suggested that when a cell divides, the telomere becomes a little shorter. The telomere is now thought of as a kind of molecular clock that keeps track of, or controls, the age of cells. It did not take long for researchers to begin to think that if the telomere could somehow be kept from getting shorter, a cell could continue to divide forever. Because their cells would never get older, the person who had such cells would also never age. In effect, an ageless telomere would become a molecular fountain of youth.
>
> Normally, the telomeres become shorter and shorter with each cycle of cell division. It is thought that a short telomere tells a cell to stop dividing. A key enzyme that can change this shortening process is telomerase, which reverses the process by adding DNA to the telomeres at the chromosome ends.
>
> In support of this hypothesis, researchers found that telomerase remains active in most immortal cell lines, such as cancerous cells that keep dividing in an uncontrolled manner. The importance of telomerase therefore has become even greater. Not only may the absence of telomerase lead to cell aging, but its presence may lead to cancer. Telomerase is also normally active in human cells that give rise to sperm and egg cells, which have to replicate.

> Will further research provide a means to keep cells from aging through the action of telomerase—yet without also dividing uncontrollably? These are significant questions to answer and an important area for future research.

41. Why is a chromosome's telomere of interest to scientists?
42. What seems to occur when a telomere gets very short?
43. How is the enzyme telomerase involved in cell aging?
44. Why is there great interest in the activity of the enzyme telomerase?
45. Why is there great interest in the results of the absence of telomerase?

Chapter 14

Genes and Protein Synthesis

Standard 5.5.12 C1 **Describe how information is encoded and transmitted in genetic material.**

GENES AND PROTEINS

Now that it has been shown that DNA is what makes up the genetic material, it is time to look more closely at genes. What is a gene? **Genes** are actually packages of information that tell a cell how to make proteins. Proteins are *polymers,* or long chains, of amino acids. As you learned already, there are 20 different types of amino acids. The order in which the amino acids are joined determines which protein is made (much like a particular order of words makes up a sentence). Every different protein has a unique sequence of amino acids. This sequence determines the shape of a protein molecule. It is the shape of the protein that allows the molecule to do its work in the cell.

Genes are specific sections of DNA molecules that are made up of linear sequences of nucleotide subunits. Proteins are made up of linear sequences of amino acids. How do cells use a linear sequence of subunits in DNA to build a linear sequence of amino acids for a protein? In all cells, except for bacteria, DNA is stored in the nucleus. Yet protein synthesis occurs outside the nuclear membrane, at the **ribosomes**. These small organelles are distributed throughout the cytoplasm. How does the genetic information in DNA within the nucleus get out to the ribosomes? A third type of molecule, *ribonucleic acid,* or **RNA**, works as a helper to transfer the information. That is, the genetic information flows from the DNA to the RNA to a protein. (See Figure 14-1.)

TRANSCRIPTION: FROM DNA TO RNA

Each gene is a portion of a chromosome; that is, it is a segment of the DNA chain. An RNA molecule called *messenger RNA* (mRNA) does the job of

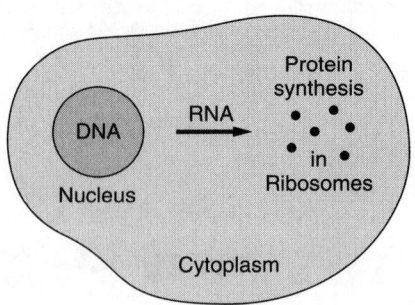

Figure 14-1 The flow of genetic information in a cell: from DNA in the nucleus to RNA to amino acids at the ribosomes.

moving the information in the base sequence out to the ribosomes. DNA is copied into mRNA by a process called **transcription**, which is similar to DNA replication. The DNA double helix opens up where a particular gene is located. Special enzymes begin to match up mRNA subunits with the complementary DNA subunits. The new mRNA molecule has the same base sequence as one strand of the original DNA. (Note, however, that RNA substitutes the base *uracil* for thymine.) This mRNA molecule then goes out of the nucleus through pores in the nuclear membrane to ribosomes in the cytoplasm. (See Figure 14-2.)

TRANSLATION: FROM RNA TO PROTEIN

So far, the genetic information, stored as a base sequence, has moved from the nucleus to the cytoplasm by using mRNA. Another problem remains: how to use the nucleotide base sequence in the mRNA to build a protein with the correct amino acid sequence. This problem involves a change of "language," from the nucleotide-base language of RNA into the amino-acid language of proteins. This process is called **translation**, and it occurs at the ribosome.

Built into every living cell in the world is a **genetic code**. It is called the *triplet code* because each different combination of three bases (like different "letters") makes up a "word," called a *codon*. Each **codon** represents a specific amino acid; each of the 20 amino acids has at least one codon, and most have more than one. (See Figure 14-3.)

In the cytoplasm, special *transfer RNA* (tRNA) molecules bind with amino acids and transport them to the ribosome. Each codon on the mRNA specifies which tRNA, with its particular amino acid, will be attached to that

Figure 14-2 The DNA sequence is copied into messenger RNA, which goes out to the ribosomes in the cytoplasm. Note that in RNA, the base uracil (U) substitutes for the DNA base thymine (T).

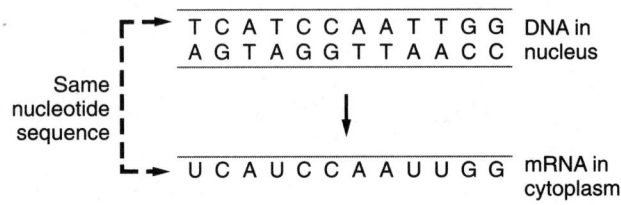

Figure 14-3 The mRNA amino acid triplet codes. Note that most amino acids are represented by more than one codon.

First Position	Second Position: U	Second Position: C	Second Position: A	Second Position: G	Third Position
U	UUU } Phe UUC } Phe UUA } Leu UUG } Leu	UCU } Ser UCC } Ser UCA } Ser UCG } Ser	UAU } Tyr UAC } Tyr UAA *Stop* UAG *Stop*	UGU } Cys UGC } Cys UGA *Stop* UGG Trp	U C A G
C	CUU } Leu CUC } Leu CUA } Leu CUG } Leu	CCU } Pro CCC } Pro CCA } Pro CCG } Pro	CAU } His CAC } His CAA } Gln CAG } Gln	CGU } Arg CGC } Arg CGA } Arg CGG } Arg	U C A G
A	AUU } Ile AUC } Ile AUA } Ile AUG Met	ACU } Thr ACC } Thr ACA } Thr ACG } Thr	AAU } Asn AAC } Asn AAA } Lys AAG } Lys	AGU } Ser AGC } Ser AGA } Arg AGG } Arg	U C A G
G	GUU } Val GUC } Val GUA } Val GUG } Val	GCU } Ala GCC } Ala GCA } Ala GCG } Ala	GAU } Asp GAC } Asp GAA } Glu GAG } Glu	GGU } Gly GGC } Gly GGA } Gly GGG } Gly	U C A G

site. The tRNA molecule has its own base triplet, called an *anticodon,* which is complementary to a codon on the mRNA. When these complementary triplets meet, the tRNA molecule releases its amino acid. The ribosome continues to move along the mRNA molecule to expose a new codon, which attaches to another tRNA molecule. In this way, amino acids are linked to form a protein. (See Figure 14-4.)

The genetic code is universal; in other words, all organisms on Earth use the same triplet codes. For example, the codon GCA stands for the amino acid alanine in all life-forms, from bacteria to trees to humans. This genetic similarity among living things is, in fact, strong evidence that all organisms evolved from a common ancestral life-form in the distant past.

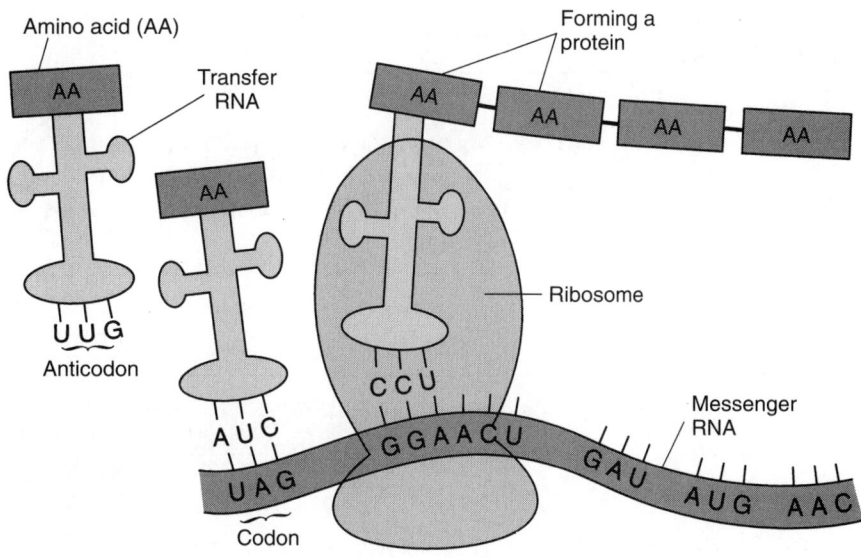

Figure 14-4 Model of translation at a ribosome, forming a polypeptide.

Mutations: A Closer Look

In Chapter 13, we defined *mutation* as a change in the base sequence of a DNA molecule. The possible effects of a mutation can now be explained in terms of what you know about protein synthesis.

The order of bases in DNA determines the order of amino acids in proteins. In certain cases, a mutation in one subunit will change the triplet code, which in turn may make a change in the amino acid. If this change occurs in a body (somatic) cell, then all other cells that are produced by mitosis from that cell in the organism's body will have the same change. It is more important, however, if the mutation occurs in the DNA of a sex cell, or *gamete*. If that gamete fuses with another gamete in sexual reproduction, then the mutation will be inherited. The genetic change will be passed on to the next generation. This will become an inherited condition: the new organism—if it survives—will have that mutation, and so will its offspring. If the mutation is harmful, the individual and its offspring will have what is known as a *genetic disease*.

GENE EXPRESSION AND CELL DIFFERENTIATION

Chromosomes contain extremely long DNA molecules. Many genes are stretched out along these molecules. For example, it is thought that there are about 35,000 different genes in human cells. After fertilization, every cell

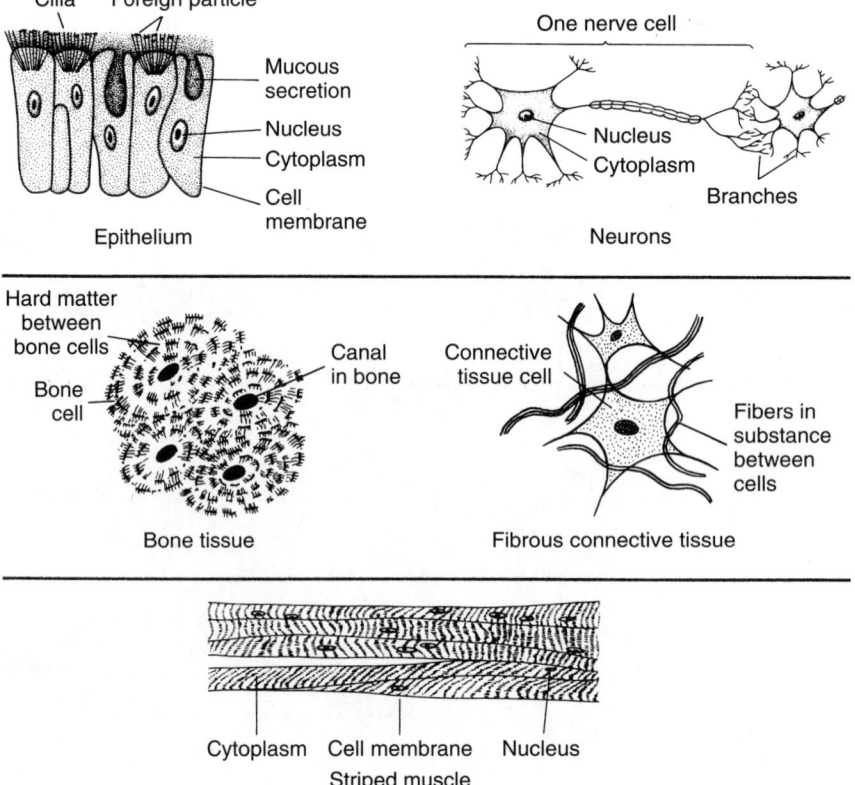

Figure 14-5 Many different types of cells make up the human body. This cell differentiation results from differences in gene expression—particular genes are "turned on" to make the specific proteins needed for each cell type.

of a growing organism arises from the mitotic cell division of other cells. Through mitosis, every cell in our body has the same 46 chromosomes, with the same DNA as the original fertilized egg cell.

You have learned that there are different types of cells in our bodies. We have skin cells, muscle cells, bone cells, nerve cells, blood cells, and so on. If all of these cells have the same DNA, why are they so different from each other? The answer is that only certain genes are used in certain cells. The use of specific information from a gene is called gene **expression**. A cell's proteins are synthesized only from genes that are being *expressed,* or "turned on." All other genes in the cell are kept silent, or "turned off." This gives each cell type its own structure, enzymes, functions, and physical characteristics. A muscle cell contracts, a nerve cell transmits an impulse, and a skin cell helps form a flat, protective layer. The process by which special types of cells are formed through controlled gene expression is called cell **differentiation**. This is an essential process of life. Without cell differentiation, we could not survive because our bodies would be made up of only one type of cell. While the exact process is not known for certain, it is thought that environmental factors—both inside and outside the cell—influence gene expression. (See Figure 14-5.)

Chapter 14 Review

Multiple Choice

1. Genes can best be described as
   A. directions for making DNA
   B. directions for making proteins
   C. the subunits of proteins
   D. directions for making RNA

2. Which path correctly describes the flow of information in cells?
   A. DNA → RNA → protein
   B. protein → RNA → DNA
   C. protein → DNA → RNA
   D. RNA → DNA → protein

3. The type of genes that an organism has is determined by the
   A. kinds of amino acids in its cells
   B. size of sugar molecules in its organs
   C. sequence of nucleotide bases in its DNA
   D. shape of protein molecules in its organelles

4. A change in the DNA base sequence that codes for a respiratory protein will most likely cause
   A. the production of a starch that has a similar function
   B. a change in the sequence of amino acids determined by the gene
   C. the digestion of the altered gene by enzymes
   D. the release of antibodies by certain cells to correct the error

5. The role of messenger RNA is to
   A. prevent genetic mutations during DNA replication
   B. match ribose-containing subunits to DNA subunits
   C. move the information in a base sequence out to the ribosomes
   D. translate the DNA base sequence into RNA at the ribosomes

6. RNA receives information from DNA by
   A. binding with a double helix as a third strand
   B. matching with subunits of a single strand of DNA
   C. making an exact copy of the DNA molecule
   D. accepting proteins through pores in the nuclear membrane

7. What happens at the ribosome?
   A. The DNA strands separate.
   B. RNA matches up with DNA strands.
   C. Genetic information is mutated.
   D. RNA is translated into amino acids.

8. The diagram below represents a process that occurs within a cell in the human body. This process is known as

   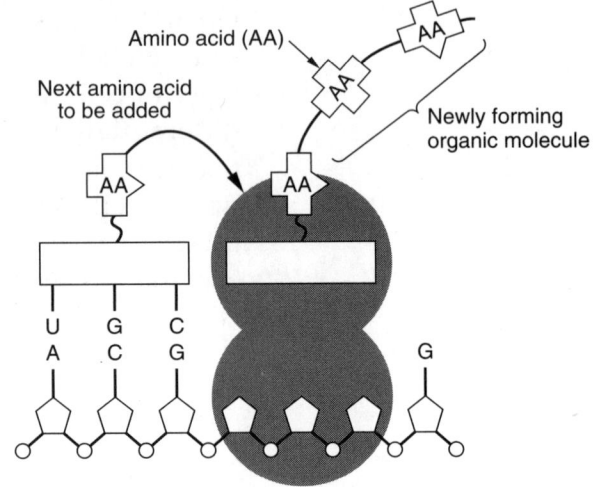

   A. digestion by enzymes
   B. energy production
   C. protein synthesis
   D. replication of DNA

9. How many nucleotide bases make up a codon?
   A. one   C. three
   B. two   D. four

10. What does a codon represent?
    A. a specific amino acid
    B. a specific base
    C. an RNA molecule
    D. an enzyme

11. The normal genetic code is
    A. different for every organism
    B. the same for all organisms
    C. constantly unstable
    D. impossible to identify

12. The sequence of amino acids in a protein is determined by the
    A. speed at which translation occurs
    B. size of the cell that produces it
    C. number of ribosomes in a cell
    D. sequence of bases in the DNA

13. The diagram below provides some information concerning proteins. Which phrase does the letter *A* represent?

    [A] Influences → Protein shape → Determines → Protein function

    A. Sequence of amino acids
    B. Sequence of starch molecules
    C. Sequence of simple sugars
    D. Sequence of ATP molecules

14. A genetic mutation is inherited if it
    A. occurs in a gamete used in sexual reproduction
    B. occurs in a cell that undergoes mitosis only
    C. gives the organism a better chance for survival
    D. endangers the organism's chances for survival

15. People with cystic fibrosis inherit defective genetic information, so they cannot produce normal CFTR proteins. Scientists have used gene therapy to insert normal DNA segments that code for the missing CFTR protein into the lung cells of people with this genetic disease. Which statement does *not* describe a possible result of this therapy?
    A. Genetically altered lung cells can produce the normal CFTR protein.
    B. The normal CFTR gene may be expressed in the altered lung cells.
    C. Altered lung cells can divide to produce other lung cells with the normal CFTR gene.
    D. Offspring of someone with altered lung cells will inherit the normal CFTR gene.

16. In total, about how many genes are contained on the 46 chromosomes in each human body cell?
    A. 500 to 1000
    B. 5000 to 10,000
    C. 30,000 to 40,000
    D. 50,000 to 80,000

17. The cells that make up a person's skin have some functions that are different from those of the cells that make up the person's liver. This is because
    A. all types of cells developed from a common ancestor
    B. environment and past history have no influence on cell function
    C. different cell types have completely different genetic material
    D. different cell types use different parts of the genetic instructions

18. The term *gene expression* refers to the fact that
    A. completely different DNA is found in different cells
    B. genes are passed through the nuclear membrane
    C. only some genes are turned on in each type of cell
    D. some cells have genes and other cells do not

19. Scientific studies have shown that identical twins that were separated at birth and raised in different homes may vary in height, weight, and intelligence. The most probable explanation for the differences is that
    A. the original genes in the twins increased in number as each of the twins developed
    B. the environments in which they were raised were different enough to affect their gene expression
    C. one twin received genes only from the mother while the other twin received genes only from the father
    D. the environments in which they were raised were different enough to change their genetic makeup

20. After a series of cell divisions, an embryo starts to develop different types of body cells such as muscle, nerve, and skin cells because
    A. the genetic code changes as the embryonic cells divide
    B. different genetic instructions are created to meet the needs of new cell types
    C. different segments of the genetic code are used to produce different cell types
    D. some sections of the genetic material are lost as a result of fertilization

Analysis and Open Ended

21. What is so important about the sequence in which amino acids are joined?

22. The following diagram shows two different structures, 1 and 2, that are present in many single-celled organisms. Structure 1 contains protein *A,* but not protein *B;* and structure 2

contains protein *B*, but not protein *A*. Which statement is correct concerning protein *A* and protein *B*?

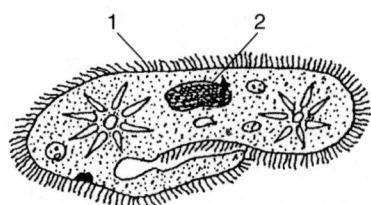

A. Proteins *A* and *B* have different functions and different amino acid chains.
B. Proteins *A* and *B* have the same function but a different sequence of DNA bases.
C. Proteins *A* and *B* have different functions but the same amino acid chains.
D. Proteins *A* and *B* have the same function and the same sequence of DNA bases.

23. Hemoglobin is a complex protein molecule found in red blood cells. Hemoglobin with the normal amino acid sequence can effectively carry oxygen to all body cells. In the disorder known as sickle-cell anemia, one amino acid is substituted for another in the hemoglobin. A characteristic of this disorder is poor distribution of oxygen to the body cells. How could one change in the amino acid sequence of this protein cause such a result?

24. In what way are the structures of DNA and protein similar?

25. Why might the location of DNA within the nucleus be a problem for protein synthesis? How does RNA solve this problem for the cell?

26. Arrange the following terms in order, going from the largest structure to the smallest one: *gene; DNA molecule; chromosome; nucleus; cell.*

27. Explain how the process of copying DNA into messenger RNA is similar to DNA replication. Your answer should include the following terms: *double helix; enzymes; subunits; base sequence.*

28. Why must the bases be grouped into triplets in order to represent amino acids?

29. In what way do ribosomes help in the process of translation?

Base your answers to questions 30 to 32 on the table below, which provides the DNA codes for several amino acids.

Amino Acid	DNA Code Sequence
Cysteine	ACA or ACG
Tryptophan	ACC
Valine	CAA or CAC or CAG or CAT
Proline	GGA or GGC or GGG or GGT
Asparagine	TTA or TTG
Methionine	TAC

30. A certain DNA strand has the following base sequence: TAC ACA CAA ACG GGG. What is the sequence of amino acids that would be synthesized from this code (if it is read from left to right)?

31. Suppose the DNA sequence undergoes the following change: TAC ACA CAA ACG GGG → TAC ACC CAA ACG GGG. How would the sequence of amino acids be changed as a result of this mutation?

32. The original DNA sequence undergoes the following change: TAC ACA CAA ACG GGG → TAC ACA CAA ACG GGT. Explain why this mutation produces *no change* in the structure or function of the protein that will be synthesized from this code.

Refer to the flowchart below to answer questions 33 and 34.

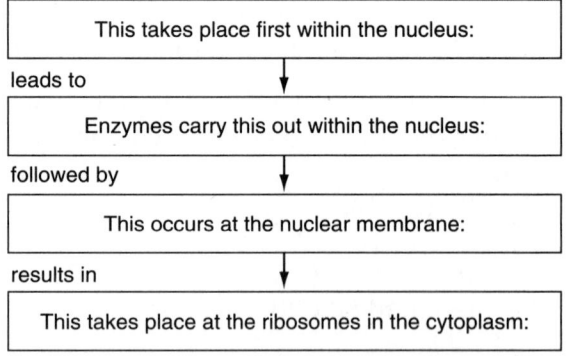

33. Use the following phrases to complete the steps listed in the flowchart boxes: a new RNA molecule moves out through pores; RNA base sequences translate into amino acid sequences; the DNA double helix opens up and unwinds; RNA subunits match up with DNA subunits.

34. The best title for this flowchart probably would be
    A. How RNA Is Made
    B. How DNA Is Made
    C. How Proteins Are Made
    D. How Ribosomes Are Made

35. Describe how RNA allows translation from the nucleotide base sequences to the amino acid sequences.

36. Explain why the genetic code is called *universal*. What is the evolutionary significance of this fact?

37. How might a mutation in a DNA molecule result in a different protein being produced by a cell?

38. Use the following diagram to answer these questions. Suppose these two cells are from the same body. (a) How is gene expression related to cell differentiation? (b) Do these cells have the same proteins? (c) Do they have the same DNA?

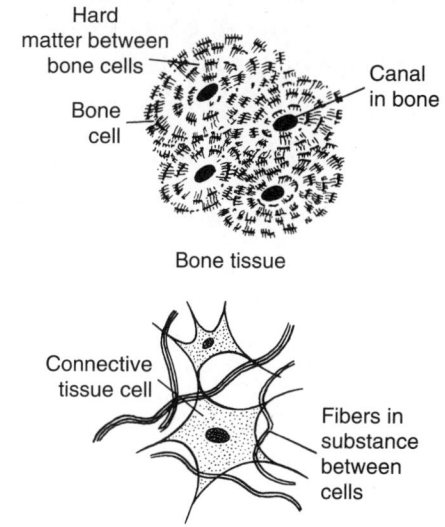

Reading Comprehension

Base your answers to questions 39 to 41 on the information below and on your knowledge of biology. Use one or more complete sentences to answer each question.

The first big surprise to arise from the decoding of the human genome was a matter of numbers. Where were all the genes? Rather than 100,000 or more genes, as scientists had predicted for years, the Human Genome Project has revealed that humans have perhaps 35,000 or fewer genes. This can be compared to the fruit fly's 12,371 genes or the 19,098 genes of the tiny roundworm *Caenorhabditis elegans*. But it is not just a matter of human pride for people to consider themselves more complex. We *are* more complex. The little worm *C. elegans,* with over 19,000 genes, has a body of only 959 cells, of which 302 are neurons—the worm's "brain." The human body, built by perhaps only 50 percent more genes than the worm, has 100 trillion cells, with the brain alone containing 100 billion cells. So where does the complexity come from?

There are two main ways that scientists think human complexity has arisen. The first way concerns proteins. Proteins are the working parts of every cell and it turns out that proteins themselves have different sections or domains in them. Ninety-three percent of the protein domains in humans are also in the worm and the fly. However, it seems that a lot of mixing and matching of these domains has occurred. Dr. Francis S. Collins, director of the genome institute at the National Institutes of Health said, "Maybe evolution designed most of the basic folds that proteins could use a long time ago, and the major advances in the last 400 million years have been to figure out how to shuffle those in interesting ways." The second ingenious way that evolution seems to have increased complexity is by dividing the genes themselves into several different segments, and using these segments in different arrangements to make different proteins.

> There are many different ways of thinking about human complexity. One scientist has compared people to the machines created by them. Dr. Jean-Michel Claverie of the French National Research Center writes, "In fact, with 30,000 genes, each directly interacting with four or five others on average, the human genome is not significantly more complex than a modern jet airplane, which contains more than 200,000 unique parts, each of them interacting with three or four others on average."

39. What is the first unexpected finding from the decoding of the human genome?
40. How do proteins contribute to the complexity of the human organism?
41. How do genes contribute to the complexity of the human organism?

Chapter 15
Asexual Reproduction and Mitosis

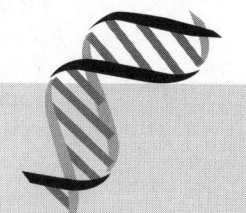

Standard 5.5.12 C1 **Describe how information is encoded and transmitted in genetic material.**

THE CONTINUITY OF LIFE

You have learned what it takes for an individual to remain alive in a constantly changing environment. Although staying alive is important for every individual organism, it is not sufficient to maintain life on Earth. No individual organism lives forever. Every organism has a typical **life span**—the length of time between when its life begins and when it ends. The continuity of life requires **reproduction**, the ability of individuals within a species to produce more of their own kind. Individuals are members of populations. It is reproduction within populations that allows species to survive. It is reproduction that allows life on Earth to continue.

THE LIFE CYCLE OF A CELL

Every cell has a life of its own. This is as true for single-celled organisms as it is for each of the billions of cells that make up the bodies of plants and animals, including ourselves. Each cell has a beginning (with a period of growth), a middle stage, and then an ending—a process known as the *cell cycle*. In the first stage, the cell begins to grow in size. Organic materials, such as amino acids and sugars, and inorganic materials, such as water, are moved into the cell. The cell increases in size by adding these materials to itself. The cell also increases its number of parts. For example, its mitochondria divide in two to make more mitochondria. If it is a plant cell, the same thing happens to its chloroplasts. (See Figure 15-1 on page 186.)

During the next stage in its life cycle, the cell stops getting larger. At this point, the genetic material in the cell—that is, the set of instructions received from the previous cell—duplicates. The genetic material is the building plan,

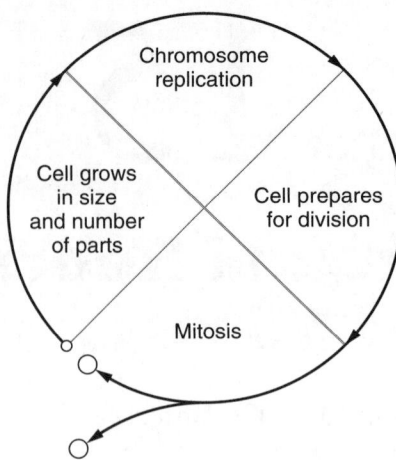

Figure 15-1 The cell cycle is the series of events that occurs in the life of a cell—from its beginning to its ending (in mitosis or death).

similar in some ways to the set of blueprints used to build a house; it contains all the information about how the cell is to be built and how it functions. As you have read, the genetic material is made up of the chemical called **DNA**, or *deoxyribonucleic acid*.

In reproducing cells, DNA is found in "packages" known as **chromosomes**. The number of chromosomes varies among different organisms. In other words, the chromosome number is specific for each type of organism. For example, a fruit fly has eight chromosomes in each body cell, a cabbage plant has 18, and a human has 46 (the 23 chromosome pairs). The exact chromosome number must be maintained for the species to continue. This means that as cells reproduce, the new cells must have the same number of chromosomes as the original cells.

The duplication of the cell's genetic material, during this middle stage in the cell's life cycle, is called **replication**. This is the most important stage in preparation for reproduction of the cell. Following this stage, some additional cell growth occurs. What is growing here is material needed for the final stage in its life cycle, called *cell division*. This is how a single cell reproduces; it divides into two new cells. (See Figure 15-2.)

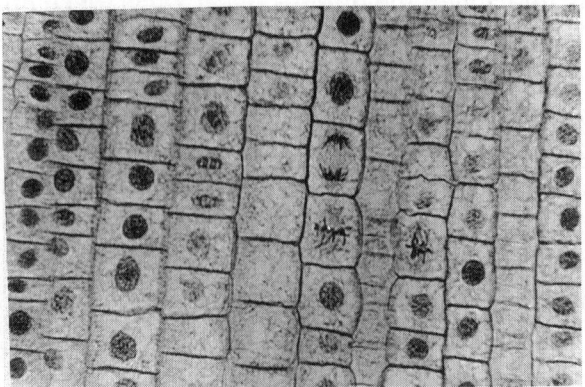

Figure 15-2 Several plant cells can be seen here preparing for cell division; chromosomes have duplicated and are dividing into two groups.

Figure 15-3 The stages of mitosis in an animal cell: The chromosomes have replicated; they pull apart; and the cytoplasm is divided to form two new cells.

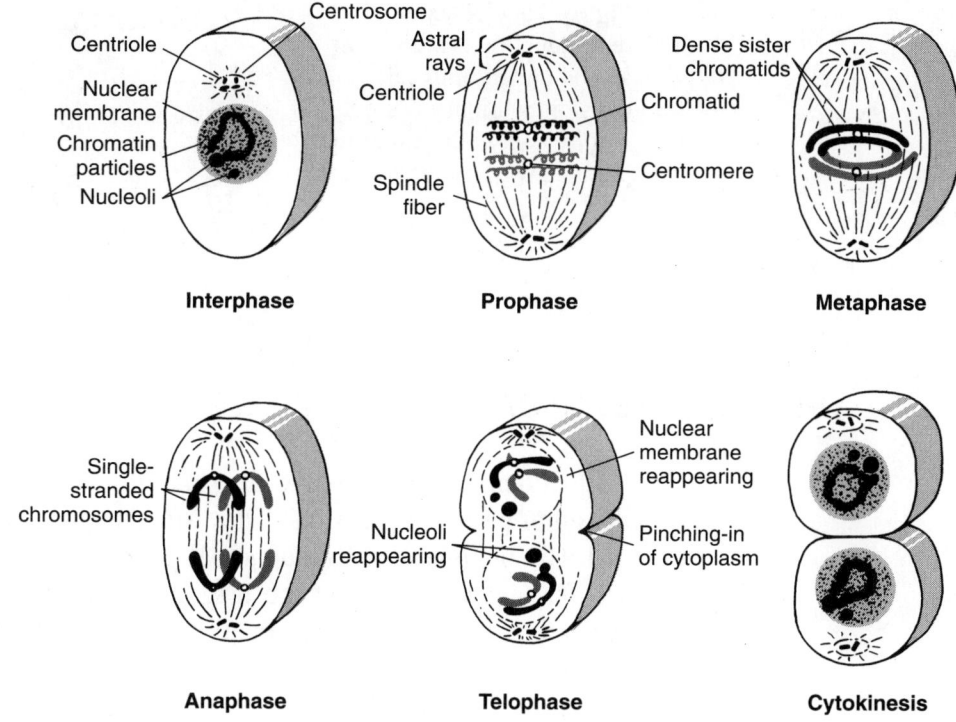

Mitosis and Cell Division

During **cell division**, the genetic material must be equally divided. When a cell divides, it must send one copy of each of its chromosomes to each of the new cells. In addition, the cytoplasm and other cell parts must be divided between the two cells.

The division of the chromosomes occurs first. This division happens during a sequence of events called **mitosis**. During mitosis, the chromosomes of a cell are divided into two equal groupings. Following mitosis, the cytoplasm of the cell divides, a process called *cytokinesis*. After that occurs, each of the new cells has a complete set of chromosomes, just like the original cell. The two new cells are called *daughter cells*. The cell that they came from, which no longer exists, is called the *parent cell*. (See Figure 15-3 and Table 15-1.)

Table 15-1 The Stages of Mitosis

Interphase	Prophase	Metaphase	Anaphase	Telophase
Growth of the cell; replication of its genetic material	Nuclear membrane breaks down; chromosomes become visible	Spindle fibers appear; chromosomes line up across middle of cell	Double-stranded chromosomes separate; identical chromatids move apart	Chromosomes move to opposite ends; nuclear membranes reappear; cell pinches in

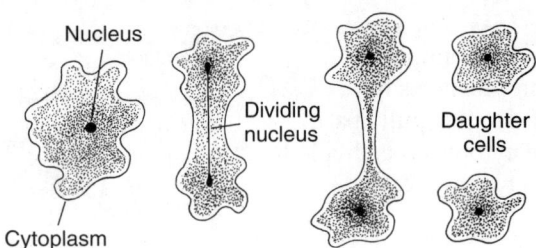

Figure 15-4 Asexual reproduction, as seen in an ameba undergoing cell division by binary fission, requires only one parent to produce two new organisms.

ASEXUAL REPRODUCTION: MAKING NEW INDIVIDUALS

Cell division produces two new daughter cells from one parent cell. The daughter cells are identical. They are also genetically identical to the parent cell. But have new individuals really been produced?

The answer is *yes,* if the original parent was a single-celled organism. An ameba, through a type of cell division called *binary fission,* splits in half to become two new, identical organisms. (See Figure 15-4.) Reproduction in an ameba involves only one parent. This is an example of **asexual reproduction**. (For reproduction to be called *sexual,* it must involve two parents.)

Plants have a variety of types of asexual reproduction. In each type, a plant or a part of the plant reproduces itself through mitosis. As a result, the offspring are identical to the parent plant. For example, strawberry plants send out horizontal stems, called *runners,* above the ground. When these runners touch the surface of the soil at another spot, an entirely new, identical plant with roots and leaves begins to grow there. The production of identical genetic copies, or *clones,* of a parent plant is called **cloning**. (See Figure 15-5.)

In some cases, animals also can reproduce asexually. For example, if an arm of a sea star is broken off, that arm can sometimes grow into a whole new sea star; and the sea star that lost the arm will grow, or *regenerate,* another arm. (See Figure 15-6.) Other types of organisms, such as yeasts, sponges, and hydra, can produce offspring asexually by **budding**. During the process of budding, a new small individual begins to grow out of the side of the parent organism. The cells that form this new individual, or *bud,* result from mitotic cell division. The bud breaks free of the parent organism when it is large enough to live on its own. (See Figure 15-7.)

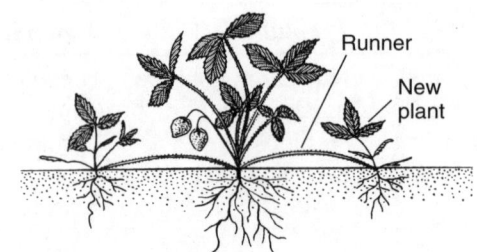

Figure 15-5 Strawberry plants can reproduce asexually by means of runners; this method of producing identical offspring is called cloning.

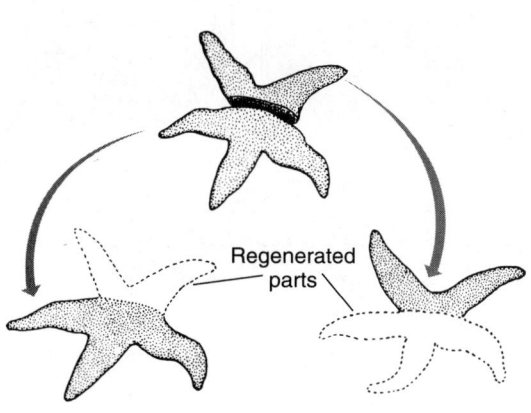

Figure 15-6 Some animals, such as the sea star (or starfish), can reproduce asexually by regrowing lost body parts when cut in two.

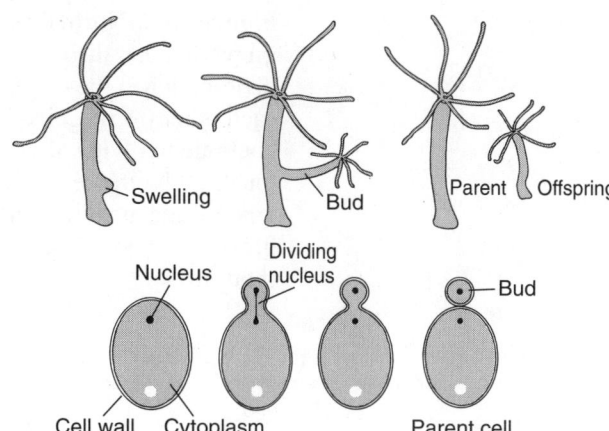

Figure 15-7 The hydra and yeasts can reproduce asexually by budding. The offspring, or bud, is smaller than the parent cell or organism.

The Rate of Cell Division

When does a cell divide? How long does it take for one segment of cell division to begin and end? Do all types of cells divide at the same rate? The answers are very important to the process of growth and development in organisms.

You read in Chapter 14 that multicellular organisms are made up of various types of cells and tissues. For example, the human body contains blood, skin, muscle, bone, and nerve tissues, among other tissues. Controlling the rate at which cells of each particular kind of tissue divide is a necessary part of homeostasis. Red blood cells have a relatively short life span, and we need millions of them, so the cells that develop into red blood cells divide quickly. (See Figure 15-8.) Bone cells, on the other hand, divide much more slowly. Skin cells normally take about 20 hours to complete one cell division; but their rate of division speeds up if you cut yourself.

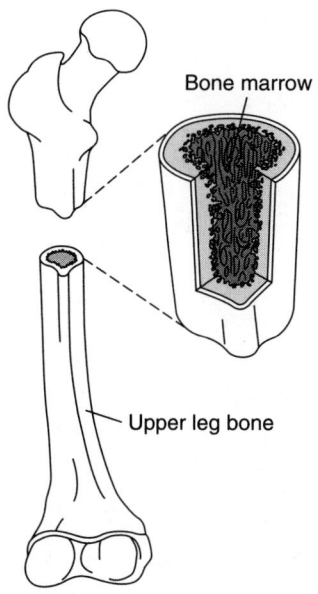

Figure 15-8 The cells of different types of tissues divide at different rates, depending on their functions. Millions of red blood cells, which divide quickly, are made in the bone marrow each day. By contrast, the bone tissue cells divided much more slowly.

Cancer: Cell Division out of Control

Cancer is a disease that results from uncontrolled cell division. Cancer cells do not seem to follow the rules or recognize the signals that control normal cell division. Uncontrolled cell growth can occur in many different types of cells. As a result, there are different types of cancer, such as skin cancer, breast cancer, prostate cancer, lung cancer, and many others. There does not seem to be just one single cause for all types of cancer. (See Figure 15-9.)

Even though there are differences, it is quite certain that all types of uncontrolled cell division involve the cells' genetic instructions, which are made of DNA. Factors that cause cancer do so by damaging or changing the DNA. These factors include the exposure of cells to certain chemicals and to radiation.

Chapter 15 / Asexual Reproduction and Mitosis **189**

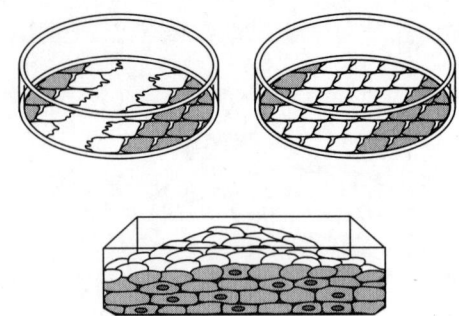

Figure 15-9 Normal cells grown in a lab dish stop dividing when they touch each other (top). Cells that continue to divide after they touch each other exhibit the kind of uncontrolled cell growth seen in cancer (bottom).

One of the most important ways to reduce the risk of cancer is to maintain good health habits. The immune system constantly attacks not only invading cells but also abnormal, cancerous cells from our own body. Most of the time, the immune system is successful. It destroys cancer cells before they can develop and cause problems. It is no surprise then that many people who suffer from **AIDS** (**a**cquired **i**mmuno**d**eficiency **s**yndrome) actually die from some type of cancer. The patient's damaged immune system is not able to protect the person from cancerous cells. Therefore, a healthy immune system is one of the best protections against cancer.

Chapter 15 Review

Multiple Choice

1. An organism's typical life span is its
   A. body length, from head to tail
   B. time between birth and death
   C. average age when it reproduces
   D. reproducing population size

2. The survival of a species depends on
   A. an environment that never changes
   B. a continuously increasing life span
   C. reproduction within its populations
   D. a limit of no more than two populations

3. During the first stage of the cell cycle, a cell
   A. grows in size
   B. divides in half
   C. duplicates genetic material
   D. makes a copy of itself

4. How does a cell grow in size?
   A. It takes in other cells.
   B. Its genetic material replicates.
   C. It divides into two parts.
   D. It takes in organic and inorganic materials.

5. The genetic material of the cell is most like
   A. the blueprints for a house
   B. the tracks for a train
   C. an advertisement for a store
   D. a fence around a house

6. A cell's DNA is located within structures known as
   A. mitochondria
   B. chromosomes
   C. chloroplasts
   D. cytoplasm

7. The number of chromosomes in each body cell
   A. is specific for each type of organism
   B. is the same for every type of organism
   C. decreases from the parent to offspring
   D. increases from the parent to offspring

8. Before cell division, the genetic material must undergo a process called
   A. reduction
   B. restoration
   C. replication
   D. reproduction

9. During the process of mitosis, the chromosomes
   A. are cut in half twice
   B. are equally divided
   C. form a circle in the cell
   D. spread through the cell

10. The diagram below represents the chromosomes in a cell. Which of the following diagrams best illustrates the daughter cells that result from the normal division of this cell?

    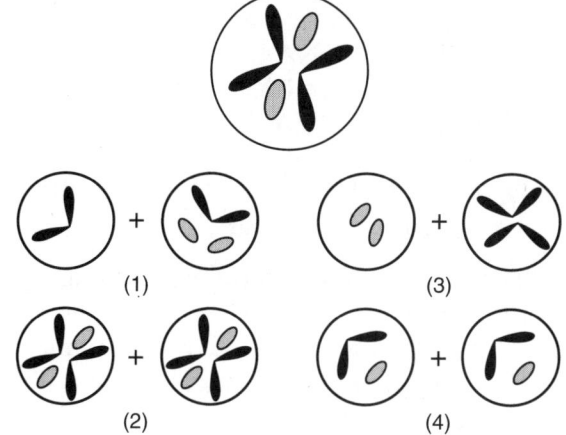

    A. diagram 1   C. diagram 3
    B. diagram 2   D. diagram 4

11. What happens *after* mitosis has occurred?
    A. The cell doubles in overall size.
    B. The genetic material replicates again.
    C. The genetic material forms a parent cell.
    D. The cytoplasm of the cell divides in two.

12. Compared to the parent cell, each daughter cell that results from the normal mitotic division of the parent cell contains
    A. the same number of chromosomes, but different genes from those of the parent cell
    B. half the number of chromosomes, but different genes from those of the parent cell
    C. the same number of chromosomes and identical genes to those of the parent cell
    D. twice the number of chromosomes and identical genes to those of the parent cell

13. In asexual reproduction, the genetic material is supplied by
    A. one daughter cell
    B. one parent cell
    C. two daughter cells
    D. two parent cells

Chapter 15/Asexual Reproduction and Mitosis

14. The diagram below represents a cell process. Which statement regarding this process is correct?

    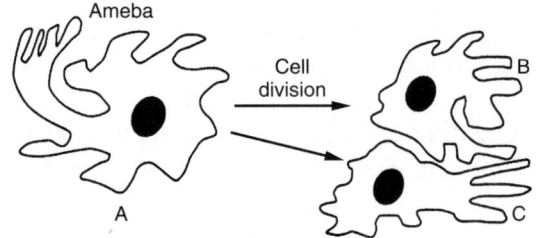

    A. Cell B contains the same genetic information as cells A and C.
    B. Cell A has DNA that is only 75 percent identical to cell B.
    C. Cell C has DNA that is only 50 percent identical to cell B.
    D. Cells A, B, and C each contain different genetic information.

15. The DNA of a plant produced by asexual reproduction would be
    A. identical to that of the parent plant
    B. similar, but not identical, to that of the parent plant
    C. totally different from that of the parent plant
    D. a combination of genetic information from several plants

16. A researcher determines that all the members of a certain population of plants on a lawn are genetically identical. The best explanation for this is that the plant reproduces
    A. sexually, by cloning
    B. sexually, by budding
    C. asexually, by cloning
    D. asexually, by budding

17. A new hydra can be produced from groups of cells that enlarge and stay attached to the parent hydra for a time before breaking off and becoming independent. This method of reproduction is called
    A. sporulation
    B. cloning by runners
    C. binary fission
    D. budding

18. One way to produce many genetically identical offspring is by
    A. using radiation to change their genes
    B. using chemicals to change their genes
    C. cloning them, so they have the same genes
    D. inserting a new DNA section into their genes

19. All of these statements describe cells cloned from a carrot *except* that they
    A. are genetically identical
    B. have the same DNA codes
    C. were reproduced sexually
    D. have identical chromosomes

20. Which statement about the rate of cell division is true?
    A. All the cells of all types of organisms divide at the same rate.
    B. The rate of cell division is related to a cell type's function.
    C. All the cells within an organism divide at the same rate.
    D. The rate of cell division is random in every organism.

21. Damage to a cell's DNA can cause cancer, which results from
    A. a slower than normal cell division
    B. a complete stop to all cell division
    C. an uncontrolled type of cell division
    D. no changes in the genetic instructions

Analysis and Open Ended

22. Why does the survival of a species depend more on its populations than on the life spans of its individual organisms?

23. Briefly explain the main purpose of the genetic material in a cell.

24. Use the following terms to replace the definitions given within the boxes in the following concept map: *Cell division; Cell growth; Mitosis; Replication.*

    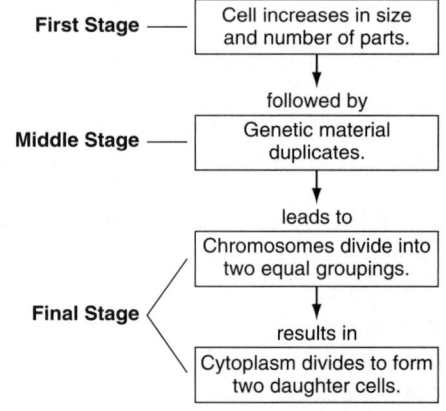

25. The best title for the concept map shown in question 24 probably would be
    A. Functions of a Cell
    B. Life Cycle of a Cell
    C. Regeneration of a Cell
    D. Genetics of the Cell

26. How does the chromosome number of one species compare with that of another species?

27. Explain why the duplication of chromosomes is necessary for the process of cell division.

28. Explain why cell division is necessary for all living things. Your answer should include the following:
    - why it is important for one-celled organisms
    - why it is important for multicelled organisms
    - *one* example of each type of organism discussed

29. During cell division, both the genetic material and the cytoplasm have to be equally divided. Which process occurs first? Which process is called *mitosis*?

Refer to the following two diagrams to answer questions 30 and 31.

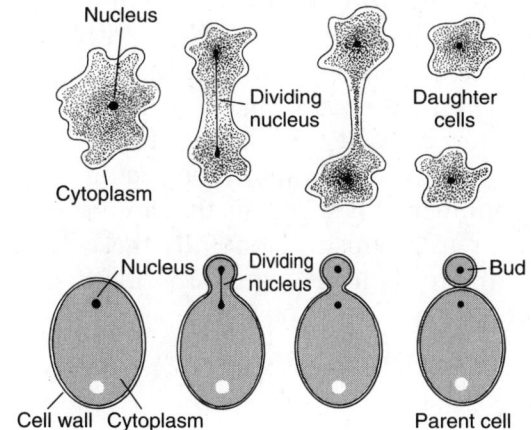

30. What type of reproduction is illustrated in these two diagrams?
    A. sexual reproduction only
    B. asexual reproduction only
    C. sexual and asexual reproduction
    D. neither type of reproduction

31. In what way is reproduction in the ameba (top) the same as reproduction in the yeast (bottom)? In what way is it different? What is the specific term for the yeast's method of reproducing?

32. Which of the following diagrams represents asexual reproduction (by mitosis)? Explain why. (*Note:* The "n" stands for the number of chromosomes in each cell; sex cells are $1n$ and body cells are $2n$.)

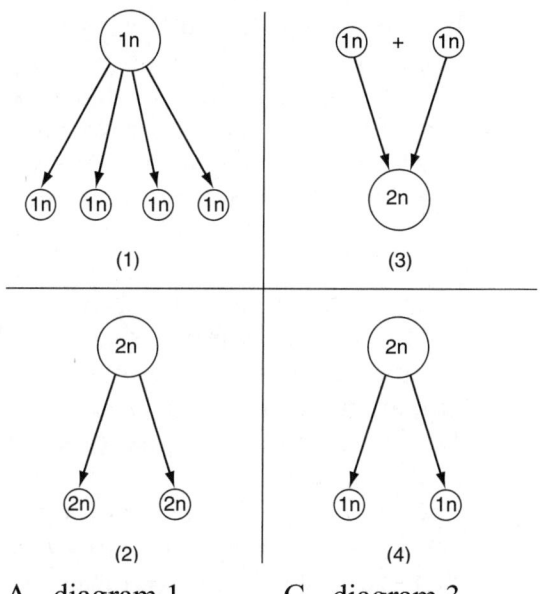

   A. diagram 1   C. diagram 3
   B. diagram 2   D. diagram 4

33. Does the rate of cell division differ from one tissue type to another within an organism? Why is this important for homeostasis? (Give *one* example.)

34. Explain why exposure to radiation and certain chemicals can cause uncontrolled cell division. What type of disease can this process lead to?

Reading Comprehension

Base your answers to questions 35 to 37 on the information below and on your knowledge of biology. Use one or more complete sentences to answer each question.

> Non-Hodgkin's lymphoma (NHL) is a cancer of the lymph system. This system collects intercellular fluid from throughout the body, returning it in tubes to the bloodstream. The tiny lymph vessels join together to eventually form large ones that empty into veins in the neck. Enlargements along these lymph vessels are known as lymph nodes. These nodes, or glands, are involved in the body's defenses against diseases. However the lymph system is also the site for NHL cancer—one of the few cancers that is occurring with greater frequency. No one knows why the incidence of NHL is increasing, but it now accounts for more than 4 percent of cancer deaths in this country.
>
> Chemotherapy, the traditional use of drugs to treat cancers such as NHL, was developed during the twentieth century. These anti-cancer drugs use a variety of methods to attack cancer cells: by attacking DNA; by shutting down protein synthesis; or by stimulating the immune system. In the mid-1990s, trials began for the use of a very different type of drug—monoclonal antibodies. These drugs are actually designer-made antibodies that have been produced to find and attack cell-surface targets that exist only on cancer cells. The monoclonal antibody drugs are therefore referred to as *targeted drugs;* they search out the cancer cells. The cutting-edge capability of these twenty-first-century drugs is to attach radioactivity or some other cancer-fighting drug to the monoclonal antibody. The targeted drug will go find the cancer cells, deliver its deadly payload, and then kill the cancer cells. This treatment is now being used against NHL with some success.
>
> Doctors currently stress that the best approach is to use both methods—twentieth-century and twenty-first-century cancer treatments. Well-respected experts are optimistic about the chances for real progress in the years ahead. For example, Dr. Andrew Zelenetz, chief of the lymphoma services at Memorial Sloan Kettering Cancer Center in New York City has said, "This is a very exciting time. We didn't have new important agents for the treatment of lymphoma for many years. Now we're seeing the emergence of these targeted therapies that are very exciting, and in fact, we're starting to see the emergence of other chemotherapeutic agents that actually have activity in lymphoma. We're entering a new era where we have both the traditional tools as well as these new targeted tools, and we're going to be seeing more of them coming down the pike. There are a number of new agents that are in development that are being tested that I think have real promise." Hopefully these new cancer-fighting agents will be developed in time to fight the increase in incidence of NHL and other potentially deadly cancers.

35. Compare the lymph system with the circulatory system.
36. How do traditional anti-cancer drugs work?
37. How do targeted drugs work in fighting cancer?

Chapter 16

Sexual Reproduction and Meiosis

Standard 5.5.12 C2 **Explain how genetic material can be altered by natural and/or artificial means; mutations and new combinations may have a positive, negative, or no effect on organisms or species.**

SEXUAL REPRODUCTION AND CHROMOSOMES

For almost all types of animals, two parents are needed for reproduction: a male and a female. This is **sexual reproduction**, because it involves parents of both sexes. Most plants use this method of reproduction to make more of their own kind, too. Sexual reproduction is very important for the survival of living things. It also plays a significant role in the process of evolution. To understand this, we must look at individual cells and examine the chromosomes within them. (See Figure 16-1.)

It Is All About the Chromosomes

Each of our cells contains **chromosomes**. The chromosomes contain the inherited information that has been passed along since the beginning of life on Earth. It is this information that determines an individual's characteristics. The chromosomes also contain the "know how" that keeps our cells functioning correctly.

Figure 16-1 A photograph of human chromosomes.

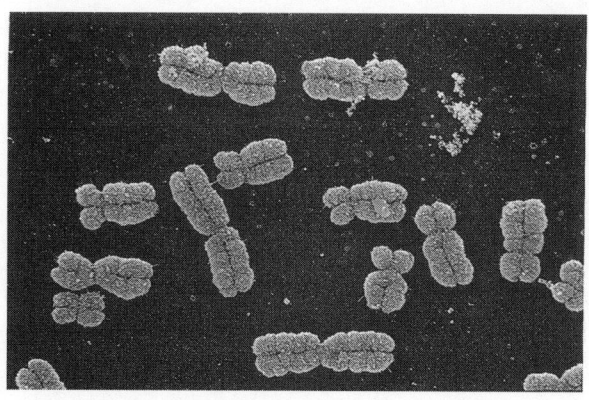

Why is sexual reproduction all about chromosomes? When a **sperm** cell and an **egg** cell unite during sexual reproduction, it is the nucleus from each cell that joins. What does the nucleus contain? Most importantly, it contains the chromosomes. So, sexual reproduction is about the combining of chromosomes from two individuals, a male (the father) and a female (the mother).

Each human body cell contains 46 chromosomes. The first cell from which each of us came—the cell that resulted from the combination of a sperm cell and an egg cell—had 46 chromosomes. Every body cell in you has 46 chromosomes. The question is: How did a sperm cell and an egg cell combine to make a new cell—that first cell of a new human—with just 46 chromosomes?

There is only one way. Both the sperm and the egg must have had only 23 chromosomes each, half the normal number of 46 chromosomes that are found in all body cells. And indeed this is the case. A special type of cell division produces sperm and egg cells, each with that reduced number of chromosomes.

Gametes: The Sex Cells

The sperm and egg cells, or *sex cells,* are also called **gametes**. In the process of sexual reproduction, the nuclei of the gametes join together. This fusion of the nuclei is called **fertilization**. The resulting cell, a fertilized egg cell, is called a **zygote**. (See Figure 16-2.) Each gamete, as noted previously, has exactly half the normal number of chromosomes. The zygote and all body cells that come from the mitotic division of the zygote contain two sets ($2n$) of chromosomes in them, one from each parent. Gametes are produced by a special type of cell division called **meiosis**, which reduces the chromosome number by one-half. This gives the sex cells just one set ($1n$) of chromosomes each. Thus, when fertilization occurs (and the nuclei fuse), the normal number of chromosomes for the species is maintained.

A Closer Look at Chromosomes

Our 23 chromosomes exist in pairs. Every body cell has two copies of each chromosome. But where does each of these two chromosomes come from? The answer is: one from each parent. Beginning with a normal body cell, which has the double set ($2n$) of chromosomes, gametes must be produced through meiotic cell division. Each gamete contains a single set ($1n$) of chromosomes; and it must be an exact set, meaning one and only one chromosome from each of the 23 chromosome pairs.

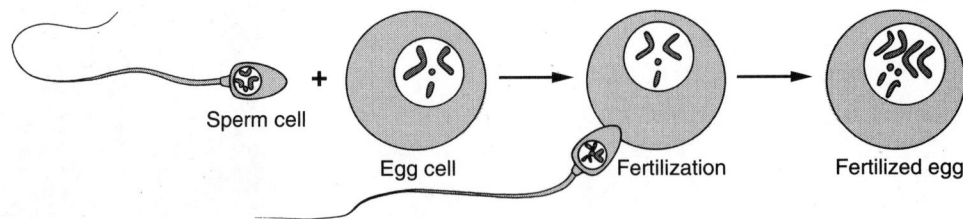

Figure 16-2 Sexual reproduction involves the joining of chromosomes from a sperm cell and an egg cell in the process called fertilization.

MEIOSIS: REDUCING THE CHROMOSOME NUMBER

Mitosis and meiosis take place before the cell division that produces gametes; and in some ways these two processes are similar. In both cases, the chromosomes replicate before either process begins. However, the results of mitosis and meiosis are very different. When mitosis is completed, each new cell's chromosome number remains the same as in the original parent cell ($2n$). When meiosis is completed, each sex cell's chromosome number is half ($1n$) the original number.

Meiosis actually involves two separate cell divisions, and these occur one right after the other. During Meiosis I, **reduction division** occurs: the diploid ($2n$) cell divides and produces two new cells, each containing the haploid ($1n$, also called *monoploid*) number of the *double-stranded* chromosomes. In Meiosis II, four new cells are produced, each containing the haploid number of the *single-stranded* chromosomes. Meiosis is called a *reduction division* because the final number of chromosomes is reduced (by half). (See Figure 16-3.)

Meiosis: The Source of Our Differences

With the exception of identical twins, children in the same family are never exactly alike. Differences can occur in eye color, hair color, height, nose shape, ear size, and many other characteristics. Why is this so, if the children were born of the same parents? The explanation arises from one of the two important jobs of meiosis. The first job of meiosis, as we have said, is to maintain the normal species chromosome number by producing gametes that have single sets of chromosomes.

The second important job of meiosis is to increase genetic variability among offspring by **recombining** genes in the eggs and sperm. (See Figure 16-4.) Genes may get exchanged between chromosomes during meiosis (in a process called *crossing-over*). Also, chromosomes may get resorted into new groupings. Because of genetic recombination during meiosis, sexual

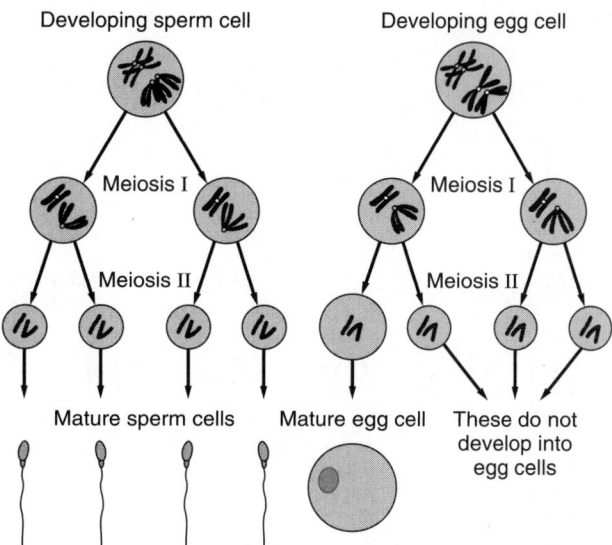

Figure 16-3 As a result of meiosis, sperm cells and egg cells have half the normal number of chromosomes for their species.

Figure 16-4 During meiosis, genetic recombination occurs when chromosomes overlap and exchange pieces.

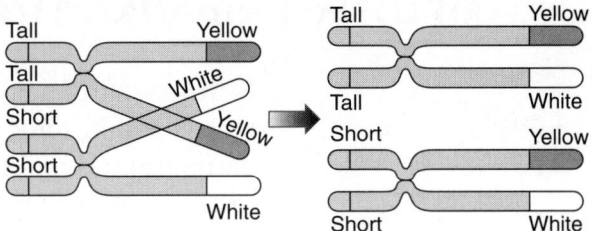

reproduction results in offspring that are different from each other and from their parents. (See Figure 16-5.) This **genetic variation** among offspring is what natural selection acts on. A greater variety of characteristics in offspring increases the chances that some individuals will be better suited than others to survive in a particular place and time. As natural selection acts on the varied offspring within a population, generation after generation, the species evolves.

Unusual Meiotic Events

The sorting of chromosomes that occurs in cell division, especially during meiosis, is part of a complex sequence of events. However, it does not always proceed correctly. A gamete may have an extra chromosome because it receives both members of a pair of chromosomes, instead of only one; or a gamete may be one chromosome short, having received neither member of a pair. If a gamete with either abnormality fuses with another gamete, problems may occur. In most instances, the zygote fails to develop. However, in some cases, the zygote does survive and it develops into an individual with an abnormal chromosome number.

The Sex Life of Flowering Plants

Of the many types of plants on Earth, flowering plants are the group that has evolved most recently. Most types of plants reproduce sexually. As in animals, the male gamete (sperm from the male reproductive organ) joins with the female gamete (egg in the female reproductive organ) to produce a zygote (the seed). The zygote grows into a new plant when external conditions are suitable. What is special about flowering plants is that the place where fer-

Figure 16-5 Each time meiosis occurs, the chromosomes line up in a different arrangement; this results in variability among resulting offspring.

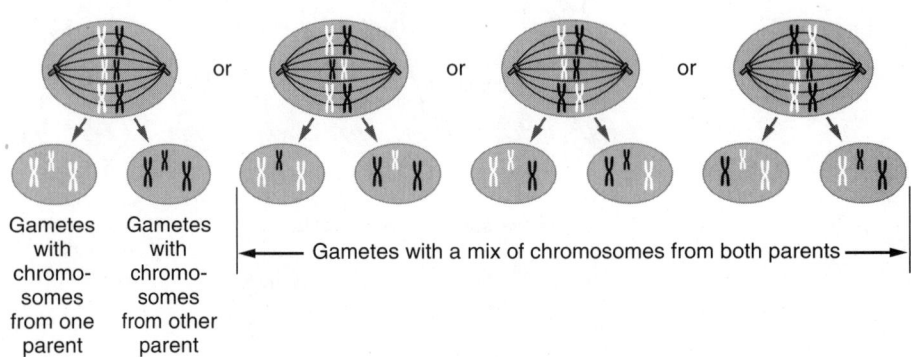

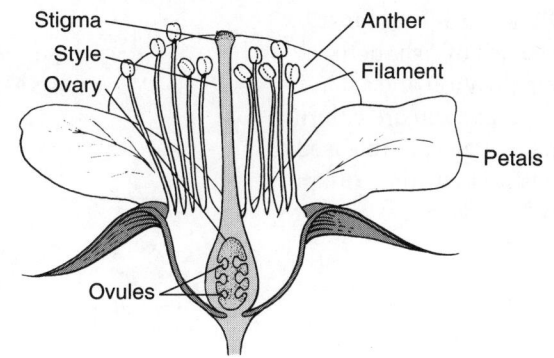

Figure 16-6 In flowering plants, sexual reproduction takes place in the flowers, which contain the reproductive structures. The female parts—stigma, style, and ovary with ovules—make up the pistil; the male parts—anther and filament—make up the stamen.

tilization occurs is very visible. The location is often brightly colored, beautifully shaped, and sweet smelling. In other words, *flowers* are the parts of plants where sexual reproduction occurs. In fact, the parts that make up a flower include the sex organs of the plant. (See Figure 16-6.)

TYPES OF FERTILIZATION AND DEVELOPMENT

Various methods of sexual reproduction occur in plant and animal species. Yet, whatever the method, sexual reproduction always involves fertilization (the fusion of nuclei from two gametes) and **development**, the growth of the zygote into a new individual. One of the main differences in the types of reproduction involves the location of the events. Both fertilization and development may occur either inside or outside the bodies of the reproducing organisms.

EXTERNAL FERTILIZATION AND EXTERNAL DEVELOPMENT

For many aquatic plants, the sex cells meet in the open water; fertilization occurs and the zygote begins to develop. Many aquatic invertebrates simply release their gametes into the water, where fertilization and development then occur. Two groups of vertebrates, the fish and the amphibians, also reproduce in water. In most species of fish, the eggs and sperm are released directly into the water, where fertilization and development of the zygotes then occur. These events, which occur in the outside environment and not inside the organism, are known as **external fertilization** and **external development**. External fertilization and development are risky; the zygotes are exposed to predators. Therefore, large numbers of eggs are released to increase the chances that some of the offspring will survive. (See Figure 16-7.)

Internal Fertilization and External Development

In the case of vertebrates that reproduce on land, the gametes still need moisture to meet and fuse. Reptiles and birds make use of the fluids inside their

Figure 16-7 In most species of fish, both fertilization and development are external; many eggs are released to ensure that some offspring will survive.

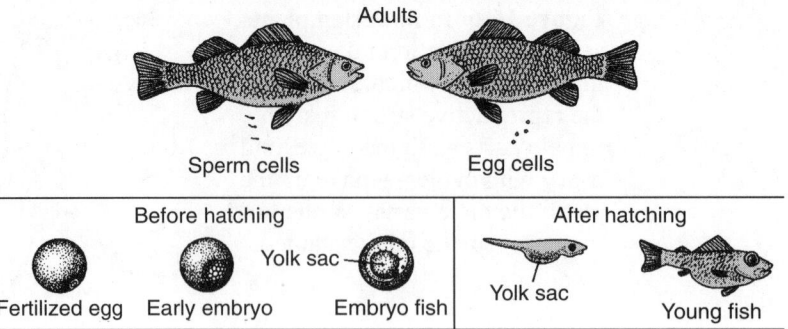

bodies for fertilization. The male and the female must mate so that the sperm can be deposited inside the female, a process known as **internal fertilization**. Then the zygote is prepared for its development on land: a watertight membrane and a protective shell form around the zygote; the egg is laid (usually within a nest), and development of the new organism occurs externally. Internal fertilization increases the chances of reproductive success and survival. Fewer eggs are produced; but some parental care is provided to help protect the developing zygotes. (See Figure 16-8.)

Internal Fertilization and Internal Development

One final pattern of sexual reproduction takes place, occurring mainly in mammals. Fertilization occurs internally, but the big difference from most other animal groups is that development of the zygote occurs within the female's body, too. Thus, mammals have **internal development**. The food for the developing **embryo** comes entirely from the body of the mother. A structure called the **placenta** has evolved in mammals to bring nutrients to the developing baby and to remove its wastes. (*Note:* The exceptions to this are the *marsupial* mammals, which complete their embryonic development within a protective pouch, and the egg-laying mammals.) After birth, the baby mammal continues to receive nourishment by nursing on milk provided by the mother's mammary glands. Although mammals' embryos are typically fewer in number, they have the most complete form of protection, since they develop within the mother's body and then receive more parental care after their birth.

Figure 16-8 In birds, fertilization is internal. The embryo is surrounded by a watertight membrane (the amnion) and then covered by a protective shell. The egg is laid and development is external.

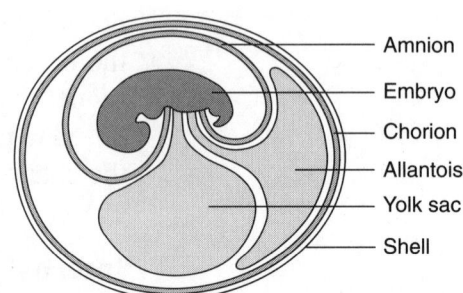

Chapter 16 Review

Multiple Choice

1. During sexual reproduction, the chromosomes of
   A. two separate individuals are combined together
   B. one individual are transferred to another
   C. one parent only are copied for its offspring
   D. two separate individuals are split apart

2. If each human body cell has 46 chromosomes, how many were in your very first body cell?
   A. 23
   B. 46
   C. 92
   D. 100

3. Most cells in the body of a fruit fly contain eight chromosomes. How many of these chromosomes were contributed by each parent fruit fly?
   A. 2
   B. 4
   C. 8
   D. 16

4. Sperm cells of the Russian dwarf hamster contain 14 chromosomes. What is the total number of chromosomes that would be found in each body cell of a normal baby hamster?
   A. 7
   B. 14
   C. 28
   D. 42

5. The gamete (sex cell) for any species should *always* contain
   A. an even number of chromosomes
   B. the normal number of chromosomes
   C. twice the normal number of chromosomes
   D. half the normal number of chromosomes

6. The diagram below represents some events in a cell that is undergoing normal meiotic cell division.

   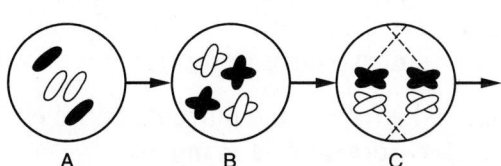

   Which of the following diagrams most likely represents the next cell that would result from the process shown?

   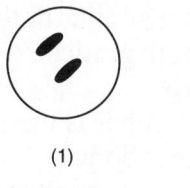
   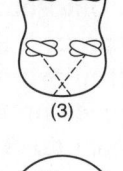

   A. diagram 1
   B. diagram 2
   C. diagram 3
   D. diagram 4

7. Compared to human cells resulting from mitotic cell division, human cells resulting from meiotic cell division should have
   A. twice as many chromosomes
   B. one-half as many chromosomes
   C. the same number of chromosomes
   D. one-quarter as many chromosomes

8. During fertilization, the parts of the sex cells that join are the
   A. membranes
   B. nuclei
   C. ribosomes
   D. vacuoles

9. Which of these items is formed during fertilization?
   A. an egg cell
   B. a sperm cell
   C. a zygote
   D. a gamete

10. In humans, most cells contain 46 chromosomes. However, some cells contain only 23 chromosomes, as a result of
    A. mitotic cell division
    B. embryonic differentiation
    C. meiotic cell division
    D. internal fertilization

11. Which statement best explains the significance of meiosis for the survival of a species?
    A. Meiosis produces egg cells and sperm cells that are completely alike.
    B. Meiosis ensures the continuation of a species by asexual reproduction.
    C. Meiosis produces equal numbers of egg cells and sperm cells in animals.
    D. Meiosis results in genetic variation among the offspring that are produced.

12. Mitosis and meiosis are similar in that
    A. the chromosomes are replicated before either process starts
    B. the chromosome number is the same when each process is done
    C. two separate cell divisions occur during both processes
    D. each process combines genetic material from two individuals

13. Which diagram correctly represents part of the process of sperm formation in an organism that has a normal chromosome number of eight?

    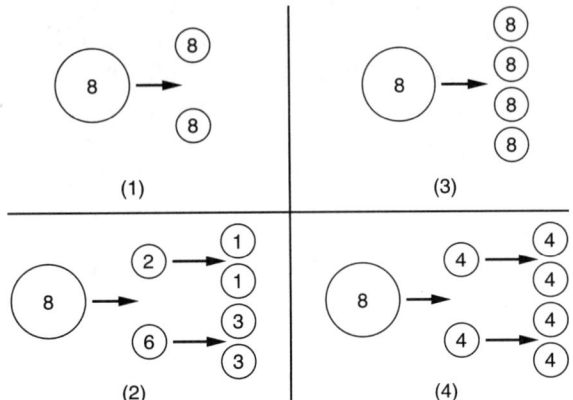

    A. diagram 1   C. diagram 3
    B. diagram 2   D. diagram 4

14. The great variety of possible gene combinations in a sexually reproducing species is due in part to the
    A. sorting of genes as a result of gene replication
    B. pairing of genes as a result of differentiation
    C. sorting of genes as a result of meiosis
    D. pairing of genes as a result of mitosis

15. During meiosis, genes may be exchanged between chromosomes. This genetic recombination usually results in
    A. overproduction of gametes
    B. variation within the species
    C. fertilization and development
    D. formation of identical offspring

16. The following diagram shows a process that can occur during meiosis. The most likely result of this process is

    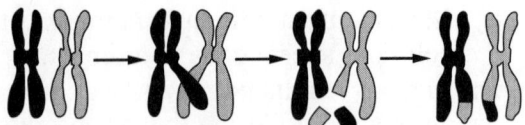

    A. a new combination of inheritable traits that can appear in the offspring
    B. a loss of genetic information, which produces a genetic disorder in the offspring
    C. an inability to pass either of these chromosomes along to any offspring
    D. an increase in the chromosome number of the organism in which this occurs

17. Mitosis produces new body cells and meiosis produces
    A. new body cells, too
    B. body cells and sex cells
    C. sex cells, only
    D. red blood cells

18. Which of the following is a characteristic found *only* in sexual reproduction?
    A. cell division
    B. cell growth
    C. fertilization
    D. chromosomes

19. Sexual reproduction in flowering plants occurs within the
    A. roots
    B. stems
    C. leaves
    D. flowers

20. An animal that has external fertilization will produce more eggs than an animal that has internal fertilization, because
    A. the siblings help raise each other without any parental involvement
    B. an animal can reproduce externally only once during its lifetime
    C. it increases the chances that some of the offspring will survive
    D. it helps the parent animal locate some eggs after they are fertilized

21. In terms of reproduction, how do mammals *differ* from most other animals?
    A. The gametes are formed internally.
    B. Fertilization takes place internally.
    C. The zygote is formed externally.
    D. The embryo develops internally.

22. What is the role of the placenta in the embryonic development of a mammal?
    A. It forms a hard, protective barrier around the developing embryo.
    B. It brings nutrients to and removes wastes from the developing baby.

C. It provides the location for fertilization of the egg to occur.
D. It provides a method of nourishing the baby after it is born.

Analysis and Open Ended

23. Briefly state what chromosomes contain and what they determine in organisms.

24. In what way is sexual reproduction basically "all about the chromosomes"?

25. Use the words *gametes, zygote,* and *fertilization* in one sentence to explain sexual reproduction.

26. Why are gametes essential to sexual reproduction, in terms of their chromosome number?

Base your answers to questions 27 and 28 on the following diagram and on your knowledge of biology.

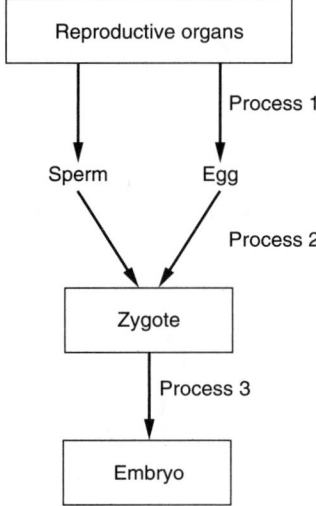

27. Explain why Process 2 is necessary in sexual reproduction.

28. State *one* difference between the cells produced by Process 1 and the cells produced by Process 3.

29. Explain why a mutation in a gamete may contribute to genetic changes in offspring, while a mutation in a body cell will not.

30. The paramecium (a single-celled organism) usually reproduces asexually. But some paramecia have developed a method by which they actually exchange genetic material with each other. State *one* advantage this simple form of sexual reproduction would have over asexual reproduction for the survival of these single-celled organisms.

31. In terms of chromosome number, what is the main difference between the results of mitosis and the results of meiosis?

Refer to the diagram below to answer questions 32 to 34. (Note: The "n" stands for the number of chromosomes in each cell.)

$1n$ + $1n$ ⟶ $2n$
A   B   C

32. What reproductive process does the diagram represent?
    A. gamete forms   C. fertilization
    B. cell division   D. recombination

33. Which of the structures in the diagram represents a gamete?
    A. structure *A* only   C. structures *A* and *B*
    B. structure *B* only   D. structure *C* only

34. Which of the structures in the diagram represents a zygote?
    A. *A* only   C. *C* only
    B. *B* only   D. *A* and *B*

35. Which of the following diagrams represents meiosis? (*Note:* The "n" stands for the number of chromosomes in each cell.)

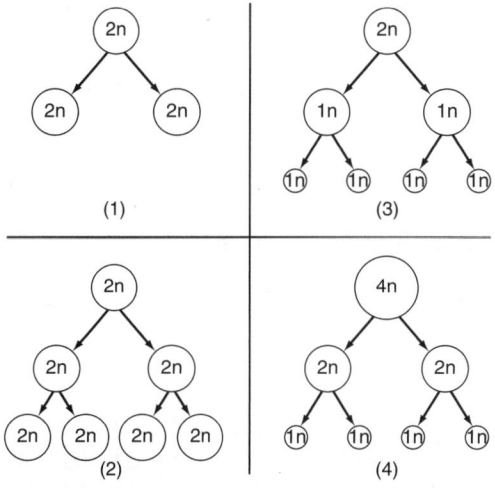

    A. diagram 1   C. diagram 3
    B. diagram 2   D. diagram 4

36. Why are offspring of organisms that reproduce sexually *not* genetically identical to their parents?

37. Compare asexual reproduction to sexual reproduction. Be sure to answer the following:
    - Which type of reproduction results in offspring that are genetically identical to the parents? Why?
    - Which type of reproduction results in offspring that are genetically different from the parents? Why?

Refer to the illustration below, which shows an important event that occurs during meiosis, to answer questions 38 and 39.

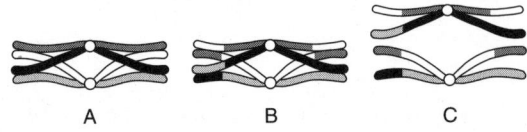

38. What is occurring in this process from steps A to B to C? How are the chromosomes in step C different from those in step A?

39. Why is this process significant in terms of the offspring that are produced?

40. Briefly describe the process of sexual reproduction in flowering plants.

41. Which characteristic of sexual reproduction has specifically favored the survival of animals that live on land?
    A. the fusion of gametes in the outside environment
    B. male gametes that may be carried by the wind
    C. fertilization within the body of a female parent
    D. female gametes that are enclosed in a hard shell

42. Which process normally occurs at the placenta?
    A. Oxygen diffuses from fetal blood to maternal blood.
    B. Maternal blood is converted into fetal blood.
    C. Materials are exchanged between fetal and maternal blood.
    D. Digestive enzymes pass from maternal blood to fetal blood.

Base your answers to questions 43 and 44 on the paragraph below and on your knowledge of biology.

Three groups of animals in which most species lay eggs for reproduction are amphibians, reptiles, and birds. Most female amphibians lay hundreds of eggs in the water, which are then fertilized by sperm from the male. Many reptiles lay up to 200 eggs at a time, often in nests on land. The eggs have a leathery shell. Birds usually lay between one and four eggs at a time in nests on land. Wild birds' eggs usually have shells similar to those of the domestic chicken. Most mammals bear live young. Some of these mammals, such as humans, usually give birth to just one live offspring at a time.

43. State *one* reason why individuals of some species must lay hundreds of eggs in order for some of the offspring to survive.

44. Explain why fertilization in reptiles and birds must be internal even though development is usually external. How are the offspring protected during their external development?

Reading Comprehension

Base your answers to questions 45 to 47 on the information below and on your knowledge of biology. Use one or more complete sentences to answer each question.

> Joseph and his parents had been referred by their family doctor to the genetics clinic at a local hospital. The doctor had made this suggestion after testing the levels of sex hormones in Joseph's blood and ordering a chromosome analysis. Joseph, a healthy 16-year-old, was doing well in school. However, even though he was almost two meters tall, he showed no signs of sexual maturity. He had not developed additional body hair, and his voice retained the high pitch of a much younger person. A long talk occurred in the geneticist's office.

> The counselor explained that Joseph has one more X chromosome than is usually found in a male. This is due to an error that occurred during meiosis when the egg or sperm from which he developed formed. Joseph and his parents learned from the counselor that about one in every thousand males has this XXY chromosome makeup—a total of 47 chromosomes instead of the normal 46. The fact that this condition is called Klinefelter syndrome was not important to Joseph and his family. However, learning about the characteristics of the syndrome was very important. The counselor told them that the characteristics of men with this syndrome usually included tall stature, some minor birth defects, small testicles, and sterility. Sometimes, but not in Joseph's case, mental retardation begins at birth.
>
> As a result of the genetics counseling, Joseph realized that he could most likely look forward to a normal life. Indeed, in a few months, his beard began to grow. More important, Joseph learned about possible options for having a family of his own someday.

45. Why did Joseph's doctor order hormone and chromosome tests for him?

46. Describe the genetic explanation for the symptoms experienced by Joseph.

47. How did the genetics counseling that Joseph and his family received benefit them?

Chapter 17

Patterns of Inheritance

Standard 5.5.12 C2 Explain how genetic material can be altered by natural and/or artificial means; mutations and new combinations may have a positive, negative, or no effect on organisms or species.

GREGOR MENDEL: THE FOUNDER OF GENETICS

How characteristics are passed from parents to offspring was a question that puzzled people for thousands of years. A set of experiments completed more than 100 years ago by Gregor Mendel helped our understanding of **inheritance**, how traits are passed from one generation to the next. Mendel, an Austrian monk, conducted hundreds of experiments using thousands of pea plants. By applying careful mathematical analysis to his work (something that had rarely been done in biology before), Mendel discovered much about the way heredity works. The study of inheritance really began with Mendel; because of this, he can rightly be called the "founder" of the science of genetics. (See Figure 17-1.)

Mendel discovered that a trait, or **hereditary** information, is passed from parents to offspring in individual units that he called *factors,* which we know today to be genes. He also discovered that the factors were passed on in specific, predictable patterns from both parents. These patterns of inheritance are described in Mendel's laws.

Mendel's Ideas About Inheritance

Mendel proposed several ideas that explained his results. These ideas were correct, even though Mendel knew nothing about chromosomes, genes, or DNA. Since that time, scientists have been able to combine Mendel's ideas with what has been learned about genetics.

Mendel's first main idea was that each characteristic, or **trait**, exists in two versions. These two versions of a gene are called **alleles**. For example, in pea plants, there are two alleles for the gene for flower color: one allele for purple flowers and one allele for white flowers. Genes exist at specific locations on chromosomes. So, one chromosome may have an allele for purple

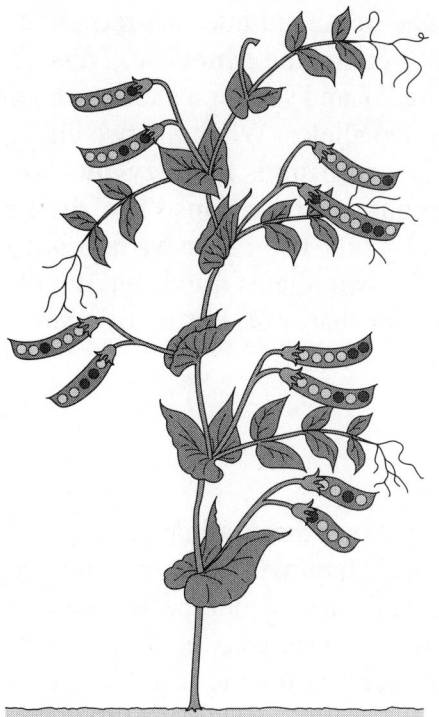

Figure 17-1 Gregor Mendel investigated the inheritance of traits from one generation to the next by carefully studying characteristics of the common pea plant.

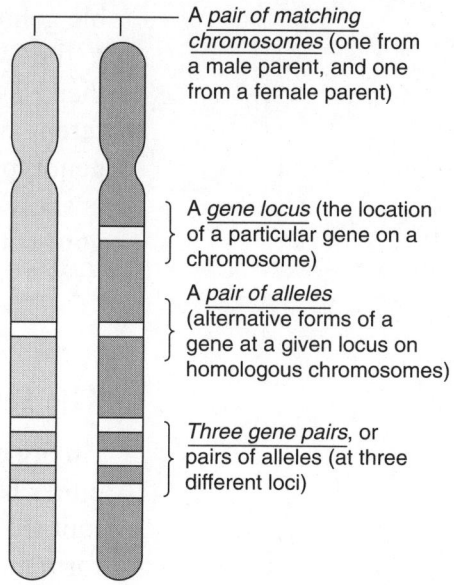

Figure 17-2 The gene for one trait, such as flower color, exists at the same place on each member of a pair of matching chromosomes. Each gene has two versions, or alleles, of that trait; for example, the allele is either purple or white for flower colors.

at the location for flower color, while the other chromosome may have an allele for white at the same location. The DNA at these locations consists of a sequence of subunits. At one allele, the DNA subunits code for proteins that result in the color purple. At the other allele, a different sequence of DNA subunits codes for proteins that result in the color white.

Mendel's second main idea was that for each characteristic, an individual inherits two copies (that is, two alleles), one from each parent. All offspring—plant or animal—that result from sexual reproduction have a double set of chromosomes, made up of the chromosome pairs. Each pair of chromosomes consists of one chromosome from the mother and one chromosome from the father. Since corresponding genes (such as for flower color) occur at the same place on each chromosome in a pair, any particular gene exists twice in each cell. Thus, every cell has two alleles for each gene. (See Figure 17-2.)

Important Genetic Terms

All the genes present in an organism make up its **genome**. The physical appearance of a trait in an organism is called the **phenotype**. In pea plants, plant height is determined by a gene with two possible alleles. The allele for tall (*T*) is *dominant* over the allele for short (*t*), which is *recessive*. For example, the phenotype of a pea plant is tall whether it has two dominant alleles, *TT,* or one dominant and one recessive allele, *Tt.* The gene combination of a trait is called the **genotype**. Thus, for a tall plant, the two possible genotypes are *TT* and *Tt.* For a short plant, the only possible genotype is *tt,* the two recessive alleles. When the two alleles of a gene are the same (either *TT* or *tt*), the organism is **homozygous** for that trait. There are two different homozygous combinations. The **dominant homozygous** has the genotype *TT* and is tall. The **recessive homozygous** has the genotype *tt* and is short. When the two alleles are different (*Tt*), the organism is **heterozygous**, or *hybrid,* for that trait, but it still shows the dominant (tall) phenotype. (See Figure 17-3.)

The Testcross

An organism that shows the dominant phenotype for a particular trait may be either homozygous dominant (*TT*) or heterozygous dominant (*Tt*) for that trait. To determine the true genotype of such a plant or animal, scientists perform a **testcross**. In a testcross, the organism is crossed with one that is homozygous recessive (*tt*) for the trait. If the test organism is homozygous dominant (*TT*), then all the offspring will be heterozygous (*Tt*) and show the dominant phenotype. If any of the offspring show the recessive (*tt*) phenotype, then the test organism must be heterozygous dominant (*Tt*). (See Figure 17-4.)

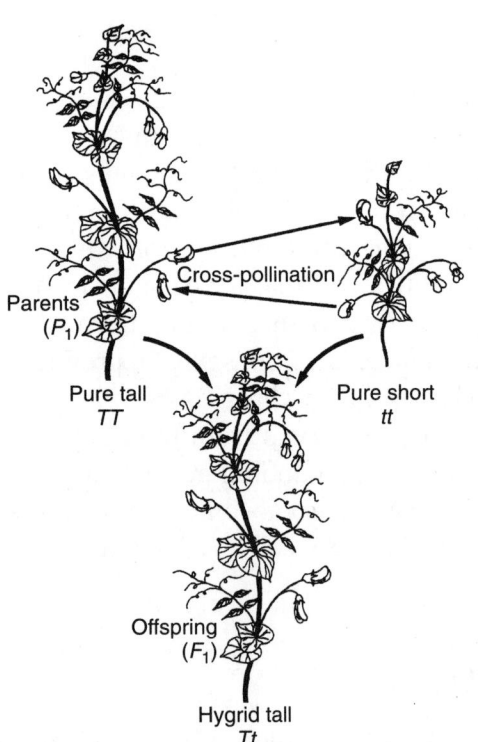

Figure 17-3 A homozygous (dominant) tall pea plant is crossed with a homozygous (recessive) short pea plant. All offspring are heterozygous (hybrid) tall; that is, they show the dominant phenotype.

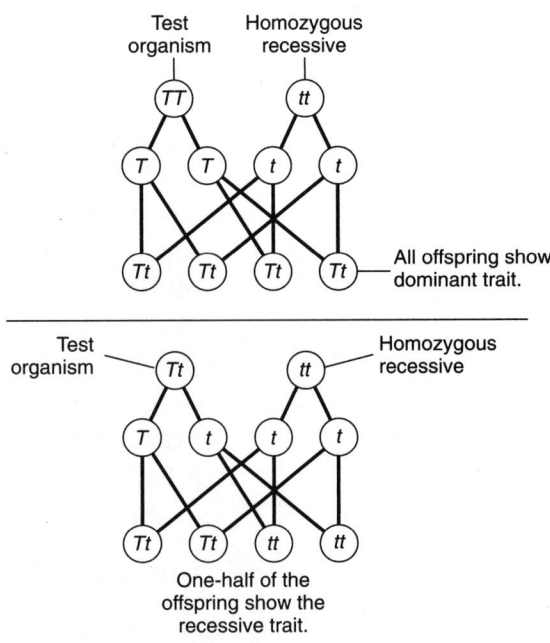

Figure 17-4 Use of a testcross to determine an organism's genotype.

MENDEL'S LAWS OF INHERITANCE

Mendel developed the following three laws to describe how an organism's various traits are inherited and expressed.

- *The Law of Dominance.* When two different alleles for a trait are inherited, only one is expressed, while the other remains "hidden." The allele that is expressed (shows its trait or characteristic) is called the **dominant** allele. The allele that is hidden is the **recessive** allele. A recessive trait will be expressed only when the organism has two copies of the recessive allele; that is, it has inherited one from each parent organism.

- *The Law of Segregation and Recombination.* The different forms of a gene (the alleles) randomly separate, or *segregate,* during the formation of gametes (in meiosis). During fertilization, they come together and new combinations of alleles are randomly formed. Mendel's experiments showed that when he let the hybrid (*Tt*) plants he had bred self-pollinate, the plants in the resulting generation showed a 3-to-1 (75 percent to 25 percent) ratio of the dominant (tall) phenotype to the recessive (short) phenotype. We know that the genotypes for these offspring were actually 1 *T* : 2 *Tt* : 1 *tt*. (See Figure 17-5.)

- *The Law of Independent Assortment.* Genes for different traits sort independently of one another during gamete formation (in meiosis). Each trait is inherited independently of all other traits. For example, in pea plants, a gamete that receives a dominant (yellow) allele for pea color can receive either a dominant (round) or recessive (wrinkled) allele for pea shape. (See Figure 17-6 on page 210.) However, if two genes are located on the same chromosome, they are inherited together (see "Linked Traits" in the text that follows).

POLYGENIC INHERITANCE AND MULTIPLE ALLELES

Mendel's experiments provided the foundation for the study of inheritance. Much of Mendel's success was due to the traits he studied. The pea plants he worked with were either tall or short, and their seeds were either yellow or green, smooth or wrinkled, and so on. But are people either tall or short? Are there only two skin colors? Do people have only blond or black hair? Of course not. In humans, and in most organisms, almost all traits are not as

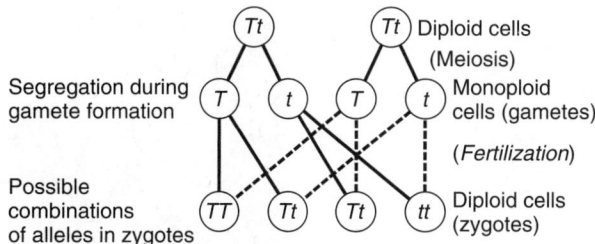

Figure 17-5 Segregation and recombination of alleles.

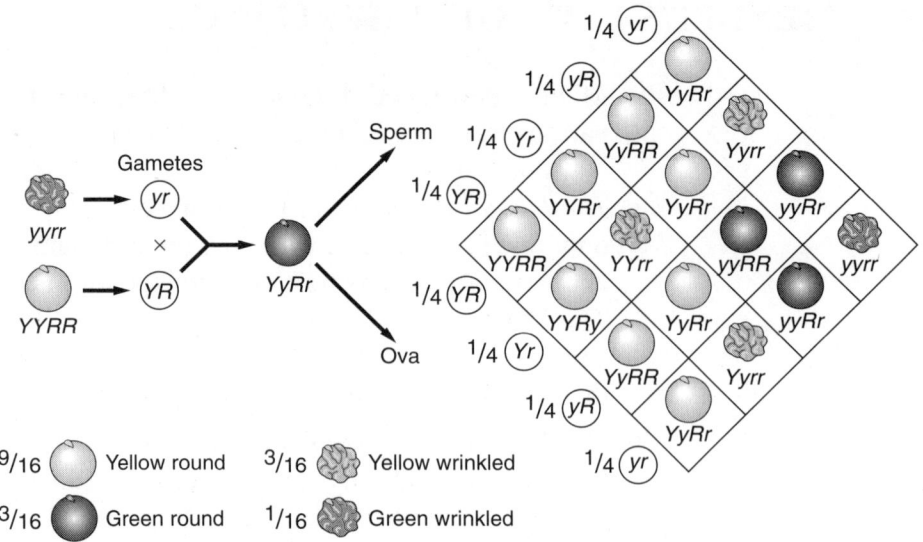

Figure 17-6 Independent assortment is illustrated by crossing organisms that are hybrid for two traits. Each trait is inherited independently and thus shows a 3 : 1 ratio among the offspring; overall, there is a 9 : 3 : 3 : 1 ratio.

clearly defined as the traits Mendel studied in pea plants. Yet Mendel was correct in his explanations.

The traits Mendel picked to study were, luckily for him, each determined by single genes. The height of pea plants is due to a single gene that occurs in just two different versions, or alleles. Human height, however, is a different story, because height in humans is determined by several genes. Sections of DNA on different chromosomes code for the proteins that affect human height. A trait that is determined by more than one gene is said to be the product of **polygenic inheritance**. (*Note:* The term *poly* means "many.") If a large group of people were arranged according to height, those with average height would be the most numerous. There would be fewer extremely short and fewer extremely tall people. (In fact, grouping individuals in this way produces a bell-shaped curve.) There are other traits that can be determined by more than one gene. For example, the many variations in human skin color show that multiple genes are involved in this trait, too. (See Figure 17-7.)

Figure 17-7 When a group of men are arranged by height, a bell-shaped curve is produced, which shows variation within a sample population.

Another type of inheritance, determined by *multiple alleles,* occurs when one gene can be represented by more than two alleles. For example, there are three different alleles (I^A, I^B, and i) that can be found at the gene location for human blood groups. Two of these alleles (I^A and I^B) are equally dominant, or *codominant.* Depending on which two alleles are present, a total of six genotypes are possible, resulting in a total of four different blood groups that can be inherited. *Codominance* is the equal effect of two alleles on the phenotype of a heterozygous trait. Since I^A and I^B are codominant, a person with that genotype has the type AB blood phenotype. Thus, this type of inheritance does not completely follow Mendel's First Law of Dominance.

INTERMEDIATE TRAITS AND LINKED TRAITS

Sometimes, rather than one allele being dominant over, or even *codominant* with, another allele, a different type of inheritance occurs. In such cases, the heterozygous (hybrid) offspring shows an intermediate phenotype for a trait, as compared to that trait in either of its parents. This is called **incomplete dominance**, and it is seen in some plants and animals. For example, when Japanese four o'clock plants that have red flowers (R) are crossed with those that have white flowers (W), the heterozygous offspring has pink flowers (RW). The hybrid pink flowers, when crossed, show the 1:2:1 ratio of red (R): pink (RW): white (W) flowers among their offspring, indicating that Mendel's Second Law of Segregation and Recombination still applies. (See Figure 17-8.)

In some cases, the genes for one type of trait are inherited along with the genes for another particular type of trait. Such traits are said to be **linked** because they are located on the same pair of homologous chromosomes. Many human traits are linked because their genes are on the same chromosome. For example, why do people with red hair also have freckles? The reason is that the genes for red hair and freckles are located on the same chromosome. These genes are linked, so they tend to be inherited together. Therefore, this inheritance pattern does not follow Mendel's Third Law of Independent Assortment.

SEX-LINKED TRAITS AND PEDIGREES

In human inheritance, many genes found on the X chromosome have no corresponding alleles on the smaller Y chromosome. These are called **sex-linked genes**, because the traits they control are determined by the female parent. Recessive sex-linked traits appear more often in males (XY) than in females (XX), because in the females there is usually a normal, dominant allele on the other X chromosome, so that the phenotype is normal. However, in males, there is no second allele; so the presence of just one recessive gene results in the recessive phenotype.

Both *hemophilia* and *color blindness* are sex-linked disorders that occur more frequently in males than in females. Hemophilia is a disorder in which

Figure 17-8 An example of incomplete dominance in Japanese four o'clock flowers: The traits are "blended" in the hybrid offspring.

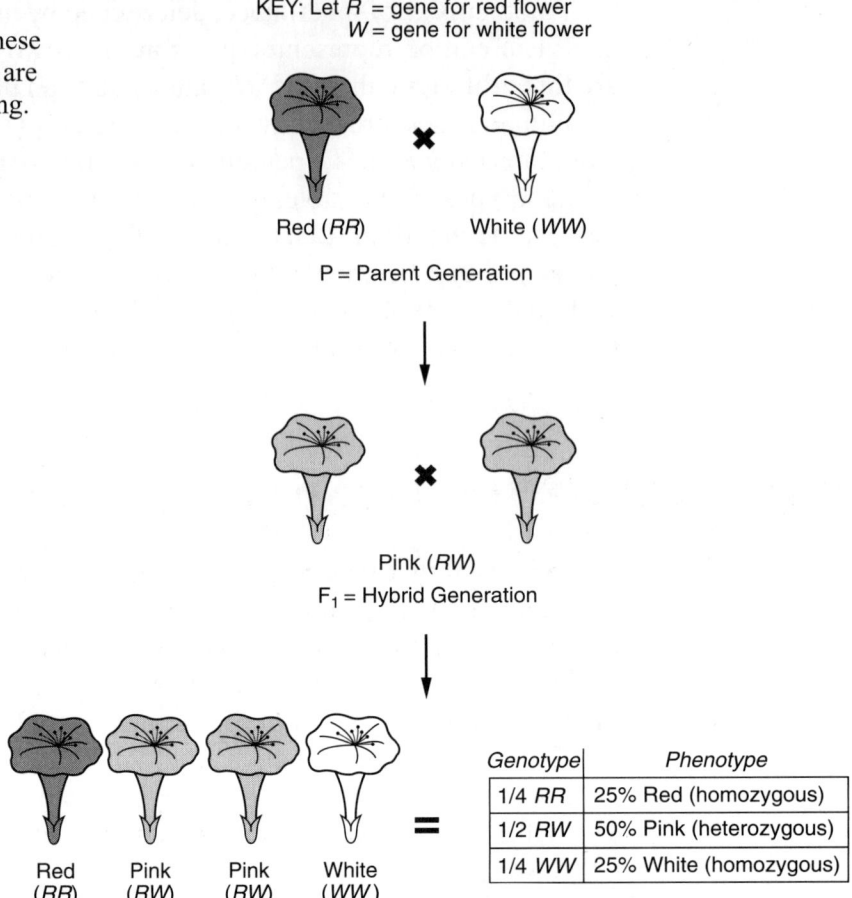

the blood does not clot properly. Color blindness is the inability to see differences between certain colors. The genes for normal blood clotting and normal color vision are dominant. The genes for hemophilia and color blindness are recessive. For a female to show either of these disorders, she must have recessive genes (alleles) on both X chromosomes. If a female has one normal, dominant gene and one recessive gene, she is a *carrier* of the disorder; she does not have the disorder, but she can pass it on to her offspring. If a male inherits one recessive gene, he will have the disorder. (See Figure 17-9.)

Figure 17-9 Inheritance of color blindness, a sex-linked trait.

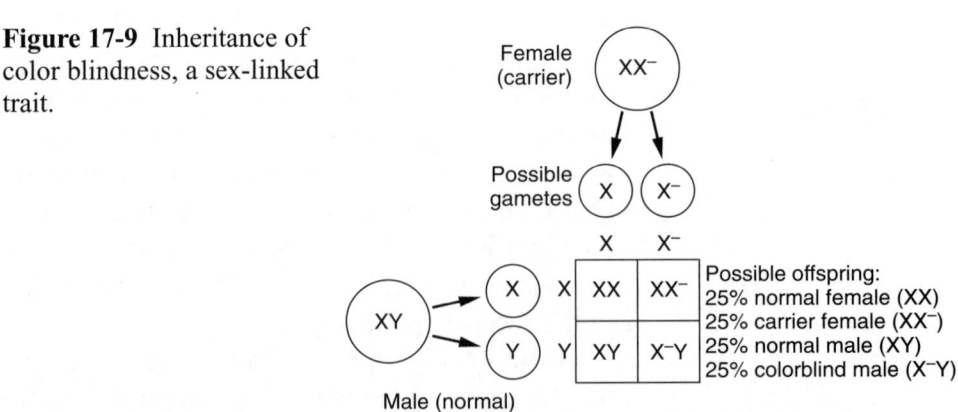

Other disorders and diseases, such as cystic fibrosis and Tay-Sachs disease, occur only if both parents have the gene. Inheritance patterns of certain traits (such as sex-linked disorders) can be traced in families for a number of generations. The patterns can be illustrated in **pedigree** charts that show the presence or absence of these genetic traits in each generation. Pedigree charts are useful because they can also identify the carriers of recessive genes and show how the genetic disorder is being inherited. A pedigree chart uses symbols to represent the parents and children in a family over several generations; squares represent males and circles represent females.

USING PUNNETT SQUARES

Probability is the likelihood that a chance event will occur. The probability of an event is usually expressed as a ratio or a fraction. For example, when throwing dice, the probability of getting one dot facing up on a die (which has six sides) would be one in six chances, written as the ratio 1 : 6 or the fraction ⅙.

Geneticists use a box-shaped chart called a **Punnett square** to show all the possible combinations of gametes and their probabilities when two organisms are crossed. (*Note:* The ratios predicted by a Punnett square may be slightly different from the results of an actual cross.) As was shown in the example of pea plants, a capital letter is used to represent the dominant allele for a trait; and the same letter, but lowercase, is used to represent the recessive allele for the same trait. Thus, for height in pea plants, a capital "*T*" represents the dominant (tall) allele, and a lowercase "*t*" represents the recessive (short) allele. The Punnett square can be used to determine the possible genotypes for height in the offspring of two pea plants whose genotypes are known. For example, when a heterozygous (tall, *Tt*) plant is crossed with a recessive homozygous (short, *tt*) plant, the resulting offspring are 50 percent tall (*Tt*) and 50 percent short (*tt*). (See Figure 17-10.)

Figure 17-10 Use of a Punnett square to determine possible genotypes of offspring.

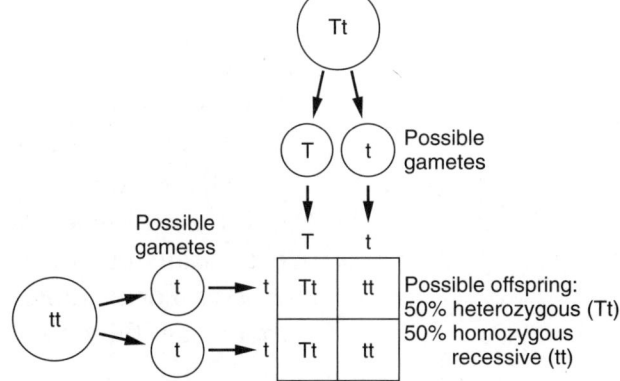

Chapter 17/Patterns of Inheritance   **213**

Figure 17-11 The expression of the gene that codes for dark-colored fur in Siamese cats depends on temperature. The parts of the body that have a lower (cooler) temperature allow that gene to be expressed; the rest of the body, which is warmer, has light fur.

ENVIRONMENT AND GENE EXPRESSION

You already know that the genes of an organism determine its physical characteristics. Yet the environment in which an organism lives can also affect the way its genes are expressed. The fur coloration pattern of Siamese cats is an example of this interaction of genes and the environment. Siamese cats have a gene that codes for an enzyme that produces dark fur. However, this enzyme works only at cool temperatures. Most of a cat's body is too warm for this enzyme to work. But you can easily identify the cooler areas where the enzyme does work: the dark ears, paws, face, and tail that are typical of a Siamese cat. Even though all parts of a cat's body have the same combination of genes (genotype), the way that the genes are expressed differs from one part to another because of different environmental influences. Thus, the environment frequently affects the final appearance (phenotype) of an organism. (See Figure 17-11.)

PLANT AND ANIMAL BREEDING

For centuries, people have chosen to breed plants and animals that have desirable traits. This is called **selective breeding**, or *artificial selection*. Over time, by allowing only those preferred organisms to breed, people have altered such traits as their size, shape, color, and even behavior.

The breeding of plants and animals has been greatly helped by the discoveries of Mendel and later geneticists. Often, breeders can identify the exact traits in which they are interested and then arrange the best genetic crosses. For example, plant breeders have produced new crops that can grow more plentifully, are resistant to diseases, and look and taste better. (See Figure 17-12.)

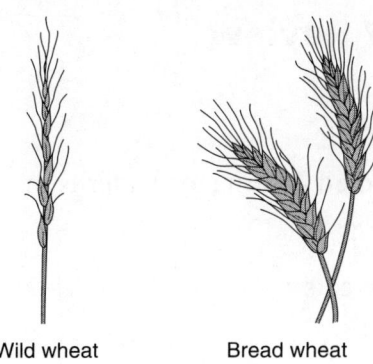

Figure 17-12 As a result of centuries of selective breeding, the small kernels of wild wheat have been transformed into the large kernels of bread wheat—a more useful crop for people.

People have bred domestic animals for many different purposes, too. This selective breeding has produced sheep with thicker wool, chickens that lay larger eggs, cows that produce more milk, faster-running horses, and dogs that perform a variety of tasks, such as herding sheep.

Chapter 17 Review

Multiple Choice

1. Mendel studied inheritance patterns in
   A. pink roses
   B. fruit flies
   C. Siamese cats
   D. pea plants

2. Mendel is credited with
   A. discovering the structure of DNA
   B. beginning the science of genetics
   C. recognizing the function of RNA
   D. starting animal-breeding programs

3. An allele is best defined as a
   A. version of a gene
   B. specialized enzyme
   C. subunit of DNA
   D. three-base code

4. One of Mendel's ideas was that
   A. alleles are responsible for mitosis
   B. alleles can cause mutations
   C. there are two alleles for each trait
   D. each gene exists as only one allele

5. In Mendel's plants, one allele produces purple flowers, while the other produces white flowers because
   A. it depends on how the sunlight reflects off the flowers' petals
   B. the DNA subunits at those two alleles code for different proteins
   C. the DNA subunits at those alleles code for different carbohydrates
   D. it depends on how hot or cold the environment is when they bloom

6. According to Mendel, for each trait inherited, offspring receive
   A. just one allele per cell
   B. one allele from each parent
   C. two alleles from each parent
   D. several pairs of alleles

7. A breeder crossed a horse that was homozygous dominant for a particular trait with a horse that was homozygous recessive for the same trait. What percentage of the foals will be heterozygous dominant?
   A. 10 percent   C. 75 percent
   B. 25 percent   D. 100 percent

8. Which of Mendel's laws states that during meiosis the different forms (alleles) of a gene are distributed to the gametes in a random fashion?
   A. Law of Dominance
   B. Law of Independence
   C. Law of Segregation
   D. Law of Mutation

9. The offspring of a mating between two heterozygous black guinea pigs, in which black coat color is dominant over white coat color, would probably have a genotype ratio of
   A. 1 BB : 2 Bb : 1 bb
   B. 3 Bb : 1 bb
   C. 2 BB : 2 bb
   D. 2 BB : 1 Bb : 1 bb

10. If a breeder wanted to determine whether a black guinea pig was homozygous (BB) or heterozygous (Bb) for coat color, the animal in question would have to be crossed with one that has the genotype
    A. BB
    B. Bb
    C. bb
    D. $BbBb$

11. Mendel's Law of Independent Assortment applies to traits whose genes are found only on
    A. homologous chromosomes
    B. the sex chromosomes
    C. the same chromosomes
    D. non-homologous chromosomes

12. Many traits such as height or skin color are determined by several genes, located on more than one chromosome. This type of inheritance is known as
    A. homologous
    B. codominant
    C. polygenic
    D. incomplete

13. A flower breeder decides to cross a red-flowered plant with a white-flowered plant (of the same species). All of the offspring produce pink flowers because their alleles for this trait show
    A. complete dominance
    B. incomplete dominance
    C. complete recessiveness
    D. incomplete recessiveness

14. Traits that are controlled by genes located on the *X* chromosome are said to be
    A. incomplete
    B. sex-linked
    C. codominant
    D. homozygous

15. If a color-blind man marries a woman who is a carrier for color blindness, it is most probable that
    A. all of their sons will be color blind
    B. half of their sons will be color blind
    C. all of their daughters will be color blind
    D. all of their sons will have normal color vision

16. A color-blind woman marries a man who has normal color vision. What are their chances of having a color-blind daughter?
    A. zero percent
    B. 25 percent
    C. 50 percent
    D. 100 percent

17. How are an organism's traits related to the environment?
    A. An organism inherits different genes depending on the environment.
    B. The genetic information is never affected by the environment.
    C. The environment can affect the expression of some genetic traits.
    D. The environment affects genetic traits only in wild organisms.

18. In a particular variety of corn, the kernels turn red when exposed to sunlight. In the absence of sunlight, the kernels remain yellow. Based on this information, it can be concluded that the color of these corn kernels is due to
    A. a different type of DNA that is produced when sunlight is present
    B. the effect of sunlight on the number of chromosomes inherited
    C. a different species of corn that is produced only in sunlight
    D. the effect of the environment on gene expression in the corn

19. In Siamese cats, the fur on the ears, paws, tail, and face is usually black or brown, while the rest of the body fur is almost white. If a Siamese cat stays indoors, where it is warm, it may grow fur that is almost white on its extremities. In contrast, if a Siamese cat mostly stays outside, where it is cold, it will grow fur that is quite dark on its extremities. The best explanation for these changes in fur color is that
    A. an environmental factor influences the expression of this inherited trait
    B. skin cells that produce pigments have a higher mutation rate than other cells
    C. the location of pigment-producing cells determines the DNA code of the genes
    D. the alleles for fur color are mutated by interactions with the environment

20. The diagram below represents the change in a sprouting onion bulb when sunlight is present and then when sunlight is no longer present. Which statement best explains this change?

    A. Onion plants need carbon dioxide to survive.
    B. Plants produce hormones to make leaves grow.
    C. Onion plants need abundant water to grow tall.
    D. Plant traits can be affected by the environment.

21. Fruit flies with the gene for curly wings will develop straight wings if kept at a temperature of 16°C during development and curly wings if kept at 25°C. The best explanation for this change in wing shape is that the
    A. genes for curly wings and straight wings are on different chromosomes
    B. outside environment affects the expression of the genes for this trait
    C. curly wings help the fruit fly develop faster in warmer environments
    D. lower outside temperature always produces the same genetic mutation

22. To produce large tomatoes that are resistant to cracking and splitting, some seed companies use the pollen from one variety of tomato plant to fertilize a different variety of tomato plant. This process is called

A. selective breeding
B. direct harvesting
C. DNA sequencing
D. crop-plant cloning

23. Mendel experimented by carrying out selective breeding and
    A. natural selection
    B. mathematical analysis
    C. molecular selection
    D. animal husbandry

24. Research based on the principles of genetics has contributed to the development of new varieties of plants and animals. Which activity is an example of such a development?
    A. testing new chemical fertilizers on food crops
    B. developing new irrigation methods to conserve water
    C. selective breeding of crops that are resistant to disease
    D. importing natural predators to control insect pests

25. Which process has been used by farmers for hundreds of years to develop new animal varieties?
    A. genetic cloning
    B. DNA splicing
    C. genetic engineering
    D. selective breeding

26. When humans first domesticated dogs, there was very little physical diversity in the species. Today there are many different breeds, such as the shepherd, the poodle, and the beagle. This increase in diversity is most closely associated with
    A. cloning of selected body cells
    B. years of mitotic cell division
    C. selective breeding for desirable traits
    D. environmental influences on inherited traits

Analysis and Open Ended

27. Even though Mendel did not know about genes, he can be called the "founder" of genetics. Why?

28. In one sentence, tell what Mendel noticed about the "factors" that were passed from parents to offspring among pea plants.

Refer to the figure below to answer questions 29 and 30.

29. The two main structures represent
    A. the pods of two different pea plants
    B. one pair of genes from a pea plant
    C. a pair of matching chromosomes
    D. mutations in the genes of a trait

30. The figure illustrates that
    A. every chromosome has at least four or five genes on it
    B. an allele for a trait is on either chromosome, but not on both
    C. several pairs of alleles are necessary to determine any one trait
    D. the gene for a trait is at the same place on matching chromosomes

31. The letters in the diagram at right represent genes on a particular chromosome. Gene *B* contains the code for an enzyme that cannot be synthesized unless gene *A* is also active. Which statement best explains why this can occur?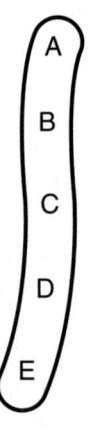
    A. A single trait can be determined by more than one gene.
    B. All the genes on a chromosome act to produce one trait.
    C. Genes are made up of double-stranded segments of DNA.
    D. The first gene on a chromosome controls all the other genes.

32. Explain how fur color in Siamese cats demonstrates an important fact about the expression of genes. In what way might this be an adaptive (beneficial) feature?

Refer to the following figures and paragraph to answer question 33.

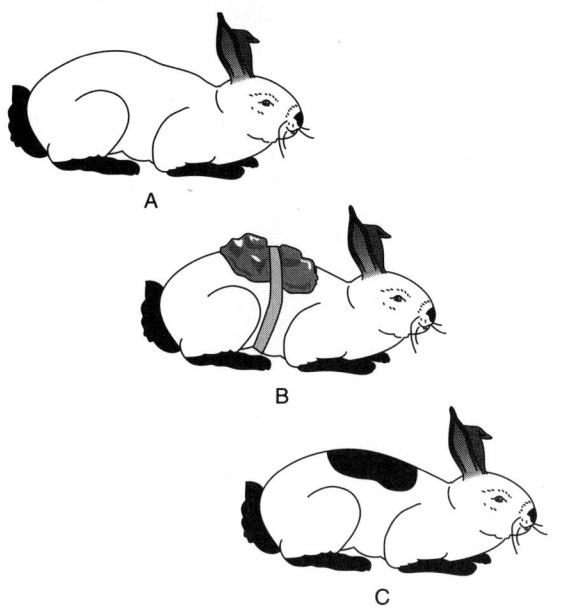

The normal color pattern of a Himalayan rabbit's fur is shown in Figure *A*. In Figure *B*, the rabbit is shown with the fur shaved from an area on its back and an ice pack applied to that area. Figure *C* shows the same rabbit after new fur has grown in the shaved area.

33. The best explanation for this change in the rabbit's color pattern is that the low temperature of the ice pack
    A. caused a genetic mutation in the fur
    B. deleted the gene for white-colored fur
    C. allowed the gene for dark fur to be expressed
    D. had no impact; it was due to a change in diet

34. Explain why hemophilia occurs more frequently in males than in females. Use a Punnett square diagram to illustrate your answer.

35. For many years, people have used different techniques to influence the genetic makeup of organisms. List five examples in which humans have altered the characteristics of plants or animals by setting up the most desirable genetic crosses.

36. When people carry out specific genetic crosses (such as those done by Mendel), are they working with organisms that reproduce sexually or asexually? Explain.

Reading Comprehension

Base your answers to questions 37 to 39 on the information below and on your knowledge of biology. Use one or more complete sentences to answer each question.

> It is not good for most organisms to breed with close relatives, a process called *inbreeding*. In corn, a very important crop, inbreeding causes the plant to become shorter, less strong, less productive, and more likely to develop diseases. The first person known to experiment with developing new strains of corn was the Pennsylvania farmer John Lorain. In 1812, he described experiments in which he crossed two different types of corn to make a hybrid that produced greater yields than either parent plant.
>
> The American botanist and geneticist George H. Shull is well known for his work on hybrid corn. Due to his research, corn yields increased 25 to 50 percent. In 1917, at Harvard University, the chemist Edward East and his student, Donald Jones, successfully combined two single-cross hybrid corn varieties to produce the first highly productive corn variety that could be grown commercially. Since the 1930s, corn hybrids have been used in countries throughout the world. Now molecular geneticists have produced even better breeds of corn by inserting genes to make the corn naturally resistant to pests. A controversy has arisen since some of this new "super-corn"

> has apparently appeared, uninvited, in Mexico where the native corn stills grows. Nevertheless, the study of genetics has helped provide more and better food for people all over the world.

37. Prepare a chart that has two columns. In the first column, list the name of each of the researchers who helped develop a better corn, in the order in which they did their research. In the second column, next to each researcher's name, describe the contribution that the person made to this research.

38. Explain the controversy surrounding genetically altered "super-corn."

39. How has the study of genetics, in this particular case, had a direct impact on people's lives?

Chapter 18

Genetics and Biotechnology

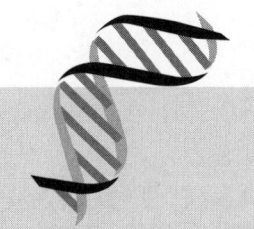

Standard 5.5.12 C3 Assess the impact of current and emerging technologies on our understanding of inherited human characteristics.

RECOMBINANT DNA TECHNOLOGY: A BRIEF DESCRIPTION

A revolution in biology began with the discovery of the structure and function of DNA, the molecule of life. This revolution has increased in importance through advances in *recombinant DNA technology.*

We know that genes are made of DNA and that they determine the characteristics of every organism on Earth. Now scientists have learned how to identify and find individual genes. Once found, these pieces of DNA can be removed and put together, or *recombined,* with other pieces of DNA. The genes can then be moved from one cell into another. The methods for doing this make up **recombinant DNA technology**.

For thousands of years, people have selectively bred certain characteristics in plants and animals to produce different crops and breeds. Now, recombinant DNA technology makes it possible to put "new" genes into organisms—that is, actually to change the genetic makeup of plants and animals. The new field of **genomics**, involving the detailed study of genetic material of different organisms, enables scientists to focus on particular genes and the traits they encode.

Through recombinant DNA technology—also called **biotechnology** or **genetic engineering**—human genes can be inserted into the genetic material of bacteria. These altered bacteria then become tiny "factories" that produce human proteins. Many other types of genes can be inserted into the genetic material of bacteria and other organisms. For example, agricultural scientists improve crops by inserting genes that make them disease-resistant, and improve livestock by inserting genes that make them grow faster or produce more milk. Perhaps most important, through advances in biotechnology, human *gene therapy* may be used to treat some genetic disorders. Although progress has been slow, someday it may be possible to remove defective

221

genes that cause certain disorders. In their place, healthy genes may be inserted that will prevent people from developing the disorder.

BASIC TOOLS OF RECOMBINANT DNA TECHNOLOGY

Scientists now estimate that humans have about 35,000 different genes in each cell. Although this figure is much lower than the 100,000 genes that had been estimated years earlier, it is still a very large number of genes. From these genes, after considerable processing, come a much larger number of different proteins in our cells. Scientists wanted to know where the genes for specific proteins were located in the DNA. They wanted to be able to move the genes from one organism and place them into another. The task seemed hopelessly complex until the 1970s, when restriction enzymes were discovered. A **restriction enzyme** recognizes a sequence of between four and six base pairs. Whenever it finds this specific sequence, the restriction enzyme cuts the DNA. The place where the cut occurs is called a *restriction site.* (See Figure 18-1.)

For a scientist, restriction enzymes are very powerful tools. With them, DNA can be cut at precise locations. In addition, the same restriction enzyme can be used to cut DNA from two completely different organisms, such as a frog and a bacterium. Pieces of DNA from one organism can then be inserted, or *spliced,* into the DNA of another organism. Remember, those "pieces of DNA" *are* the actual genes themselves. (See Figure 18-2.)

Yet, restriction enzymes are not enough. Scientists also need a way to move pieces of DNA from the cell of one organism to a cell of another organism, where the foreign DNA can replicate. So scientists have developed special molecules, called **vectors**, which can move pieces of DNA from one organism to another. (See Figure 18-3.)

Bacterial cells are used most often to receive the new segment of DNA from a vector. Bacteria reproduce quickly. Since the altered genes are passed

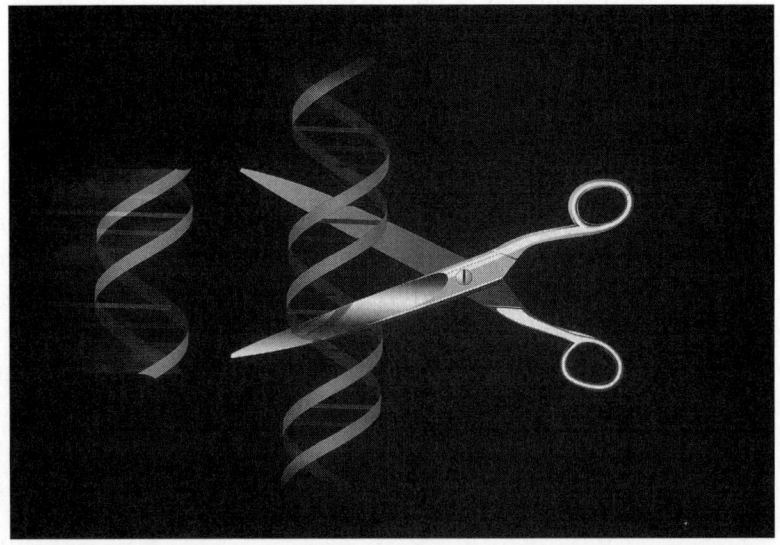

Figure 18-1 In this piece of computer artwork, the scissors represent the restriction enzyme that is used to cut a piece of DNA. Another piece of DNA, perhaps the gene to produce human insulin, would be inserted where the cut is made.

Figure 18-2 The action of a restriction enzyme called Eco RI: The original double strand of DNA is cut; then another piece of DNA that has been cut (perhaps from a different organism) is inserted to form recombinant DNA.

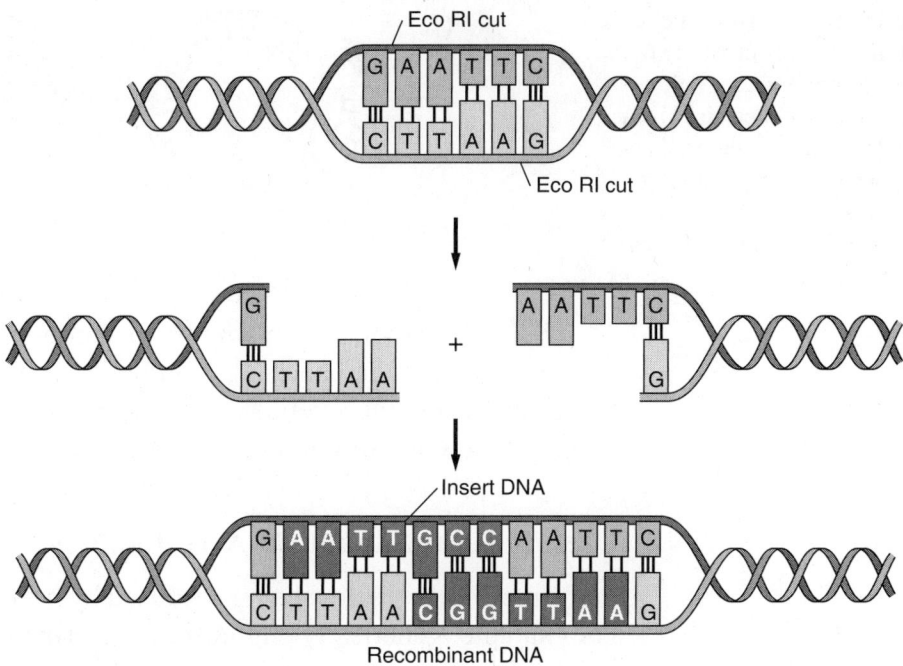

Figure 18-3 Scientists use vectors, such as circular pieces of bacterial DNA, to insert and move genes from one organism to another. The vector is usually placed within a bacterial cell, because it can reproduce quickly and make many more copies of the recombinant DNA.

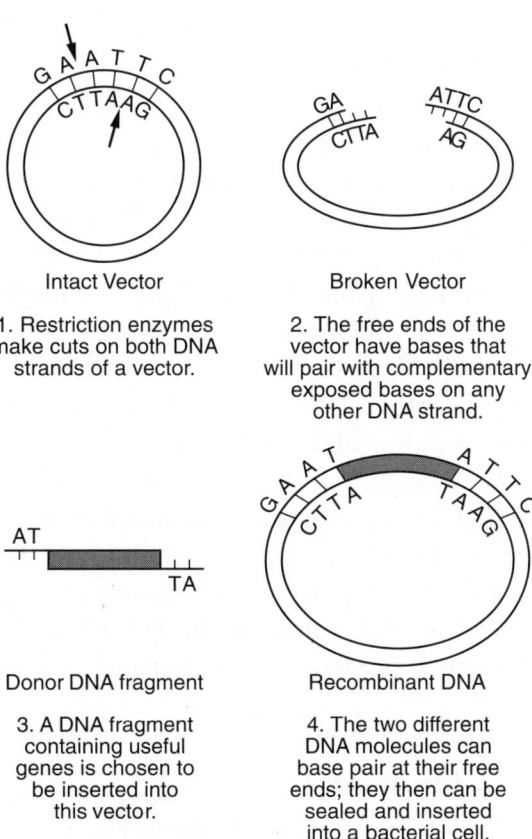

on to every cell that develops (through mitosis), soon there are thousands of bacterial cells that contain the new DNA. In other words, once the piece of new DNA is in the reproducing bacteria, the amount of it increases as the number of bacteria reproduce and increase. This is one way to make much larger quantities of the recombinant DNA.

Chapter 18 / Genetics and Biotechnology

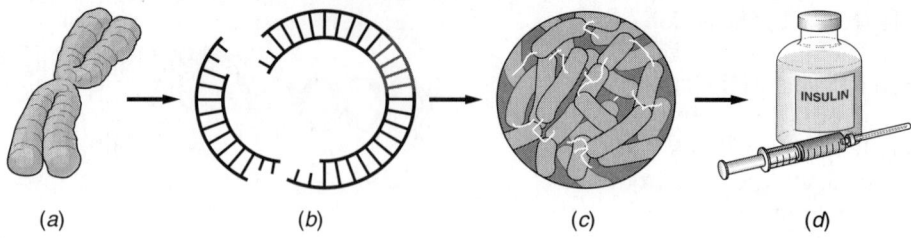

Figure 18-4 The insulin gene from a human chromosome (*a*) is inserted into a vector (*b*), a circular piece of bacterial DNA. The bacteria that receive this vector then contain the gene to produce insulin (*c*), which is collected for use by people who have diabetes (*d*).

There is another reason why the bacteria are encouraged to reproduce. The new DNA in the bacteria is coding for the production of a specific protein of interest to the scientists. The more bacterial cells there are living and reproducing, the more protein that is produced.

Uses of Recombinant DNA Technology

Some medicines are now being produced by bacteria with recombinant DNA. For example, genetically engineered bacteria use a human gene to produce insulin. It is, therefore, pure human insulin. (See Figure 18-4.) In the past few decades, biotechnology research teams have devised ways to mass-produce medically or industrially useful proteins by modifying the DNA of various other organisms. Some animals now produce specific proteins in their cells, milk, urine, or eggs. Even silkworms have been genetically altered to produce a partial form of human collagen—an important structural protein—in their silk. This material could be used in artificial skin grafts.

Other areas of medicine are benefiting from the use of biotechnology. Dozens of human genetic disorders can now be diagnosed with recombinant DNA technology. Often, individuals can be diagnosed with a disease even before they show any symptoms. This is because the gene that causes the disease can be identified in their DNA. Such identification can even be performed on a fetus while still in its mother's womb. The process, called *DNA fingerprinting,* can be used to identify traits that run in families. It is even used as evidence to determine the guilt or innocence of individuals in criminal trials. (See Figure 18-5.) DNA fragments are separated by means of **gel electrophoresis** to compare DNA from one person with that of another person. If the two DNA samples are very similar, they will produce many similar-sized fragments that line up next to each other on the gel. (Refer to Figure 2-11 on page 32.) Biotechnology is used to help clean up the environment, too: Genes are inserted

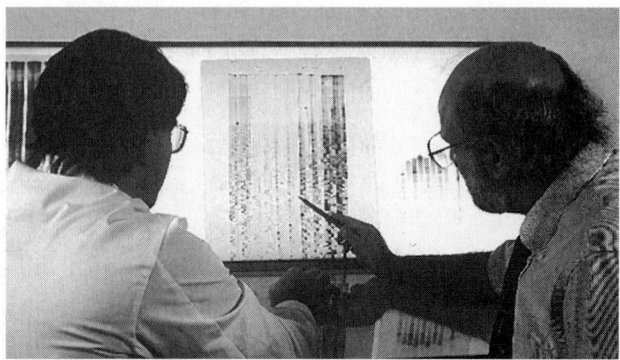

Figure 18-5 Results from the analysis of DNA fingerprints can be used to argue the guilt or innocence of individuals in criminal trials.

into bacteria to give them the ability to remove hazardous substances. As mentioned before, farm animals and crop plants also receive helpful genes.

Stem Cells

Human embryonic **stem cells** are unique because they are the cells that have not yet developed into mature cells that do a specific job in the body. Therefore, they have the ability to become almost any kind of tissue. Researchers hope they can use stem cells to produce specific tissues such as heart, lung, or nerve. The tissues grown from stem cells may offer cures to people who suffer from conditions such as diabetes, Alzheimer's disease, Parkinson's disease, and spinal-cord injuries. However, there is controversy about using stem cells from human embryonic tissue. So, researchers are trying to find other methods to help them find cures, such as using adult human stem cells, as well as embryonic stem cells.

The Human Genome Project

At one time, determining the sequence of nucleotides in a particular type of DNA was difficult and time-consuming. In the 1960s, it took seven years just to determine the sequence of a DNA molecule with only 77 nucleotides. Now, like many other tasks, the analysis of DNA is automated. Laboratory equipment analyzes DNA quickly, and computers tabulate the results. Because of these technological advances, in the late 1980s molecular biologists began to plan for what they considered the most important biology investigation of all time: determining the entire nucleotide sequence of human DNA. In 1992, a worldwide effort, called the **Human Genome Project**, began to analyze the three billion base pairs of human DNA. Molecular biologists the world over worked on this project; the age of "human genomics" had truly begun.

In February 2001, the first working draft of the entire human genome was published. By April 2003, through the combined efforts of dozens of scientists, the Human Genome Project had completed its "mapping" of the human genome, two years ahead of its proposed deadline.

Scientists agree that the Human Genome Project, described by some people as the effort to read the "book of humankind," is just the beginning of human genetic research. Only through this effort will we be able to understand ourselves on the molecular level, the most basic level of all. In fact, scientists plan to eventually understand all life-forms on this level. Since 1990, researchers around the world have begun to map the genetic codes of such diverse organisms as a protozoan, various fungi, the honeybee, the sea urchin, the chicken, the cat, the horse, and the chimpanzee.

FAMILY HISTORY AND GENETIC DISORDERS

Sadly, genetic disorders occur in some infants. These disorders are not caused by infectious microorganisms. Instead, genetic disorders result from inborn errors that are caused by defects in genes. Now that the human genome has been decoded, new and powerful information-processing tools are being used to find the genes that cause human genetic disorders. As noted above, it is even possible to find out if a baby has a genetic disorder before it is born.

As we learn more about genetic disorders, we realize that there are risks in having children. To assess the risks, people must have information about their family's medical history. A man and a woman who want to become parents can go to a trained genetics counselor for help in determining their risks of having a child with a genetic disorder. A genetics counselor prepares a pedigree chart showing the occurrence of any genetic disorders in past generations of the couple's families. Such a chart, showing a person's family history for a particular trait, can be analyzed and patterns of inheritance can be determined. This information helps prospective parents make informed decisions about having children.

Genetic Screening: Detecting Disorders

As noted above, some genetic disorders can be detected before birth. Biochemical tests can be done to show the presence of a genetic disorder. More often today, the DNA of a fetus is studied directly to see if something is abnormal in its genes. Scientists can take a photograph of the chromosomes in a fetal cell by using a camera attached to a microscope. The chromosomes are paired up and then numbered from largest to smallest. The picture that results is called a **karyotype**. When scientists study a normal human karyotype, they find 23 pairs of chromosomes: the 22 perfectly matched chromosome pairs (called *autosomes*) and one pair of *sex chromosomes,* either XX or XY, for a total of 46 chromosomes. (See Figure 18-6.)

Genetic disorders can result not only from a defective gene, but also from an abnormal number of chromosomes. Unusual events can happen during meiosis as the paired chromosomes separate, leading to either one missing or one extra chromosome in a gamete. Such abnormal chromosome numbers cause disorders that can be detected prior to or at birth. For example, Down syndrome is due to an extra chromosome number 21. (See Figure 18-7.)

Sickle-cell anemia is a disorder that is caused by a single-gene defect, not by an abnormal chromosome number. A mutation in the DNA base sequence of the gene for hemoglobin causes this disease, which reduces the ability of red blood cells to carry oxygen. (See Figure 18-8.)

Phenylketonuria is one of the most studied genetic disorders. The allele for this condition prevents a newborn from producing the enzyme that breaks down the amino acid *phenylalanine*. As a result, phenylalanine builds up in

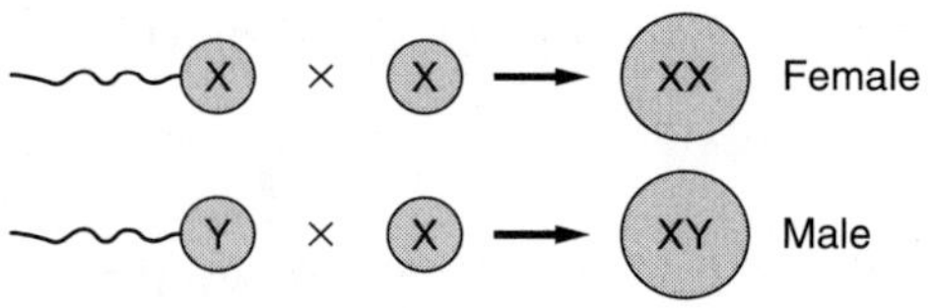

Figure 18-6 When scientists study a normal human karyotype, they find 23 pairs of chromosomes: the 22 perfectly matched chromosome pairs, called autosomes, and one pair of either XX or XY chromosomes. The X and Y chromosomes, which do not match each other, are called the sex chromosomes.

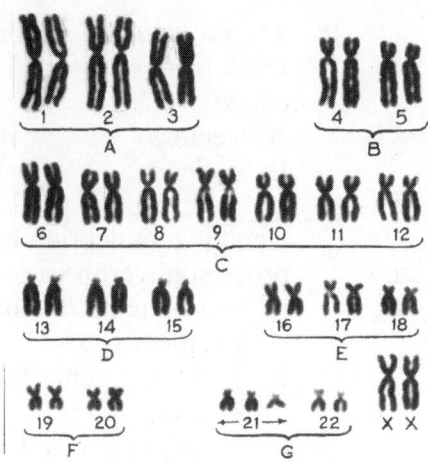

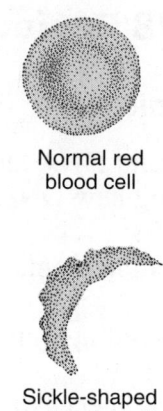

Figure 18-7 The karyotype of a person with Down syndrome shows that the disorder is caused by inheritance of three copies of chromosome 21.

Figure 18-8 The sickle-shaped red blood cell is characteristic of sickle-cell anemia, a genetic disorder caused by a mutation in the base sequence of the gene for hemoglobin.

the baby's blood, which interferes with the development of the brain, causing mental retardation. Fortunately, some genetic disorders can be treated. Routine tests for this disease are now done on newborns. With such early detection, the baby's diet can be changed to prevent the disease's effects from developing, and the child can lead a normal life.

Chapter 18 Review

Multiple Choice

1. Recombinant DNA technology involves
   A. creating new DNA from their molecular subunits
   B. inbreeding plants or animals with similar DNA
   C. interbreeding plants or animals with different DNA
   D. splicing pieces of DNA into other sections of DNA

2. Terms that describe the methods by which scientists change the genetics of organisms include all of the following *except*
   A. biotechnology
   B. genetic engineering
   C. agricultural engineering
   D. recombinant DNA technology

3. Humans produce more than 30,000 different proteins in their bodies. To make these proteins, we must have between approximately
   A. 100 and 500 genes
   B. 5,000 and 15,000 genes
   C. 30,000 and 40,000 genes
   D. 50,000 and 100,000 genes

4. Scientists use restriction enzymes to
   A. limit the length of DNA molecules
   B. stop parts of DNA from replicating
   C. prevent certain genes from being expressed
   D. cut specific base-pair sequences out of DNA

5. The diagram below represents some steps in a scientific procedure. The letters *X* and *Y* represent the

   A. hormones that stimulate the replication of bacterial DNA
   B. hormones that trigger rapid mutation of genetic information
   C. enzymes that aid the insertion of new genes into an organism
   D. enzymes that cause mutations in the X and Y sex chromosomes

6. The molecules that can move cut pieces of DNA from one organism to another are called
   A. vectors   C. transformers
   B. splicers   D. combiners

7. Genetic engineering has been used to improve some crop varieties by
   A. reproducing the old genes for wild characteristics
   B. removing genes that cause them to get diseases
   C. inserting genes that make them disease resistant
   D. adding animal genes that make them grow faster

8. Why do scientists insert human genes into bacteria?
   A. to give bacteria some human features
   B. to make large amounts of human proteins
   C. to dispose of our defective genes
   D. to find out what the bacteria will do

9. When a human gene is inserted into a bacterial cell to become part of its DNA, the process is an example of
   A. DNA fingerprinting
   B. biotechnology
   C. karyotyping
   D. reproduction

10. Bacterial cells are very useful for recombinant DNA technology because they
    A. reproduce quickly and increase in number
    B. carry out sexual reproduction like humans
    C. are almost identical to human body cells
    D. can be placed safely within a human body

11. The production of certain human hormones by genetically engineered bacteria results from
    A. inserting a specific group of amino acids into the bacteria
    B. interbreeding two different species of harmless bacteria
    C. splicing a piece of human DNA into a vector and then inserting it into bacteria
    D. deleting a specific amino acid from human DNA and inserting it into bacteria

12. The term *karyotype* refers to a
    A. group of similar alleles
    B. photograph of chromosome pairs
    C. cross between two plants or animals
    D. pair of traits that are linked

Analysis and Open Ended

13. How is recombinant DNA technology different from the traditional practice of selective breeding of plants and animals?

14. Describe three examples of how recombinant DNA technology is being used today. Your answer must include at least *one* example each for (a) plants; (b) animals; and (c) humans.

15. Explain the unique nature of human embryonic stem cells.

16. Why are stem cells of great interest to medical researchers?

Base your answers to questions 17 and 18 on the passage below and on your knowledge of biology.

For a number of years, scientists at Cold Spring Harbor Laboratory in New York have been working on the Human Genome Project to map every known human gene. By *mapping,* researchers mean that they are trying to find out on which of the 46 chromosomes each gene is located and exactly where on the chromosome the gene is located. The scientists want to be able to improve the health of people. By locating the exact positions of defective genes, scientists hope to cure diseases by replacing defective genes with normal ones, a technique known as *gene therapy.* Scientists can use specific enzymes to cut out the defective genes and insert the normal genes. They must be careful to use the enzyme that will splice out only the target gene, since different enzymes will cut DNA at different locations.

17. State one reason why the Human Genome Project is considered important for human health.

18. Explain why scientists must use only certain enzymes when inserting or removing genes from a cell.

19. During the 1970s, scientists discovered restriction enzymes. Why was this discovery so important for the advancement of genetic engineering?

20. What are vectors and how are they used in genetic engineering?

21. The following diagram represents a procedure used in biotechnology. Name a specific substance that can be produced by this technique and state how humans have benefited from the production of this substance.

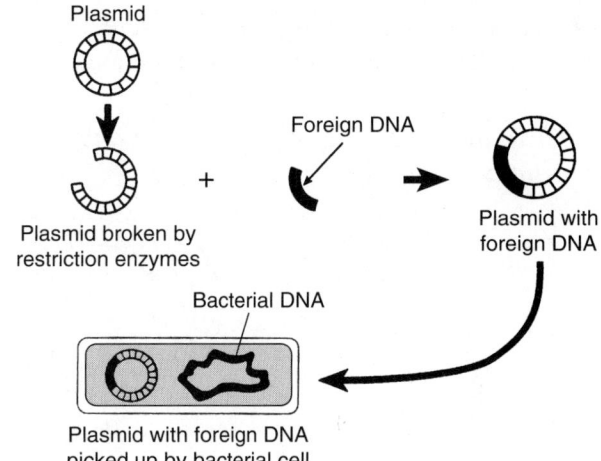

22. The diagrams below illustrate a process used in biotechnology. For each step (*a* through *d*) shown, use one of the following phrases to label what the structure represents: DNA fragment with useful genes is chosen; DNA fragment has been inserted into bacterial DNA; restriction enzyme makes cuts in bacterial DNA; free ends of bacterial DNA are exposed.

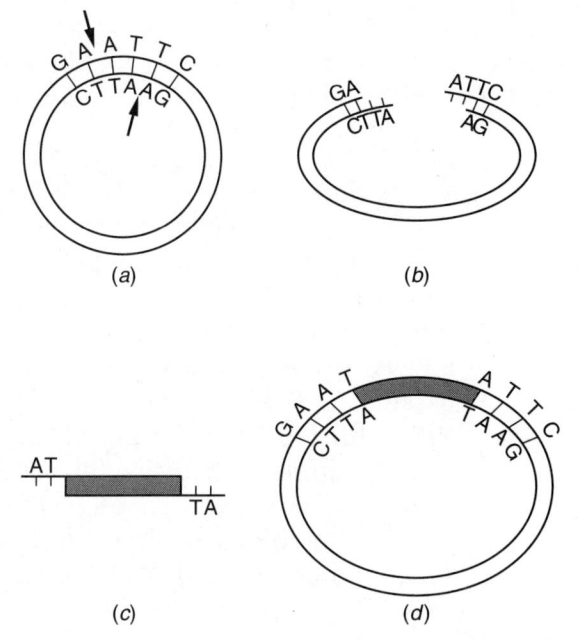

23. Biotechnology can be used to diagnose genetic disorders in people. Explain how a specific genetic disorder can be diagnosed. Your answer must include:
    - the name of *one* genetic disorder that can be diagnosed;
    - a description of *one* technique used to diagnose the disorder;
    - a description of *one* characteristic of the disorder.

24. How is biotechnology used to detect a disease even before its symptoms appear?

Refer to the set of diagrams below to answer questions 25 and 26.

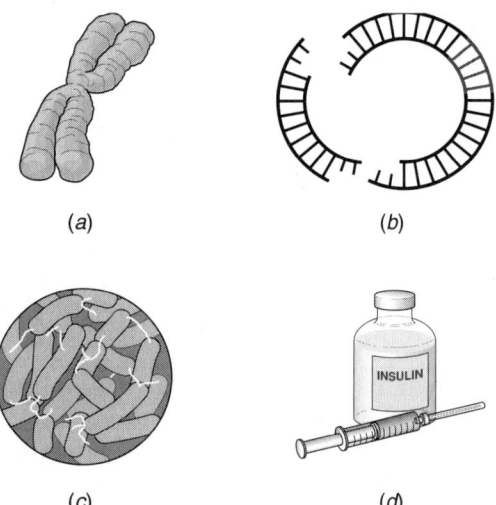

25. For each step (*a* through *d*) illustrated, write one sentence to explain which part of the genetic engineering process it represents.

26. Which one of the following titles would best describe the set of diagrams?
    A. Restriction Enzymes and Genetic Advances
    B. Agricultural Uses for Genetic Engineering
    C. Genetic Engineering and Medicine Production
    D. Genetic Engineering and DNA Fingerprinting

27. The bacteria that are used in recombinant DNA experiments have been changed so that they can survive only under special conditions in the laboratory. Why do you think scientists have included this safety precaution in their work?

28. Sickle-cell anemia is due to a single-gene defect that affects hemoglobin. Use your knowledge of genetics to explain how one defect in a DNA base sequence could upset production of this protein.

29. Why is early detection important for treating the genetic disease phenylketonuria?

30. Why might prenatal diagnosis be important for a couple that has a family history of a genetic disorder?

Reading Comprehension

Base your answers to questions 31 to 33 on the information below and on your knowledge of biology. Use one or more complete sentences to answer each question.

> Every time a prisoner awaiting a death sentence is proven innocent by DNA evidence and released, it makes the news. And it should. Nothing demonstrates the power of DNA technology better. Ray Krone owes his freedom, and probably his life, to this technology. In 2002, he was released from an Arizona prison after serving ten years. During that time, Mr. Krone, who had served in the U.S. Air Force and worked as a letter carrier with no criminal record, was tried twice for the sexual assault and stabbing murder of a bartender in 1991. Mr. Krone was in the bar where the victim worked the night of the murder. The only evidence used to convict him was the similarity between the pattern of tooth marks on the victim, where she had been bitten, and Mr. Krone's teeth.
>
> The first trial sentenced Mr. Krone to death, the second trial to a life sentence. Finally, after 10 years, DNA testing was done on saliva from bite marks found on the

victim's clothing. Not only did the DNA *not* match that of Mr. Krone, but it *did* match that of a person serving time in another Arizona prison for an unrelated sex crime. The odds were 1.3 quadrillion (1,300,000,000,000,000) to 1 that it was this other man's DNA on the victim and not that of Mr. Krone or anyone else. A judge ordered the immediate release of Ray Krone when the DNA test results were announced.

The DNA match was made possible because Arizona now has a database that contains a DNA profile of every prison inmate. In fact, every state in America now has such a database; and a national system, the National DNA Index System (NDIS), was started in 1998. By 2002, the one-millionth DNA profile had been entered into the computerized system. DNA evidence collected from any crime scene can now be quickly compared to that of any one of the million convicted offenders in the NDIS database. The system is quickly growing and the technology of DNA testing is rapidly improving. For example, a portable DNA testing kit is under development in Britain. It will be smaller than a suitcase and will be linked to the national DNA database of that country. It is expected that the crime scene evidence will be put in a solution and then placed inside the mobile unit. Silicon chip technology in the testing kit will then extract a DNA profile that will be sent to the national database via a laptop computer. The results may be returned in under an hour to the detective's palm-held computer. Saliva on discarded cigarette butts at crime scenes has already been used successfully to provide DNA profiles of suspects.

It is hoped that someday, thanks to this kind of technology, there will be no more wrongful convictions such as that of Mr. Krone, and more positive identifications of those who do deserve the jail time.

31. Compare the evidence used to convict Ray Krone in 1991 with the evidence used to release him in 2001.

32. Describe the system that has been put in place in the United States to use DNA technology to solve crimes.

33. How is the technology of DNA testing being improved for use at crime scenes?

Unit V
Environmental Systems and Interactions

STANDARD 5.10.12 A/B
Environmental Systems

All students will develop an understanding of the environment as a system of interdependent components affected by human activity and natural phenomena.

Enduring Understanding IV Living systems interact with natural occurring processes in the physical environment. Human intervention can affect the balance of natural cycles within the environment. Scientific data must be considered in the analysis of human decisions which would impact these cycles and alter the living world.

Chapter 19
Natural Systems and Interactions

Standard 5.10.12 A1 **Distinguish naturally occurring processes from those believed to have been modified by human interaction or activity.**

ECOLOGY AND ECOSYSTEMS

An aquarium is a self-contained miniature world of life. Like the living things in an aquarium, every organism on Earth lives within its surroundings, or environment. All living things interact—they affect other living things and their environment; and all living things depend on each other and their environment—they are interdependent. These relationships of *interaction* and *interdependence* between organisms and their environment are studied in the branch of biology known as **ecology**. (See Figure 19-1.)

Figure 19-1 An aquarium is like a miniature ecosystem—the living things interact with one another and with the nonliving parts of their environment.

235

Every organism has to live somewhere. The environment in which an organism lives is molded by many different *factors,* such as availability of food and water, amount of sunlight, temperature, and type of soil. Conditions that involve other living organisms are known as **biotic** factors. For a fish in the aquarium, the biotic factors could include other fish, snails, algae, and plants. Conditions that involve nonliving things are known as **abiotic** factors. For that same fish, the abiotic factors could include the water, air bubbles, gravel, acidity, temperature, and light.

The interaction and interdependence of organisms and their environment can be understood by examining specific places. All the living and nonliving factors (such as water, soil, and air) that interact in one specific place are the parts of an **ecosystem**. For example, a pond is an ecosystem. A forest is an ecosystem. Even the little aquarium is an ecosystem that can be used to study ecology. Large geographic regions that are characterized by a particular climate and ecosystem are called **biomes**.

An Organism's Habitat and Niche

Every organism is adapted through evolution to live in a particular place. Each species of organism is adapted to a specific set of conditions. The place where an organism lives is its **habitat**. The habitat of a bullfrog is a pond. The habitat of a giant anteater is open grassland. Thus, an organism's habitat is its "address."

To understand an organism's relationship to its environment, we must know more than its address, or *where* it lives. We must also know its "occupation," or what it does and *how* it lives. The occupation of an organism is called its **niche**. The niche of an organism includes how it gets food, reproduces, avoids predators, and so on. The behavioral adaptations of an organism make up its niche. As you have read, these adaptations are the result of evolution just as are its physical adaptations. The niche of an organism determines its habitat. In other words, the ways that an organism has evolved to survive will also determine where it can live. For example, a woodpecker cannot live in grasslands; the bird's niche involves finding its insect food in the trunks of trees. So woodpeckers need a habitat that has trees, and the insects that live in them, in order to survive. (See Figure 19-2.)

Figure 19-2 The niche of an organism determines where it can live. For example, the woodpecker needs to live in a forested habitat because its niche involves finding insects that live in trees.

ENVIRONMENTAL FACTORS AND ECOSYSTEMS

Since organisms have specific habitat requirements, different environmental conditions can limit where they live. These conditions are called **limiting factors**. For example, the availability of sunlight is an important limiting factor for plants. The amount of sunlight in oceans and lakes varies but does not reach below a certain depth. Below that depth, it is too dark for aquatic plants to grow. Temperature, of course, is another major limiting factor. Each species of plant or animal has a fairly narrow temperature range that it prefers. In other words, the organism can tolerate temperatures only within this range. Other environmental factors, such as chemical nutrients in the habitat, may be less obvious, but are still very important for a species' survival. In general, organisms have a tolerance range for a variety of environmental factors, sometimes a very narrow one, usually neither too low nor too high. The tolerance range determines the best conditions for a particular organism in a specific location.

All life depends on water, and organisms have a variety of adaptations that enable them to survive in very specific ranges of available moisture. For example, water constantly moves out of openings on the surface of leaves. So some species of trees, such as pines, have evolved ways to save water—they have narrow leaves, or needles, from which little moisture is lost. In areas with abundant rainfall, water loss is not a problem, so the trees have large, flat leaves.

Aquatic Ecosystems

As stated, water is essential to life; and many ecosystems occur in the water. A map of the world shows individual oceans; however, all the world's oceans are actually connected. Some ecologists consider this world ocean to be one tremendously large *saltwater ecosystem*. The main limiting factors in the ocean are saltiness, temperature, and sunlight. As the amount of salt in ocean water varies, the density of the water also changes. The temperature of ocean water also differs from place to place and this, too, affects water density. Cold water is denser than warm water, so the colder water sinks. The amount of sunlight also varies over different parts of the ocean, and it penetrates only to a certain depth. Because they need light to carry out photosynthesis, all plants and algae in the ocean live only in the top (photic) zone of the water. (See Figure 19-3.)

There are two main types of *freshwater ecosystems* on the surface of Earth: lakes and ponds, which are bodies of still water; and rivers and streams, which are running water. In the still-water ecosystems, temperature and light are the main limiting factors. In the running-water ecosystems, the temperature and light are fairly constant at any given point but vary along the length of the river or stream. For example, the water flows faster and colder at the start of a river (headwaters) than at its end (mouth). As a result, different kinds of plants and animals are adapted to survive in different parts of a river. In general, conditions (such as temperature) in water are fairly

Figure 19-3 Sunlight is a limiting factor for the growth of plants. Because sunlight penetrates only about 200 meters below the ocean's surface, aquatic plant life is restricted to that top (photic) zone.

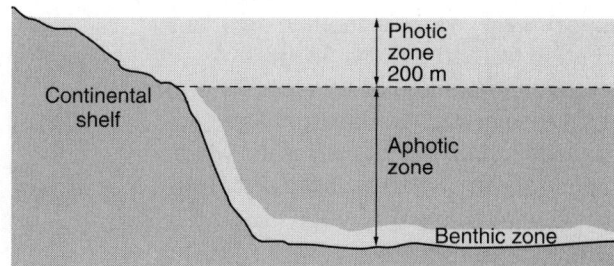

constant over wide areas. They change little over time and, when they do, they change very slowly.

Land Ecosystems

Conditions on land are very different from those in the water. From season to season, and from one part of the day to another, temperatures on land can vary widely. Variations in temperature, moisture, soil type, length of days and nights, seasons, and altitude all work together to produce many different land ecosystems.

There are three main forest ecosystems on Earth, each with its characteristic types of trees: *tropical rain forests, broad-leaved forests,* and *needle-leaved forests*. The tropical rain forests exist in a wide band north and south of Earth's equator. They have large amounts of rainfall, warm temperatures, and a stable length of daylight throughout the year. Abundant life of all kinds exists in the tropical rain forests.

Farther north and south of the equator, climate patterns change. Definite seasons occur, with variations in temperature and rainfall. This creates the *temperate forest* environment of broad-leaved, or *deciduous,* trees, such as maple and oak. Here, fewer tree species are found, and the seasonal dropping of leaves is typical. Still farther north is the *taiga,* where it is colder. Here are the great forests of needle-leaved, or *coniferous,* evergreen trees, such as spruce and pine. (See Figure 19-4.)

Farthest from the equator, the temperatures and amount of rainfall are too low for trees to grow. Only one or two months of the year are warm enough to support plant growth. Here, in the *tundra,* a thick layer of mosses, lichens, grasses, and low shrubs covers the surface.

In regions where precipitation, but not temperature, decreases, there are *grasslands,* because the limited moisture prevents the growth of trees. These areas support large populations of grazing animals, such as antelope. Finally, the driest places on Earth are the *deserts*. In a typical desert, plants include water-storing cactuses, shrubs with roots that grow deep to reach water, and wildflowers and grasses that flourish briefly after the infrequent rains. (See Figure 19-5.) Animals also show adaptations to the desert conditions. For example, the kangaroo rat lives underground for much of the day to avoid the heat, and desert predators such as coyotes and foxes are also more active at night.

Figure 19-4 Water is another important limiting factor for plants. In areas of abundant rainfall, plants such as the maple (left) have broad, flat leaves; in areas of limited rainfall, plants such as the white pine (right) have narrow, needle-shaped leaves to reduce water loss.

Figure 19-5 The giant saguaro cactus, which grows in the desert, is adapted to store water in its thick stem.

THE STUDY OF POPULATIONS

Ecologists are more interested in groups of organisms than in individuals. All the organisms of one species that live in one place at a particular time make up a **population**. No population ever lives alone. Other organisms—plants, animals, fungi, and microorganisms—are also present. For example, a field has a population of mice, but it also has populations of wildflowers, insects, mushrooms, birds, and snakes. All of the populations that interact with each other in a particular place make up a **community**. In large part, the study of ecology is about populations and communities.

Factors That Affect Population Growth

Most organisms are able to reproduce rapidly. Even if a pair of individuals produced only two offspring each year, the growth rate would be enormous if all offspring survived. Suppose a population was founded by two individuals. By doubling each year, after only 10 years there would be a population of more than 1000 individuals. Many organisms produce even greater numbers of offspring, such as fish that lay thousands of eggs per year. If these eggs all hatched and survived to reproduce, the number of resulting fish would be enormous. Trees also produce many thousands of seeds per year. With such growth rates, Earth could quickly be overcrowded with living

Figure 19-6 The carrying capacity for a population of organisms will fluctuate slightly in a stable population.

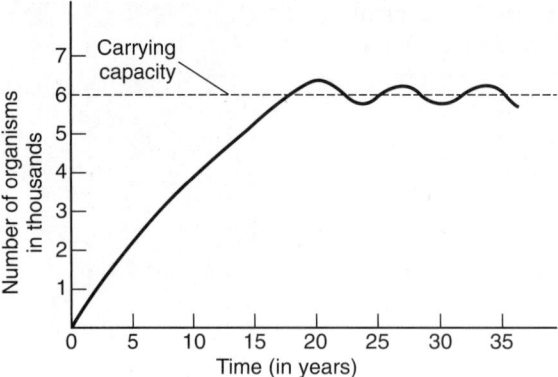

things. Of course, this does not occur. Thus, an important area in ecology is the study of what controls population growth.

A key factor in population growth is **density**, which is the number of individuals in a population in a given area. A population with a small number of individuals in a particular area has a low density. When a population's density is low, there is usually sufficient food and space for existing organisms. The birth rate increases, while the death rate drops. As a result, the density of the population begins to increase at a faster rate. But this rate of increase cannot last forever. At some point, the population gets too crowded.

The most basic needs of organisms from their habitats are food and space. However, every habitat has limits. When population density increases too much, the available food and space decrease. At that point, the population density has reached a maximum for the particular habitat. The death rate increases, while the birth rate drops. The size of a population that can be supported by any ecosystem is called the **carrying capacity**. (See Figure 19-6.) Population growth slows and may reach zero growth as the population size approaches an area's carrying capacity. Zero growth means the population size is no longer increasing—the birth rate and death rate are about equal. The rate at which a population grows is shown in Figure 19-7.

Factors that limit the size of a population include *competition* for food and space. This competition increases as population density increases. Another problem with high population density is that the more crowded a population is, the easier it is for predators to find prey. In a crowded population, there is also a greater chance for disease to spread among individuals.

COMMUNITY INTERACTIONS

Organisms that live in the same community are always interacting with each other. The survival of many plants and animals often depends on the relationships they have with other organisms. (See Figure 19-8.) Some important types of relationships are discussed below. As previously stated, **competition** is one of the main interactions between organisms. For example, if two different species of birds ate the same species of butterflies from treetops in

Figure 19-7 A typical population growth curve. A new population increases slowly at first, because there are few individuals. As the population increases, the growth rate increases rapidly as more individuals reproduce. After some time, it levels off to zero growth as the population approaches the area's carrying capacity.

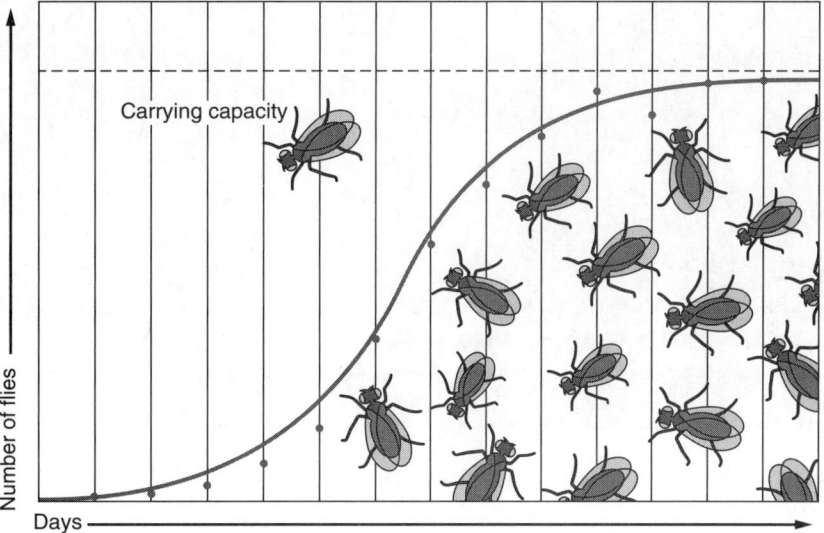

the same forest, there would be competition between them. In this case, the competition between the two bird species would be intense because of the overlap in their niches. However, if the two bird species fed on different insects in the same treetops—or on the same insects but in different parts of the trees—there would be less overlap in their niches and reduced competition between them.

The greatest competition usually occurs between members of the same species, because such individuals are more likely to share identical niches. In other words, they live in exactly the same way and compete most intensely for the same limited resources in their area. Competition results in natural selection. Recall the thousands of eggs produced by the fish—most will not survive to maturity. According to Darwin's theory, only the most fit individuals live to pass on their genes to offspring. Therefore, competition is an important force in the process of evolution.

Predation is one of the most basic interactions that occurs between organisms. In nature, most living things will either "eat or be eaten." That is, they are either **predator** or **prey**. In **predation**, members of one population are the food source for members of another population. Even grass seeds

Figure 19-8 A community, such as that illustrated living in and around this pond, is made up of many different organisms. The survival of most plants and animals depends on the relationships they have with the other organisms in their environment.

Chapter 19 / Natural Systems and Interactions **241**

Figure 19-9a The mouse is considered the predator when it preys on plants.

Figure 19-9b The snake is the predator when it preys on the mouse, which is then the prey.

can be considered prey when a mouse—the predator, in this case—eats them. Of course, the mouse becomes the prey when a snake, another predator, eats it. (See Figures 19-9a and 19-9b.)

Types of Symbiotic Relationships

As stated above, almost no organism lives entirely alone. In fact, most organisms have close relationships with at least one other type of organism. This close relationship is called **symbiosis**. In this relationship, one type of organism can live near, on, or even in another organism. Each partner in the relationship can either help, harm, or have no effect on the other partner.

A **parasite** lives on or in another organism that it uses for food and, sometimes, for shelter. The organism the parasite uses is called the **host**. In this type of interaction, one organism is helped while the other is harmed. This relationship is different from the **predator-prey** relationship. In that relationship, the prey is usually killed right away. In parasitism, the host organism usually continues to live, but it is harmed. Parasites usually evolve together with their host and have characteristics that make them specifically adapted to it. This is called **coevolution**, the process by which species change because of, or along with, changes in other species around them. As you read in Chapter 18, there are parasitic flatworms, such as flukes and tapeworms, which are adapted to live in the human body. Songbirds, too, may have many parasites that are specifically adapted to live inside their bodies. (See Figure 19-10.)

In some *symbiotic relationships,* one organism benefits while the other organism is unaffected; this type of symbiotic relationship is called *commensalism.* For example, when a Cape buffalo walks through the African plains to browse, it disturbs insects in the grass. Birds called cattle egrets gather around the buffalo and feed on the insects near it. The birds are helped by finding food, and the Cape buffalo remains unaffected.

Finally, there are symbiotic relationships in which both parties benefit; this is known as *mutualism.* An example of this is the relationship between

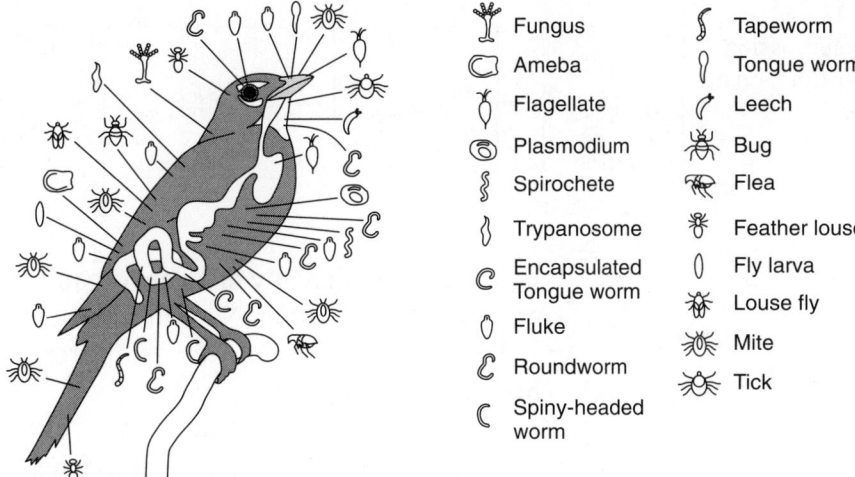

Figure 19-10 One bird may be host to many different types of parasites. Many parasites have evolved to be specifically adapted to their host, which is typically harmed, but not killed, by the parasites.

a type of acacia tree and a species of stinging ants. The trees produce hollow thorns. The ants make their nests in these thorns and feed on sugars produced by the plant. If any other insect lands on the acacia, the ants quickly surround and kill it. The ants have shelter and a source of food, and the tree is protected from other plant-eating insects.

ECOLOGICAL SUCCESSION: CHANGING COMMUNITIES

A complex variety of interactions exists within an ecosystem. Many different populations live together and affect each other. In a forest, for example, there are populations of grasses, shrubs, trees, fungi, insects, worms, bacteria, birds, reptiles, and mammals that make up the community. However, this community will probably not remain the same over time. Some populations of organisms may disappear entirely while other populations may move in from somewhere else. Existing populations may increase or decrease in number. The community may even change with the seasons. Yet, despite these kinds of changes, the forest itself—especially a very old forest—may remain essentially the same over many years. If so, it has become a stable community.

But then a sudden, dramatic change may occur. For example, a fire may destroy much of the forest. Some animals may die, while others are able to escape. What happens now? Did the fire destroy the forest? No. What scientists observe is the process of ecological **succession**.

Usually, succession is a series of slow changes that occurs until a stable *climax community* is reached. *Primary succession* occurs when a new environment appears for the first time, such as when an island emerges or a volcano produces new rock. The gradual succession of communities that occurs in these places can tell us a great deal about how ecosystems on Earth evolved during ancient times. The first plants or plantlike organisms to populate such an area are called *pioneer organisms;* these are usually lichens or algae. This progresses to mosses, grasses, shrubs, and then, finally, a stable

Chapter 19/Natural Systems and Interactions **243**

Figure 19-11
Ecological succession occurs, over time, on a new island.

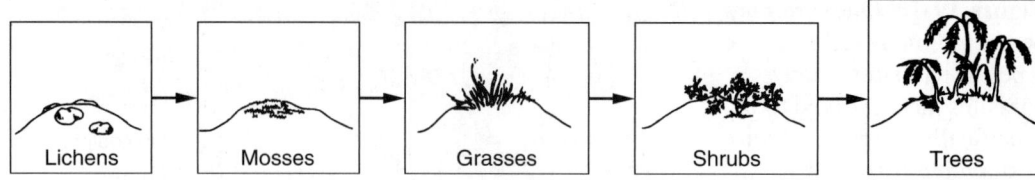

forest community, with many different organisms in it. Also, over time, ponds and small lakes tend to fill in. Natural materials accumulate in and around the water until there is a gradual succession from an aquatic to a land community. (See Figure 19-11.)

Most naturally occurring successions take much longer than a person's lifetime. However, after a sudden disturbance, such as a forest fire, the community quickly begins to go through a series of changes. This process is known as *secondary succession* and it, too, continues until a stable climax community is reached. These changes often follow similar patterns wherever the same kind of disturbance has occurred.

ECOLOGY AND FIELDWORK

Scientists who study ecology—that is, populations, communities, and ecosystems—are called *ecologists*. They gather information by conducting **fieldwork**, the study of living things in their natural habitats. The information can be used, for example, to determine an animal's population size, diet, territorial boundaries, and behavioral patterns. Fieldwork also includes the study of the physical factors in the environment, such as the soil, rocks, air, and water. The research often includes laboratory studies that follow up on data gathered in the field.

Sampling Populations

One aspect of fieldwork is *sampling* a population of organisms. The technique of **sampling** uses a representative portion of a population to determine one particular characteristic of the entire population. Sampling can be used to estimate the number of individual organisms of one species that live in an area. By knowing how many of each species are in an ecosystem, scientists can begin to understand how organisms interact.

Within a small area, trees and shrubs can be counted directly because they cannot run away or hide. Over a large area, it takes too long to count each tree. That is when ecologists can use indirect sampling methods. They count samples from a small portion of the area, and then these numbers are *extrapolated*, or projected, to arrive at a possible number for the whole area. For example, suppose that there are 10 oak trees in a 1-hectare portion of a 200-hectare forest. To find out about how many oaks would be in the whole forest, you can set up the following proportion:

$$10 \text{ oaks}/1 \text{ hectare} = x/200 \text{ hectares}$$
$$10 \text{ oaks}/1 \text{ hectare} \times 200 \text{ hectares} = x$$
$$2000 \text{ oaks} = x$$

Therefore, there are approximately 2000 oak trees in the 200-hectare forest. The result of indirect sampling is accurate only if the area sampled is representative of the whole area.

Large animals can usually be counted by sight, unless they are very elusive (hard to find). But smaller animals, such as field mice, are harder to count. So, a population of small animals is trapped, marked, and then released. The animals are marked with spots of paint or, with larger animals, tagged on their ears. A second trapping done at a later date provides information that is used to predict the total population size. For example, 25 field mice can be trapped, marked, and released in a given area. Several days later, 20 field mice are collected in a second trapping. Of these 20 mice, 5 had been marked on the first trapping. If you assume both samples were random, then the first trapping must have contained five-twentieths of the total population of field mice in that area. By extrapolating, you can see that the total population of field mice in this area is about 5×20, or 100 mice.

Chapter 19 Review

Multiple Choice

1. Ecology can best be described as the study of
   A. all the plants in a certain environment
   B. living organisms and their environment
   C. living factors that affect an organism
   D. nonliving factors that affect an organism

2. A *biotic* factor in a snake's environment would be
   A. sunlight   C. sand
   B. water   D. a mouse

3. An *abiotic* factor in an eagle's environment would be
   A. a tree   C. a snake
   B. water   D. an insect

4. Which event illustrates the interaction of a biotic factor with an abiotic factor in the environment?
   A. Water temperature affects water density in the ocean.
   B. The lamprey eel survives as a parasite on other fish.
   C. Shorter daylight hours cause maple trees to lose leaves.
   D. A gypsy moth caterpillar eats the leaves of an oak tree.

5. An ecosystem is best described as the
   A. type of food that an organism eats
   B. type of home an organism builds
   C. group of organisms in a particular place
   D. living and nonliving factors in one place

6. Which of the following can be considered an ecosystem?
   A. a large rock
   B. a bird's nest
   C. a rain forest
   D. a rain cloud

7. A habitat can be described as an organism's
   A. average size
   B. main function
   C. wild behavior
   D. natural address

8. A frog's habitat would be the
   A. pond it lives in
   B. sounds it makes
   C. insects it eats
   D. color of its skin

9. An organism's niche is most similar to a person's
   A. character   C. address
   B. occupation   D. personality

10. In a forest community, a shelf fungus and a slug live on the side of a decaying tree trunk. The fungus digests and absorbs materials from the tree, while the slug eats algae growing on the outside of the trunk. These organisms do not compete with one another because they occupy
    A. the same habitat, but different niches
    B. the same niche and the same habitat
    C. the same niche, but different habitats
    D. different habitats and different niches

11. Light, temperature, and water are examples of environmental
    A. habitats
    B. niches
    C. limiting factors
    D. adaptations

12. Why do plants live only in the top zone of the ocean?
    A. There is too much salt in deeper water.
    B. They automatically float to the top.
    C. Plants cannot grow underwater.
    D. Plants need sunlight to make food.

13. The main limiting factors in freshwater ecosystems are
    A. temperature and light
    B. salt and temperature
    C. density and light
    D. altitude and depth

14. Ecosystems vary more on land than in water because
    A. there is more land than water on Earth's surface
    B. conditions on land vary more than they do in water
    C. evolution of new species does not occur in water
    D. organisms that live in water become extinct sooner

15. Earth's three main forest ecosystems vary because they
    A. experience different temperatures and rainfall
    B. are located near different major cities

C. differ greatly in overall size
D. are in different stages of development

16. An example of a population in a lake ecosystem would be all the
    A. lake trout
    B. lake trout and brown trout
    C. plants and trout
    D. soil, plants, and fish

17. A community can best be described as all of the
    A. plant species in one particular place
    B. organisms of one species in a particular place
    C. populations that interact in a particular place
    D. animals that interact in a particular place

18. Which ecological term includes everything represented in the illustration below?

    A. ecosystem   C. community
    B. population   D. species

19. When population density is low, the
    A. birth rate increases and the death rate drops
    B. death rate increases and the birth rate drops
    C. birth rate and the death rate both increase
    D. birth rate and the death rate both decrease

20. An environment can support only as many organisms as the available food and space will allow. Which term is best defined by this statement?
    A. biological feedback
    B. homeostatic control
    C. carrying capacity
    D. biological diversity

21. Competition between organisms can best be described as an interaction in which the organisms
    A. rely on the same resources
    B. work together to find food
    C. live in the same place but eat different food
    D. eat the same food but live in different places

22. Competition can occur between members of
    A. the same species only
    B. different species only
    C. both the same and different species
    D. two different communities only

23. A parasitic relationship differs from a predator-prey relationship in that a
    A. host organism is killed right away, but prey is not
    B. prey organism is killed right away, but a host is not
    C. parasite helps its host, but a predator kills its prey
    D. prey organism benefits, but a parasite's host does not

24. In the symbiotic relationship between cattle egrets and Cape buffalo,
    A. both species benefit
    B. both species are harmed
    C. one species benefits while the other is harmed
    D. one species benefits while the other is unaffected

25. Some crocodiles let small birds enter their mouths to pick bits of food from between their teeth. The crocodiles get clean teeth, while the birds get an easy meal. In this type of relationship,
    A. both animals benefit
    B. both animals are harmed
    C. only the crocodiles benefit
    D. only the birds benefit

26. The relationship between the crocodiles and birds could be described as a
    A. predator-prey relationship
    B. parasite-host relationship
    C. symbiotic relationship
    D. competitive relationship

27. In a certain ecosystem, rattlesnakes are predators of prairie dogs. If the prairie dog population started to increase, how would the ecosystem most likely regain stability?

A. The rattlesnake population would start to decrease.
B. The prairie dog population would increase rapidly.
C. The rattlesnake population would start to increase.
D. The prairie dog population would begin to prey on the rattlesnakes.

28. In the relationship between the hollow-thorned acacia trees and the stinging ants,
    A. both the trees and the ants benefit
    B. both the trees and the ants are harmed
    C. the tree benefits, while the ants are harmed
    D. the ants benefit, while the tree is unaffected

29. A new island formed by volcanic action may become populated by living communities as a result of
    A. a decrease in the amount of organic material
    B. the lack of abiotic factors in the area
    C. decreased levels of carbon dioxide in the area
    D. the process of ecological succession

30. Succession in an ecosystem is *usually* a
    A. sudden event that changes the ecosystem
    B. series of very rapid changes in the area
    C. series of slow changes that occurs in the area over time
    D. short period of rapid change followed by a stable period

31. In which of the following cases is *direct* sampling not possible?
    A. determining the number of lions in an area of South Africa
    B. counting the population of prairie dogs in a colony
    C. determining the size of a bison herd in Yellowstone
    D. finding the number of oak trees on Green Mountain

Analysis and Open Ended

32. What is the difference between biotic and abiotic factors in an ecosystem? Your answer should include the following:
    - the definition of *biotic*
    - the definition of *abiotic*
    - *two* examples of biotic factors
    - *two* examples of abiotic factors

33. Identify one *abiotic* factor that would directly affect the survival of organism *A* shown in the diagram below.

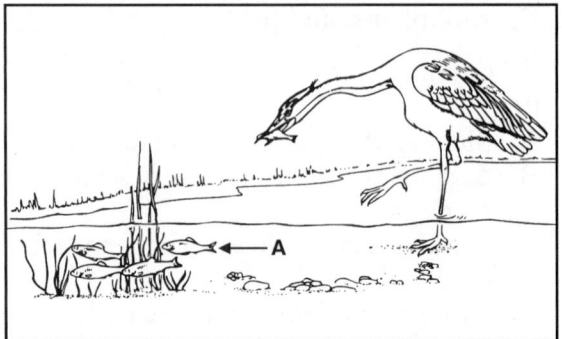

34. Why does the niche of an organism determine its habitat? Give one example.

35. Using an example such as sunlight or moisture, explain the idea of a limiting factor.

36. List the three main limiting factors in the ocean. Why do they have an effect on life in the sea?

37. Why is population growth more rapid when population density is low?

Refer to the graph below, which shows the growth curve for a population of Paramecium caudatum, *to answer questions 38 to 40.*

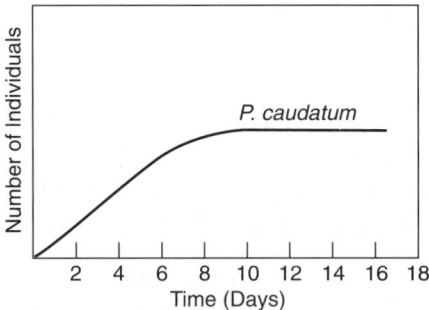

38. Why does the slope of the graph increase from the beginning to the middle?
    A. The death rate begins to increase.
    B. The growth rate slows after four days.
    C. The population grows while still below carrying capacity.
    D. There is intense competition for resources.

39. The level (flat) portion at the top of the graph indicates that the population
   A. is still growing
   B. is slowly shrinking
   C. is neither growing nor shrinking
   D. no longer exists in that location

40. How does the size of the paramecium population change as it approaches carrying capacity?

41. Why is competition for food and space usually greatest among members of the same species? How does this relate to the process of evolution?

Refer to the diagram below to answer questions 42 and 43.

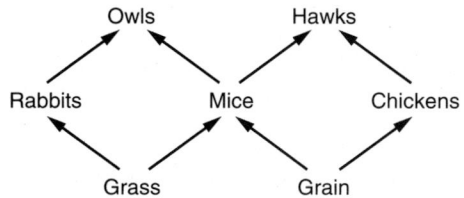

42. Based on the diagram, which of the following statements is true?
   A. Rabbits and owls compete for grass.
   B. Mice and chickens compete for grain.
   C. Rabbits and chickens compete for grass.
   D. Chickens and rabbits compete for grain.

43. Owls hunt at night, whereas hawks hunt during the day. This has the effect of
   A. reducing competition for mice because the birds occupy separate forests
   B. reducing competition for mice because the birds occupy separate niches
   C. increasing competition for mice because the birds occupy the same niche
   D. reducing competition for rabbits and chickens because the birds eat more mice

44. Choose an animal that you find of interest; then list two biotic and two abiotic factors for which it competes with other animals in its environment.

45. Compare and contrast predation and parasitism. Your answer should include:
   • the definitions of predation and parasitism
   • *one* way predation and parasitism are similar
   • *one* way predation and parasitism are different

46. As shown in the following figure, the remora has a suckerlike disk on its head by which it attaches to the underside of a shark. The remora feeds on leftovers from the shark's meals, without taking anything from the shark's body. This is an example of a symbiotic relationship in which

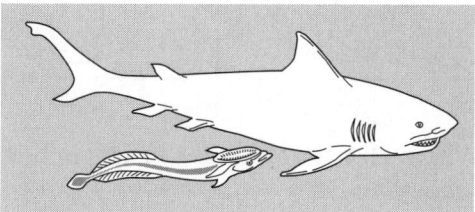

   A. both parties benefit by being able to catch more food
   B. one party benefits and the other is directly harmed
   C. one party benefits and the other is apparently unaffected
   D. both parties are harmed by not being able to swim as fast

47. The diagram below shows changes that might occur over time in an area after a forest fire. Which statement is most closely related to the events shown in the diagram?

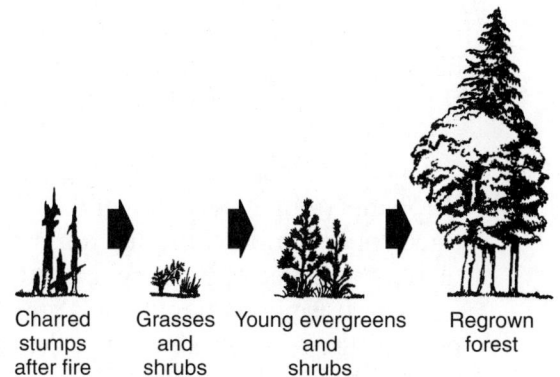

   A. The lack of animals in an altered ecosystem causes natural succession.
   B. Abrupt changes in an ecosystem result only from human activities.
   C. Stable ecosystems never become established after a natural disaster.
   D. Abrupt environmental upsets can cause long-term changes in an ecosystem.

Base your answers to questions 48 to 50 on the stages of succession shown below and on your knowledge of biology.

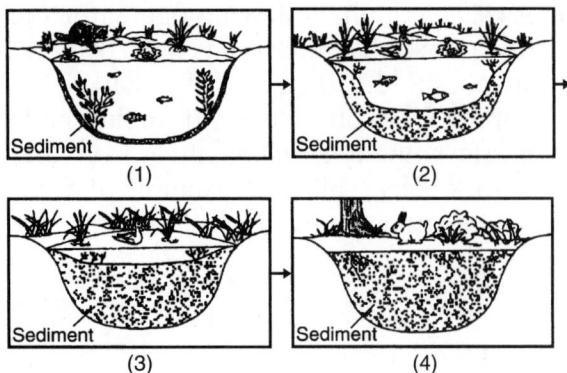

48. Which statement helps to explain this type of succession?

A. Species are replaced until a new pond ecosystem is established.
B. Species are replaced until a stable land ecosystem is established.
C. Humans replace all species over time and fill in all niches.
D. Animals control all changes in the plant species of an area.

49. Which population of organisms would be most harmed by the ecological changes occurring in this community?
A. trees C. raccoons
B. fish D. rabbits

50. Identify *one* factor that could disrupt the final stage of this ecosystem.

Reading Comprehension

Base your answers to questions 51 to 54 on the information below and on your knowledge of biology. Use one or more complete sentences to answer each question.

> In the early 1980s, an Asian elephant was born at the Bronx Zoo in New York City. Its charming antics made it a crowd favorite. It was heralded as the first elephant born in New York in more than 10,000 years (since prehistoric mastodons lived there). But a year and a half later, it died of unknown causes.
>
> In the 1990s, another Asian elephant was born in America, this time at the National Zoo in Washington, D.C. Like the New York baby, this elephant also died at about the same age. Several years later, scientists learned that the Washington baby elephant had died after being infected with a herpes virus. Later, tests of its preserved tissues revealed that the New York baby elephant had also died of a herpes virus infection.
>
> The virus that had killed the baby Asian elephants is commonly found in African elephants (which may live in the same zoos). In the African animals, the virus produces only a mild skin infection and sores. However, when the virus infects Asian elephants, especially young ones, it can produce deadly results. The reverse can also happen. It is now thought that a similar virus is found in Asian elephants, but it does not kill them. However, when this virus enters an African elephant, it may cause death. By 2002, a total of 22 baby elephants in the United States had died of the viral disease.
>
> Yet there is hope for infected baby elephants. One of the earliest symptoms of infection is a purple tongue. If this symptom is noted, treatment with the human antiviral drug famciclovir can be started. This drug can save the animals' lives if it is administered early enough. Scientists hope to develop tests to identify elephants that carry the virus and do not become ill. That way, they can try to prevent these elephants from coming into contact with, and passing the virus onto, other healthy elephants.

> Since 2002, several more baby elephants—both African and Asian—have been born in U.S. zoos. In time, scientists hope to develop a vaccine that will protect them from the herpes viruses, but that is still in the future. For now we can just hope that these young elephants remain uninfected and healthy.

51. Why was the birth of the Bronx Zoo elephant given so much attention?

52. In what way were the deaths of the Bronx Zoo and National Zoo elephant babies connected?

53. What do the facts about the herpes virus in elephants show about the relationship of the Asian elephant populations to the African elephant populations?

54. Why might a baby elephant born in a U.S. zoo today have a better chance of surviving than those born in the zoos more than 25 years ago?

Chapter 20

Human Impact on the Environment

Standard 5.10.12 B1 Assess the impact of human activities on the cycling of matter and the flow of energy through a system.

PEOPLE CHANGING THE ENVIRONMENT

Up until about 10,000 years ago, all humans hunted and gathered their food. Then, people started planting crops and domesticating animals; this marked the beginnings of agriculture. As a result of agriculture, people use the land differently from before. When people cut down and burn trees to make room to plant crops and graze livestock, wild animals are often forced to leave the area—they lose their habitat.

Advances in science and technology have produced even greater changes in the environment. About 200 years ago, developments in science and technology led to the Industrial Revolution, which greatly increased the ways that humans affect the environment.

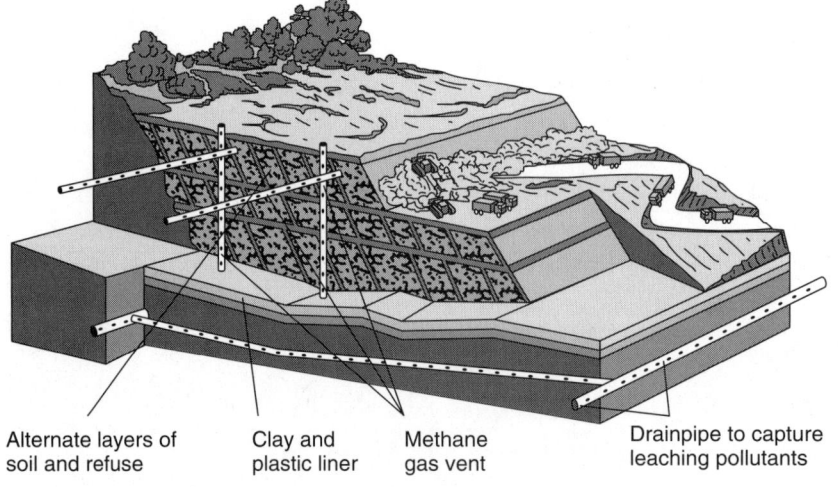

Figure 20-1 A sanitary landfill is constructed in a way that limits the effects of the waste materials on the surrounding environment.

Alternate layers of soil and refuse

Clay and plastic liner

Methane gas vent

Drainpipe to capture leaching pollutants

Figure 20-2 Some landfills contain waste products that are particularly toxic; such sites may pose a health threat to nearby communities.

Changes to the Land: Adding Wastes

All organisms produce wastes as a normal by-product of their life processes. However, since the time of the Industrial Revolution, the amount of wastes produced by humans has increased greatly. Also, since that time, the kinds of wastes have changed. Many of the wastes do not decompose, and they often contain harmful chemicals. These waste materials, called *solid wastes,* are often deposited in landfills, areas in which garbage is buried. (See Figure 20-1.)

In a sanitary landfill, attempts are made to limit the effects of the wastes on the environment. Other kinds of landfills are much more harmful to the environment, such as a *toxic* waste dump. The most dangerous of all toxic wastes are radioactive substances. (See Figure 20-2.)

Changes to the Land: Losing Soil

Although soil may be mistaken for dirt, it is actually a very valuable resource. In fact, this combination of organic and inorganic matter takes hundreds of years to form. Without good nutrient-rich soil, called *topsoil,* we could not grow food. Land ecosystems depend on this resource, too. (See Figure 20-3.)

Figure 20-3 Nutrient-rich topsoil takes a long time to form and is of great value to humans and wildlife. Crops cannot grow in soil that lacks adequate nutrients.

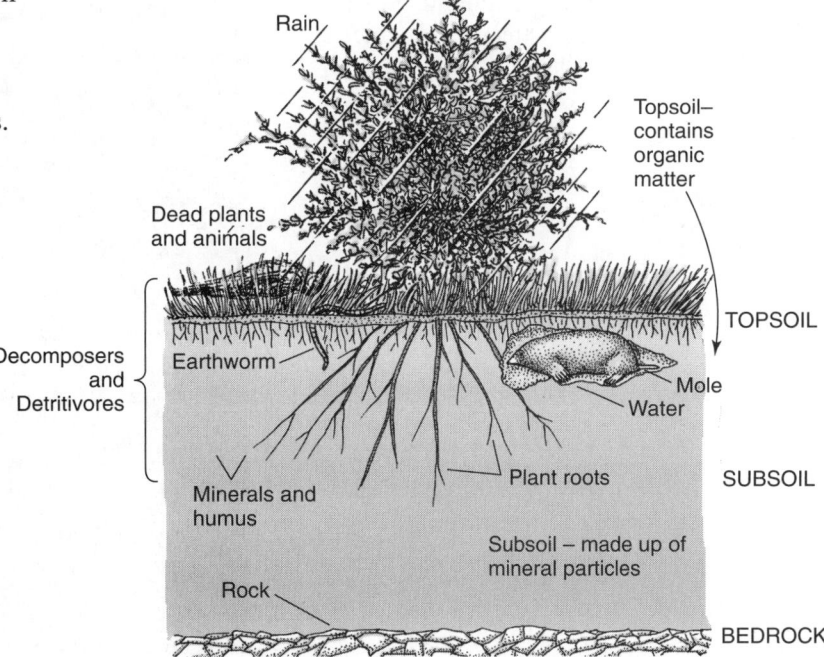

Chapter 20/Human Impact on the Environment **253**

Soil is now becoming unavailable because of human activities. For example, when toxic chemicals enter the ground, the soil becomes unusable. Poor farming practices and overgrazing by livestock can strip an area of all vegetation. The land becomes bare and, if weather patterns change and less rain falls, the land becomes a desert.

As people cut down forests or remove the plants that grow in an area, there is an increase in soil loss, or *erosion*. Both wind and water can cause erosion: rain washes away loose soil, and strong winds blow it away.

Changes to the Water

Because flowing water carries away wastes, a stream or river has always seemed the perfect place to dump garbage. Today, as the human population increases in size, many more wastes are placed into streams and rivers. There are simply more wastes introduced than the natural ecosystems can handle. The river or stream, once full of life, begins to lose its ability to support the same species of organisms as before. (See Figure 20-4.)

In addition, the types of wastes deposited have changed. Industries sometimes dump toxic chemicals into rivers. Fish that need clean, well-oxygenated water are replaced by fish that can live in water with lower oxygen levels. If the levels of pollutants keep increasing, these fish will die, too. A major advance in dealing with the problem of water pollution has been the development of sewage treatment plants. In these treatment plants, human organic

Figure 20-4 When a community dumps sewage and industrial wastes into its streams and rivers, the local aquatic ecosystems—and those of communities living downstream—become polluted.

Figure 20-5 Schematic of a city's water treatment process: Freshwater from a reservoir goes through several physical and chemical processes before it is considered clean enough to pipe into peoples' homes.

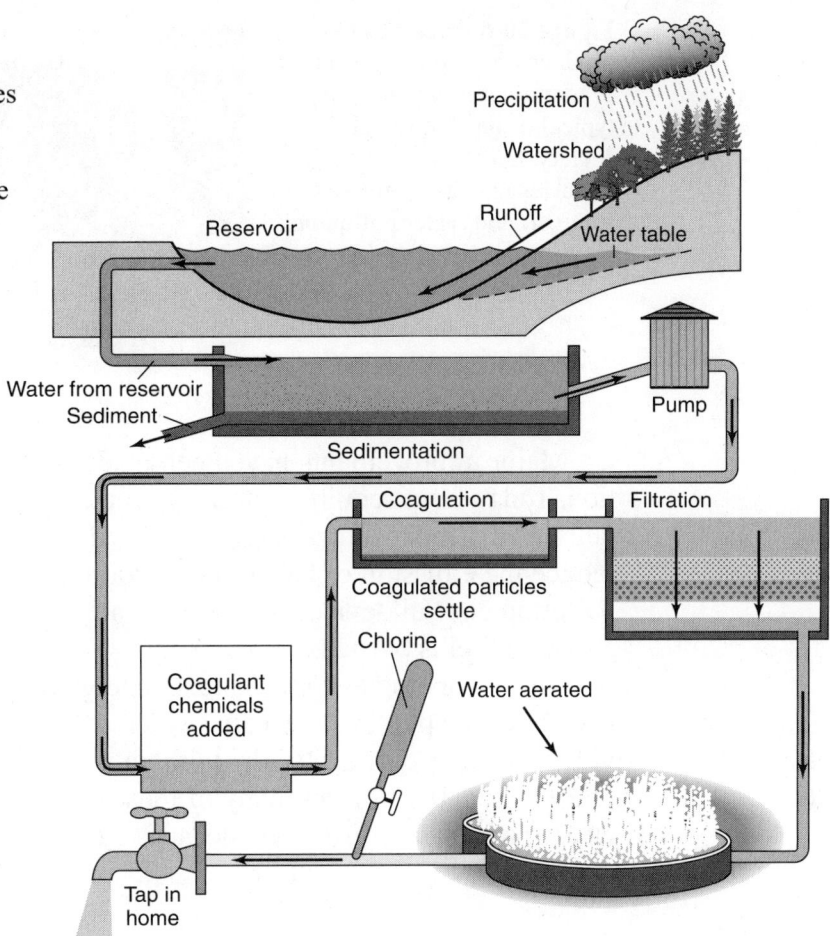

wastes are treated in large tanks. Wastes in the water are chemically digested by bacteria. The remaining solids, including dead bacteria, then settle to the bottom and are removed. Chlorine is added to the water to kill bacteria. Finally, the purified water is released into a river or stream.

Many cities obtain their water from underground wells. The water from these wells, called *groundwater,* accumulates over time and is stored naturally between layers of rock. There are above-ground reservoirs of freshwater, too. New York City relies on a system that directly transports clean water from upstate reservoirs through underground pipes. Other cities have their water treated first. (See Figure 20-5.)

Changes to the Air

No one owns the air; we all share the air, which forms a continuous blanket over Earth. If the air becomes polluted in one place, that pollution can easily spread to another place. Gases and tiny solid particles are constantly added to the air by human activities. If these substances are not normally found in the air and are harmful, they are called air *pollutants*. The burning of *fossil fuels*—coal, oil, and natural gas—to power cars, heat homes and offices, and produce electricity creates air pollution. In addition, many industries release pollutants into the air from huge smokestacks. (See Figure 20-6.)

Figure 20-6 Factories that produce enormous quantities of manufactured goods are typical of our industrial society. Unfortunately, these factories may also release some air and water pollutants.

Major improvements have been made in the efforts to reduce air pollution. Today, laws require factories to reduce or prevent the release of pollutants from smokestacks. Devices called "scrubbers" are installed, which reduce the emission of harmful compounds. Car engines, too, have built-in pollution control devices that reduce the amount of pollutants added to the air when fuel is burned.

New technologies for producing energy also have been developed. Solar collectors and photovoltaic cells, which use the sun's radiant energy, can provide us with heat or electricity without polluting the air. The Clean Air Acts of 1970 and 1977 began many of these changes. As a result, the air is now cleaner than it was just a few decades ago.

GLOBAL AIR POLLUTION PROBLEMS

Acid rain is a form of air pollution that produces far-ranging effects. Sulfur dioxide and nitrogen oxides are produced when fossil fuels are burned. Winds carry these gases high into the atmosphere and over long distances. They combine with water droplets in the air, which fall back to the ground as acid rain. Many forests and lakes in North America and Europe have been severely damaged by acid rain. (See Figure 20-7.)

Perhaps even more important are the effects of **global warming** and ozone depletion. Carbon dioxide (CO_2) in the **atmosphere** helps keep Earth

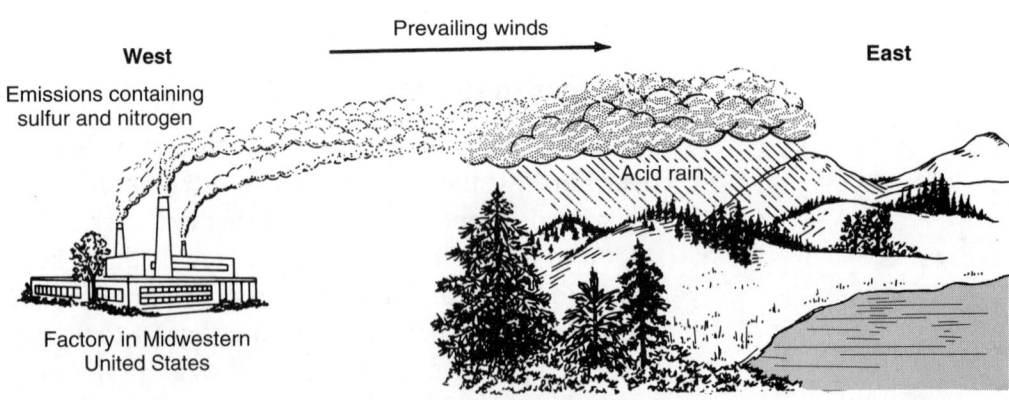

Figure 20-7 Due to acid rain—which contains chemicals from the smoke of Midwestern factories—lakes in New York's Adirondack Mountains have become more acidic, causing harm to wildlife.

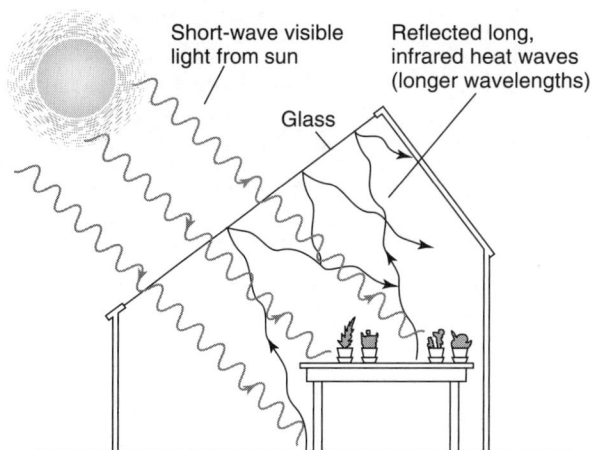

Figure 20-8 The greenhouse effect: Carbon dioxide in the air traps infrared energy, warming the atmosphere. This is similar to the way in which the glass roof of a greenhouse traps heat, keeping the plants warm.

warm by trapping heat. This is called the "greenhouse effect." But the amount of CO_2 has been increasing in the atmosphere due to the burning of fossil fuels and deforestation—in particular, the destruction of countless trees in rain forests that used to absorb CO_2. With more CO_2 in the atmosphere, more heat is trapped. Many people are concerned that, as a result, Earth's climate is getting warmer. Such a change in climate could have major effects on habitats and organisms everywhere. (See Figure 20-8.)

Scientists are also concerned about the effects of certain air pollutants that harm the **ozone shield**. This effect is known as ozone **depletion**. The layer of ozone gas that surrounds Earth high in the atmosphere blocks out harmful ultraviolet (UV) radiation. The UV rays are part of the energy that reaches Earth from the sun, and they can damage the DNA in our cells, causing skin cancer.

Chlorofluorocarbons (CFCs), found in air conditioners and refrigerators, are suspected of causing the most ozone depletion. In 1987, an agreement was signed by many countries to protect the ozone layer by limiting or banning the use of these chemicals. Progress has been made since the agreement was signed, and the ozone layer seems to be less depleted.

HUMAN POPULATION GROWTH

The most serious problem that now affects all life on Earth is the rate at which the human population is increasing. Over most of its history, the human population increased slowly. However, the rise of agriculture caused a rapid increase as people settled down with a more secure source of food. More recently, the Industrial Revolution, combined with scientific advances in farming and medicine, caused an explosion in human population size. (See Figure 20-9 on page 258.)

What is Earth's carrying capacity for humans? It is now known that the growth rate for the human population peaked in 1990 and is now declining. However, the population is still growing and it is believed it will reach a peak of about 9 billion around 2050. Others think that the population is already past Earth's carrying capacity and that serious environmental problems have

Figure 20-9 Industrial and scientific advances in the last 50 years have caused the human population to more than double in size—an increase that may put us past Earth's carrying capacity and cause environmental problems.

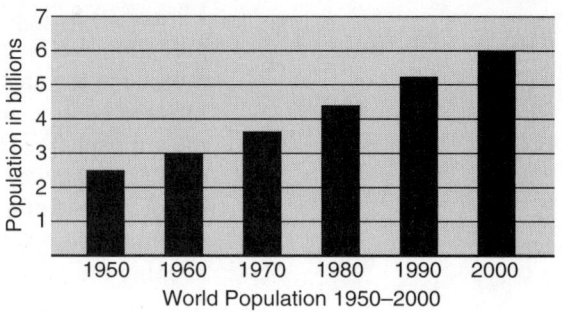

World Population 1950–2000

already begun, such as the crisis of an inadequate freshwater supply throughout much of the world.

An exploding human population may lead to the extinction of vast numbers of species. Many organisms are already endangered due to the loss of habitat and other human factors. Industrialization (which causes more air, water, and land pollution), acid rain, global warming, and ozone shield depletion are worldwide concerns. Earth—as one large, complex, intact ecosystem—is threatened by an ever-increasing human population. (See Figure 20-10.)

In 1994, participants from 160 countries at an international conference agreed that Earth's population cannot continue to grow at its current rate; yet they disagreed about how to lower the growth rate. What is clear is that our population, like that of any other organism, cannot increase forever. Either we will find a way to control human population size or nature will do it for us. Polluted soil, air, and water; lack of food and space; and widespread disease may ultimately limit human population size. However, individual choices and government planning could also limit it. People can still try to make the right decisions.

Figure 20-10 The exploding human population leads to overcrowding, pressure on limited resources, loss of wildlife habitat, increased pollution, and other problems that may threaten Earth's health as well as our own.

Chapter 20 Review

Multiple Choice

1. Over time, human populations have usually
   A. produced fewer and fewer wastes
   B. increased the amount of wastes produced
   C. prevented harmful chemicals from being produced
   D. decreased the amount of waste in landfills

2. Which organism has had the most negative impact on Earth's ecosystems?
   A. gypsy moth   C. zebra mussel
   B. human   D. shark

3. Which phrase would be appropriate for area *A* in the chart below?

Technological Device	Positive Impact	Negative Impact
Nuclear power plant	Provides efficient, inexpensive energy	A

   A. Produces radioactive wastes
   B. Provides light from radioactive substances
   C. Results in greater biodiversity
   D. Reduces dependence on fossil fuels

4. The most dangerous of all toxic wastes are
   A. solid wastes
   B. nutrient-rich soils
   C. radioactive substances
   D. plastic garbage

5. When humans cut down forests in an area,
   A. soil is lost through erosion
   B. the soil becomes richer
   C. new soil forms quickly
   D. flooding is prevented

6. When too many wastes are dumped into a river, the wastes
   A. eventually disappear through dilution
   B. gradually cause harm to the river ecosystem
   C. are carried away, where they can cause no harm
   D. are broken down immediately

7. Which is *not* a cause of increased water pollution?
   A. dumping sewage into streams and rivers
   B. addition of chlorine to treated water
   C. pouring industrial wastes into rivers
   D. fertilizers washing off land into streams

8. Dumping raw sewage into a river will lead to a reduction in the dissolved oxygen in the water. This condition will most likely cause
   A. an increase in all fish populations
   B. an increase in the depth of the water
   C. a decrease in most fish populations
   D. a decrease in water temperature

9. In a sewage treatment plant, bacteria are
   A. added to the water before it is released into a river
   B. killed by chlorine at the beginning of the process
   C. used to chemically digest wastes in the water first
   D. left in the purified water because they are harmless

10. "Natural ecosystems provide an array of basic processes that affect humans." Which statement does *not* support this quotation?
    A. Bacteria of decay help recycle materials.
    B. Treated sewage is less damaging to the environment than untreated sewage.
    C. Trees add to the amount of atmospheric oxygen.
    D. Lichens and mosses on rocks help to break down the rocks, forming soil.

11. A negative impact of technology is an increase in the
    A. development of new products
    B. availability of different foods
    C. wastes released into the environment
    D. societal awareness of the environment

12. Which type of waste will decompose most quickly?
    A. foam cup   C. glass bottle
    B. plastic bag   D. banana peel

13. Which practice will best protect the soil?
    A. removing excess trees from it
    B. planting vegetation on it
    C. allowing cattle to feed on the land
    D. adding lots of chemicals to it

14. An increase in the use of fossil fuels is an indication of which type of society?
    A. hunter-gatherer
    B. agricultural
    C. industrial
    D. horticultural

15. Acid rain forms when
    A. carbon dioxide traps heat near Earth
    B. ozone is depleted from the atmosphere
    C. gases from fossil fuels combine with water droplets in the air
    D. chlorine is added to waste water

16. Methods used by people to reduce the emission of pollutants from smokestacks are an attempt to
    A. lessen the amount of insecticides in the environment
    B. lessen the formation and harmful effects of acid rain
    C. eliminate diversity in natural habitats
    D. use nonchemical controls on pest species

17. Changes in the chemical composition of the atmosphere that may produce acid rain are most closely associated with
    A. insects that excrete acids
    B. factory smokestack emissions
    C. runoff from acidic soils
    D. flocks of migrating birds

18. Since the start of the Industrial Revolution, there has been
    A. a sharp decrease in the greenhouse effect
    B. an increase in global atmospheric cooling
    C. a decrease in CO_2 released into the atmosphere
    D. an increase in the amount of wastes produced

19. People can have a large negative impact on ecosystems when they
    A. conserve natural resources
    B. modify the environment
    C. restrict the use of chemicals
    D. pass laws to protect habitats

20. By causing atmospheric changes through such activities as extensive tree harvesting, humans have
    A. caused the destruction of habitats
    B. established equilibrium in ecosystems
    C. affected global stability in a positive way
    D. replaced nonrenewable resources

21. The effect of CO_2 and other greenhouse gases on the atmosphere can best be likened to that of a
    A. blanket
    B. balloon
    C. pitcher of water
    D. crowd of people

22. Deforestation will most directly result in an increase in
    A. atmospheric carbon dioxide
    B. wildlife populations
    C. atmospheric ozone
    D. renewable resources

23. Which human activity would have the most direct impact on the oxygen–carbon dioxide cycle?
    A. reducing the rate of ecological succession
    B. destroying large forested areas
    C. decreasing the use of water
    D. banning the use of leaded gasoline

24. Chlorofluorocarbons are harmful to the environment because they
    A. kill fish in lakes
    B. form acid rain
    C. cause ozone depletion
    D. increase the greenhouse effect

25. The ozone layer of Earth's atmosphere helps to filter ultraviolet radiation. As the ozone layer is depleted, more ultraviolet radiation reaches Earth's surface. This increase in ultraviolet radiation may be harmful because it can directly cause
    A. photosynthesis to stop in all marine plants
    B. mutations in the DNA of organisms
    C. abnormal migration patterns in waterfowl
    D. sterility in most species of mammals and birds

26. Of the four items listed below, which factor is often responsible for the other three?
    A. increase in levels of toxins in freshwater
    B. increased poverty and malnutrition
    C. increase in the human population
    D. increased depletion of finite resources

Analysis and Open Ended

27. Briefly explain how humans change the land through agriculture. Your answer should include the impact on the following:
    - topsoil
    - forests
    - wildlife

28. How has industrialization changed the types of wastes produced by humans?

29. Why might building a landfill near an aquatic ecosystem cause harm to it?

Base your answers to questions 30 and 31 on the graph below, which shows pollution from nitrogen-containing compounds (nitrates) in a brook flowing through a forested area and a deforested area between 1965 and 1968.

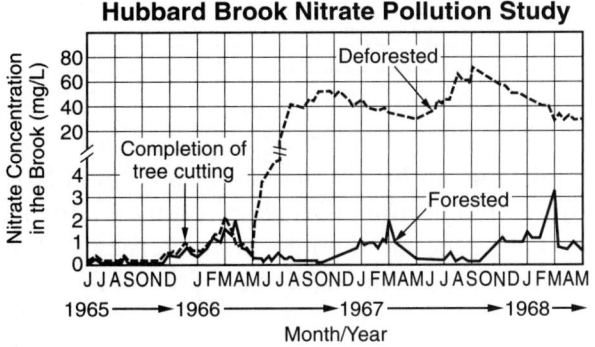

30. State how nitrate pollution in the brook changed after it flowed through the deforested area.

31. Explain how deforestation contributed to this change.

32. Explain why toxic waste dumps are most harmful to the environment.

33. Why is good topsoil considered so valuable? List three human activities that cause loss of soil.

Refer to the following graphs to answer questions 34 and 35.

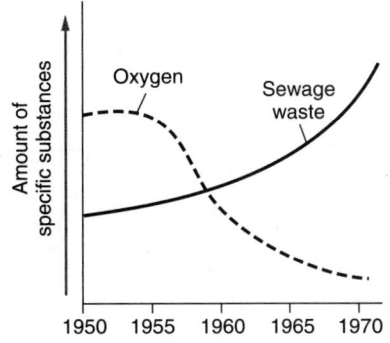

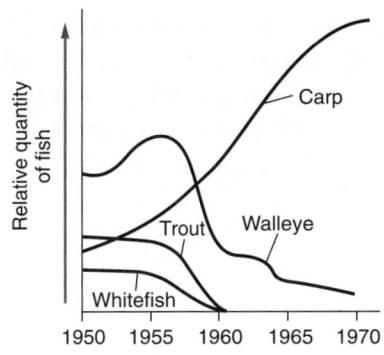

34. According to the graphs, an increase in sewage waste in a lake over time would be associated with
   A. an increase in dissolved oxygen and an increase in most fish populations
   B. a decrease in dissolved oxygen and an increase in most fish populations
   C. an increase in dissolved oxygen and a decrease in most fish populations
   D. a decrease in dissolved oxygen and a decrease in most fish populations

35. According to the graphs, the fish species that adapted most successfully to the change in oxygen content over time was the
   A. trout, because it can live in highly oxygenated water
   B. carp, because it can live in poorly oxygenated water
   C. walleye, because it can live in highly oxygenated water
   D. whitefish, because it can live in poorly oxygenated water

36. Explain how a new power plant built on the banks of the Rocky River could have an environmental impact on the Rocky River ecosystem downstream from the plant. Your explanation must include the effects of the power plant on:
   • water temperature
   • dissolved oxygen
   • local fish species

37. Describe the importance of sewage treatment to both people and wildlife.

38. Both car exhaust and factory emissions add pollutants to the air. For each case, tell *how* it adds to air pollution and *what* is being done to reduce the problem.

Chapter 20/Human Impact on the Environment **261**

39. The map below shows the movement of some air pollution across part of the United States. Which statement is a correct inference that can be drawn from this information?

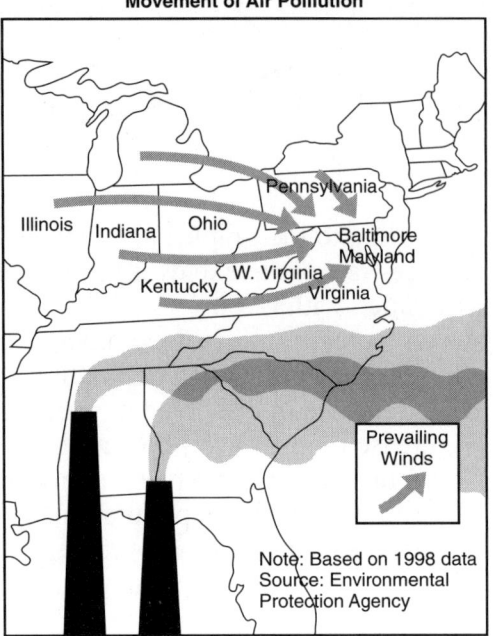

A. Illinois produces more air pollution than the other states shown.
B. The air pollution in Baltimore is increased by pollution from other areas.
C. There are no air pollution problems in the southern states.
D. The air pollution in Virginia clears up quickly as the air moves toward the sea.

Refer to the illustration below to answer questions 40 and 41

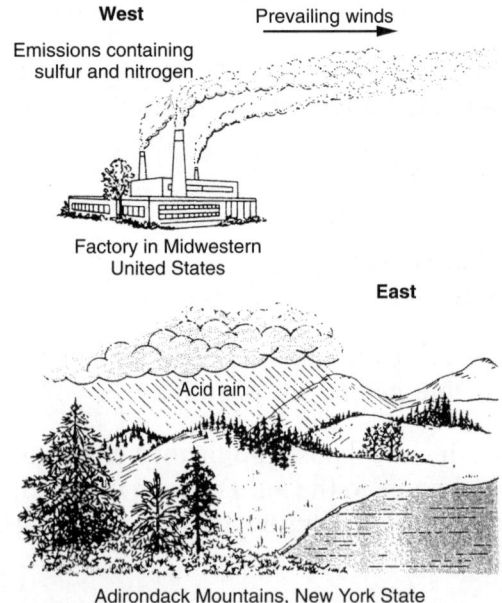

40. According to the illustration, acid rain is both an air pollutant and a water pollutant. Explain why this is true.

41. For each type of pollution (air and water), give *one* example of the kind of habitat the acid rain affects. By what means does it reach these different ecosystems?

Base your answers to questions 42 and 43 on the information below and on the following diagram.

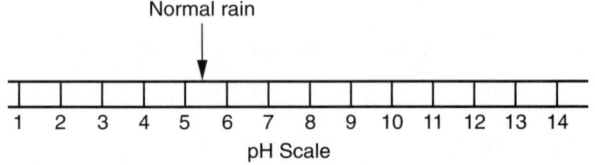

Acid rain can have a pH between 1.5 and 5.0. The effect of acid rain on the environment depends on the pH of the rain and the characteristics of the environment. It appears that acid rain has a negative effect on plants. The following scale shows the pH of normal rain. (*Note:* Lower pH numbers are more acidic.)

42. Provide the information that should be included in a research plan to test the effect of pH on the early growth of bean plants in the laboratory. In your answer be sure to:
    • state a hypothesis
    • identify the independent variable
    • state *two* factors that should be kept constant

43. Construct a data table in which you could organize the research results.

Refer to the following graph to answer questions 44 and 45.

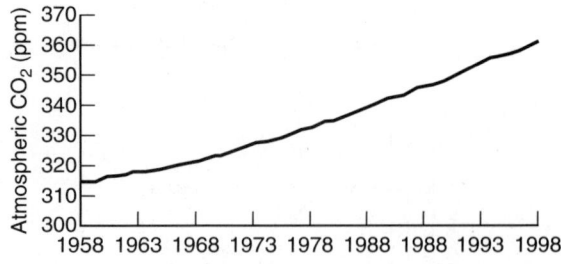

44. According to the graph, from the late 1950s to the late 1990s, the amount of CO_2 in Earth's atmosphere has been
    A. steadily decreasing
    B. steadily increasing

C. staying about the same
D. going up and down

45. Changes in the amount of CO_2 in Earth's atmosphere have been correlated with steadily increasing average global temperatures over the past 50 years. Based on this statement and the data in the graph, you could reason that as the amount of CO_2 in the air
    A. increases, the average temperature decreases
    B. increases, the average temperature increases
    C. decreases, the average temperature stabilizes
    D. decreases, the average temperature increases

46. Some scientists are urging that immediate action be taken to stop activities that contribute to global warming. Discuss the effects of global warming on the environment and describe some human activities that may contribute to it. Your answer *must* include:
    - an explanation of what is meant by the term *global warming*
    - *one* human activity that is thought to be a major contributor to global warming
    - an explanation of *how* the human activity may contribute to the problem
    - *one* negative effect of global warming if it continues for many years

47. In the early 1980s, scientists discovered "holes" in the ozone shield that surrounds Earth. What is one negative effect that this environmental change could have on humans?

48. What substance is thought to cause the depletion of Earth's ozone shield? What has been done to try to solve this problem?

49. Choose *one* ecological problem to discuss from the following list: *global warming; destruction of the ozone shield; acid rain; increased nitrogen and phosphorus in lakes; loss of biodiversity.* In your answer be sure to state:
    - the type of ecological problem you have chosen
    - *one* human action that may have caused the problem
    - *one* way in which the problem may negatively affect humans
    - *one* way in which the problem may negatively affect the ecosystem
    - *one* positive action that people can or did take to reduce the problem

50. Describe *two* specific methods that have been recently used by people to reduce the amount of chemicals being added to the environment.

51. All living organisms are dependent on a stable environment. Describe how humans have made the environment *less* stable for other organisms by:
    - changing the chemical composition of air, soil, and water
    - reducing the biodiversity of an area
    - introducing advanced technologies

Base your answers to questions 52 and 53 on the information below and on your knowledge of biology.

Amphibians have long been considered an indicator of the health of life on Earth. Scientists are concerned because amphibian populations have been declining worldwide since the 1980s. In fact, in the past decade, at least twenty species of amphibians have become extinct and many others are endangered.

Scientists have linked this decline in amphibians to global climatic changes. The destruction of many eggs produced by the Western toad is an example of this. Warmer weather during the last three decades has led to a decrease in rain and snow in the Cascade Mountain Range in Oregon. A negative result has been reduced water levels in the lakes and ponds that serve as reproductive sites for the Western toad. As a result, the toads' eggs are exposed to more ultraviolet light. This makes the eggs more susceptible to a water mold that kills the embryos by the hundreds of thousands.

52. The term that is commonly used to describe the worldwide climatic changes mentioned in the passage above is
    A. global warming
    B. deforestation
    C. mineral depletion
    D. industrialization

53. State *two* ways in which the decline in amphibian populations could disrupt the stability of the ecosystems they inhabit.

Refer to Figure 20-9 on page 258 to answer the following question.

54. By the year 2000, the worldwide human population had reached

A. three times the size it was in 1950
B. three times the size it was in 1960
C. two times the size it was in 1970
D. two times the size it was in 1960

Reading Comprehension

Base your answers to questions 55 to 57 on the information below and on your knowledge of biology. Use one or more complete sentences to answer each question.

> Dr. David Vaughan is a British scientist who has been studying glaciers for a long time. As a glaciologist, he is very interested in the ice that covers and surrounds the great landmass at the South Pole. If the climate change that is occurring on Earth causes global warming, then the ice of Antarctica will start to melt. The melting ice would, eventually, raise sea levels around the world and the results would be disastrous. There is enough ice in just the western part of Antarctica to cause a rise of five meters in sea levels. This would flood many coastlines where millions of people live. However, knowing what is happening to the ice of Antarctica is very difficult.
>
> Much attention has been given in recent years to ice shelves, floating masses of ice that surround much of Antarctica. While their melting will not directly affect sea levels—that ice is already in the sea—the loss of the ice shelves would make it much easier for the huge masses of land ice to melt. So, Dr. Vaughan and other glaciologists have been closely monitoring a series of collapses of ice shelves that began in 1995. In January 1995, a 770-square-mile section of an ice shelf along the Antarctic Peninsula broke apart suddenly. This area of ice was 35 times larger than all of Manhattan. Another even larger ice shelf that was at least 400 years old broke apart in 1998. And, in just 35 days beginning on January 31, 2002, the largest collapse to be seen in 30 years occurred. About this ice shelf, Dr. Vaughan said, "We knew what was left would collapse, but the speed of it is staggering." The area of ice that disappeared was 220 meters thick and contained 720 billion tons of ice!
>
> These ice-shelf collapses are not entirely unexpected. The temperatures in the area of the Antarctic Peninsula, a long sliver of land pointing toward South America from Antarctica, have been rising steadily since the 1940s. The average temperature is now 2.5 degrees Celsius higher than it was in 1945. This is the fastest rate of warming seen any place on Earth. With the higher temperatures comes the melting of ice. Is this change an early warning sign of global warming elsewhere? Quite possibly. For example, seasonal shrinkage of Arctic sea ice has been increasing, too. More square miles of Arctic ice have melted—and at a faster rate—in recent years than had been recorded in previous years.
>
> Scientists are determined to study ice around the world—especially in the Antarctic—even more closely to get an answer. But how can this study be done, knowing how difficult it is to get to Antarctica? Go into outer space! And this is exactly what has been done. A satellite, called *ICESat* (*I*ce *C*loud & *L*and *E*levation *Sat*ellite), launched in 2003, is now orbiting Earth to track precise changes in ice sheets around the world.

55. Why is the possible melting of Antarctic ice of such concern?

56. How is the melting of ice shelves related to the rise of sea levels?

57. How is the use of outer space helping in the study of the Antarctic?

Chapter 21

Efforts to Protect the Environment

Standard 5.10.12 B2 Use scientific, economic, and other data to assess environmental risks and benefits associated with societal activity.

PROTECTING THE BIOSPHERE

In April 1970, the first Earth Day marked the beginning of the modern environmental movement. Many environmental organizations were founded at this time, and the government passed several environmental protection laws. Can we protect the environment not only for people but for all species? The total area of land, water, and air on Earth's surface where life is found—that is, all the ecosystems—makes up the **biosphere**. (See Figure 21-1.) These ecosystems need protection. Saving the biosphere means paying attention to local, regional, and global problems.

Figure 21-1 The biosphere is the total area of Earth's land, air, and water in which life is found. Earth is the ultimate ecosystem; although environmental problems may start out as local ones, they can become regional and, eventually, have a global impact.

265

A Change in Attitude

Sometimes the most important, and difficult, changes concern accepted attitudes in our society. For example, what if we thought our lifestyle should not harm the environment and other living species? Would we be willing to make the necessary changes to accomplish this?

An industrialized society mainly views Earth as a source of valuable resources for its use. In contrast, ecology teaches us that humans are just one of many interdependent species that also need resources to live. In order for our species to survive, we must make sure that these important relationships within ecosystems also survive.

Think Globally, Act Locally

The future health of the environment will depend on people's attitudes and behaviors. It has been suggested that people should learn to appreciate the "hidden costs" of many consumer goods. In other words, the environment pays a price for the products used by people in an industrial society. Environmentalists have encouraged people to live by the "3 Rs": reduce, reuse, and recycle. To *reduce* consumption, you would use less of a product or resource; for example, fewer paper towels can be used to clean up a spill. You can also *reuse* a product; for example, paper or plastic grocery bags that you bring food home in can be taken back the next week and used again. Finally, many used products can be made into other products; in many cities, you now have to *recycle* plastic, glass, metal, and paper. These materials are used again in other products, such as benches made up of a "wooden" building material that is actually a form of recycled plastic. Such recycling helps to conserve our natural resources. (See Figure 21-2.)

Figure 21-2 Recycling of plastic, glass, and metal containers is required in some cities. As shown here, students can help in the recycling effort by sorting and recycling metal soda cans and bottles.

Figure 21-3 The mongoose was introduced to Puerto Rico to control its rat population. However, its introduction produced unexpected harmful results, such as an increase in the June beetle larvae population.

Introducing "Alien" Species

Great care must be taken to avoid the environmental damage that can occur when a species from a distant region is introduced into a new environment. Without any predators or natural controls, the new "alien" species can reproduce without limit, upsetting the stability of the ecosystem; or it can affect populations of other species. This happened, for example, when the mongoose was introduced to Puerto Rico to control the rat population. The mongoose started preying on other species, too, such as lizards. As a result, the June beetle larvae population increased, becoming a serious agricultural pest. (See Figure 21-3.) In a similar way, the harmful effects that result from the chemical control of agricultural pests can be avoided by the use of local means of natural, or biological, control. For example, native predators of insect pests can be used as a means of biological control.

RENEWABLE VERSUS NONRENEWABLE RESOURCES

Air, water, and sunlight are some of the important resources that are basic to life on Earth. Modern industrialized society requires other resources, too, such as coal, oil, and metal ores.

Resources can be considered renewable or nonrenewable. A *renewable resource* can be replaced within a generation. Enough of the resource is being made (by natural processes) to replace what is being used. For example, the wood used to build houses can be replaced if enough trees are replanted. The sun's energy and the wind can be considered renewable resources, too.

A *nonrenewable resource* cannot be replaced within our lifetime; it exists in limited amounts and takes a very long time to form. This includes such energy sources as coal, oil, and natural gas. In addition, metals such as gold, silver, iron, copper, and aluminum and nonmetals such as sand, gravel, and limestone are nonrenewable resources.

One way to protect the biosphere is to use renewable energy sources. For example, electricity can be made in a dam from the power of falling water, rather than from the burning of coal. Wind power can turn the blades on giant windmills to generate electricity. (See Figure 21-4 on page 268.) Sunlight, or **solar energy**, can be used to heat water and buildings, and to produce

Figure 21-4 Windfarms, such as this one in California, harness wind power—another renewable resource—to generate electricity.

electricity. (See Figure 21-5.) Finally, steam and hot water from deep beneath Earth's surface, or *geothermal energy,* can be used to heat buildings and make electricity, too. (*Note: geo* means "earth"; *thermal* means "heat.") All of these are renewable energy sources.

SUSTAINABLE DEVELOPMENT

If the economy is not growing, people think that something is wrong. Yet, unlimited growth is not possible. We cannot use more and more of Earth's resources indefinitely; and we cannot continue to add more and more pollution to our environment. We need to find a way to live that is sustainable—a way that does not ruin Earth's ability to support life in the future. Improving the

Figure 21-5 The diagram illustrates how energy from sunlight can be collected by a solar panel and used to heat a building.

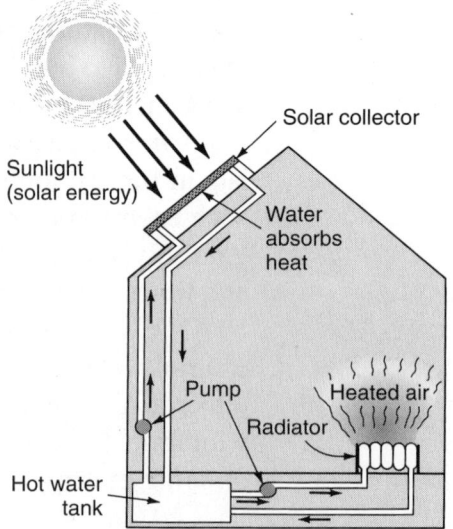

way we live without harming the environment is called *sustainable development*. By making these changes now, we may secure a healthier future for our planet.

Sustainable Development of Forests

Some of the greatest damage to the environment is done by cutting down trees. The most economical way to harvest trees is to cut down an entire forest, a method called *clear-cutting*. (See Figure 21-6.) Such deforestation causes animals to lose habitat; and the rain—no longer absorbed by tree roots—flows right into streams, carrying valuable topsoil with it. Due to soil erosion, the land cannot support plant growth; and the freshwater habitat for fish is disturbed, too.

Sustainable development in forestry means replacing every tree that is cut with a seedling and making sure that the seedling survives. A forest—with trees of all sizes and different ages, continually regrowing—would provide a healthy habitat for other woodland species and would prevent soil erosion.

Environmental Protection in a Developed Country

Like other industrialized developed countries, the United States uses a great amount of energy and resources. The environment has been seriously affected by the wasteful use of resources; that is why environmental awareness has increased. One important result of this increased awareness was the establishment of the Environmental Protection Agency (EPA), which is responsible for safeguarding the environment for future generations. Another

Figure 21-6 The wood from forests has many uses: lumber, paper pulp, and fuel. Although cutting all the trees in one area is economical for loggers, the environment pays a high price in terms of soil erosion and loss of habitats.

Figure 21-7 The spotted owl of the Pacific Northwest is an endangered species. By protecting the owl, the Endangered Species Act also protects the old-growth forest in which it lives.

response to environmental issues was passage of the Endangered Species Act, which regulates a wide range of activities that affect threatened or endangered plant and animal species. (See Figure 21-7.)

Environmental Protection in a Developing Country

People's lives in developed countries are very different from those in developing countries. In industrialized nations, the average standard of living is high and most people can expect to live for 70 or more years. But in many developing countries, most people are poor and have a much shorter life span. So, saving the biosphere means different things in rich nations and poor nations. Environmental leaders are now learning about these differences.

Environmentalists realize that it is very difficult to set aside parks for endangered animals if doing so stops the local people from obtaining food and housing. One solution that has worked—called "parks for people"—directly involves the local people in pr'srotecting their environment. For example, in part of Kenya, villagers work as guides for tourists who come to see the wildlife. (See Figure 21-8.)

Saving the Biosphere: A Worldwide Effort

In 1992, representatives from 178 countries attended the largest environmental meeting ever held, known as the Earth Summit. The main theme of

Figure 21-8 In Africa, near Kenya's Ewasu River, local guides take tourists on camel safaris. This is one way in which "parks for people" programs can help local people earn a living from their natural environment without doing it harm.

the meeting was sustainable development. The Kyoto Treaty, based on negotiations first held in 1997, was finally ratified in 2005 by 141 nations (the United States did not sign). Its goal is to limit the emissions of greenhouse gases such as CO_2 from 35 industrialized nations. At the G-8 Summit of major industrialized nations, held in July 2008, topics included climate change and funding for cleaner energy technologies to reduce greenhouse gases. Work continues, but differences among the countries interfere with more progress. While local and regional efforts to save the environment are important, the global effort matters most. The protection of Earth's biosphere for the future requires the efforts of all people.

Chapter 21 Review

Multiple Choice

1. The biosphere is the total area where life exists on or in Earth's
   A. land only
   B. water only
   C. land and water
   D. land, water, and air

2. A major reason that humans have negatively affected the environment in the past is that they
   A. often lacked an understanding of how their activities affect the environment
   B. attempted to control their population growth
   C. passed laws to protect certain wetlands
   D. discontinued the use of certain chemicals used to control insects

3. Recycling of materials such as glass, metal, and plastic helps to
   A. keep the cost of groceries low
   B. conserve our natural resources
   C. prevent natural resources
   D. build more wooden houses

4. Which human activity would be *least* likely to disrupt the stability of an ecosystem?
   A. disposing of wastes in the ocean
   B. increasing the human population
   C. using more fossil fuels
   D. recycling bottles and cans

5. An industrialized society mainly views Earth as a
   A. home for wildlife
   B. hazardous place to live
   C. source of natural resources
   D. barren landscape

6. Coal and wood are found in nature. They are both examples of
   A. enzymes   C. metals
   B. resources   D. proteins

7. An example of a renewable resource is
   A. natural gas   C. coal
   B. silver   D. wood

8. The definition of a renewable resource is that it
   A. can be replaced by nature within a generation
   B. cannot be replaced by nature within a generation
   C. is manufactured by humans
   D. is not too expensive to use

9. A nonrenewable resource is one that
   A. is replaced by nature as fast as it is used
   B. exists in a limited supply that can run out
   C. is recycled naturally by Earth's systems
   D. does not pollute ecosystems when it is used

10. Suppose that the average life span of most people is about 70 years and that there is only a 70-year supply of fossil fuel left on Earth. Then imagine that you are a member of a government panel that is deciding on how to handle the fuel situation. The best possible decision you could suggest would be to
    A. use up all the fuel in the present generation and not worry about the future
    B. find some alternative energy sources so that the fossil fuel lasts longer
    C. destroy the remaining fossil fuel so that no nations will fight over it
    D. have people return to a farming society so that they do not need the fuel

11. Which practice would most likely deplete a nonrenewable natural resource?
    A. harvesting pine trees on a tree farm
    B. restricting water usage during a period of water shortage
    C. burning coal to generate electricity in a power plant
    D. building a dam and a power plant to use water to generate electricity

12. All of the following are nonrenewable energy sources *except*
    A. coal   C. falling water
    B. gas   D. natural gas

13. Which statement is true about endangered species and nonrenewable resources?
    A. They are both living things that need to be protected.
    B. They are both nonliving things that need to be protected.
    C. They both need to be protected so they do not disappear.
    D. They both can be renewed quickly if they do disappear.

14. The goal of sustainable development is to
    A. achieve unlimited economic growth at any cost
    B. make sure the economy expands at a steady rate
    C. improve the way we live without harming the environment
    D. expand our lifestyle even if we run out of natural resources

15. Which method would you *not* use to solve environmental problems?
    A. promote global awareness
    B. cooperation among people
    C. "parks for people" programs
    D. increasing the population

16. Which action would best illustrate people's concern for the biosphere?
    A. passing game laws that limit the number of animals that may be hunted
    B. increasing the use of pesticides that may drain off farms into river systems
    C. allowing air to be polluted by only those factories that use new technology
    D. removing resources from nature at a faster rate than they are being replaced

17. One way to help provide a suitable environment for the future is to urge individuals to
    A. apply ecological principles when making decisions that have an impact
    B. agree that population controls have no impact on environmental matters
    C. control all aspects of natural environments
    D. work toward increasing global warming

Analysis and Open Ended

18. What is meant by "think globally, act locally" in terms of environmental protection?

19. Define the "3 Rs" proposed by environmentalists. Give an example of each one.

20. Why is it important for industries to be involved in recycling programs?

21. Compare and contrast renewable and nonrenewable resources. Your answer should include:
    - a definition of *renewable* and of *nonrenewable*
    - *one* example of a renewable resource
    - *two* examples of a renewable *energy* source
    - *two* examples of a nonrenewable resource

22. Recycling can extend the use of nonrenewable resources but *cannot* restore them. Humans can restore renewable resources to reduce some negative effects of increased consumption. Identify *one* resource that is renewable, and describe *one* specific way people can restore this resource if it is being depleted.

23. Use the following terms to complete the concept map below, which lists a variety of natural resources: *flowing water; copper; wind power; wood (charcoal); oil; gold; trees (lumber); limestone; geothermal energy; gravel; natural gas; coal; sunlight; sand; silver.*

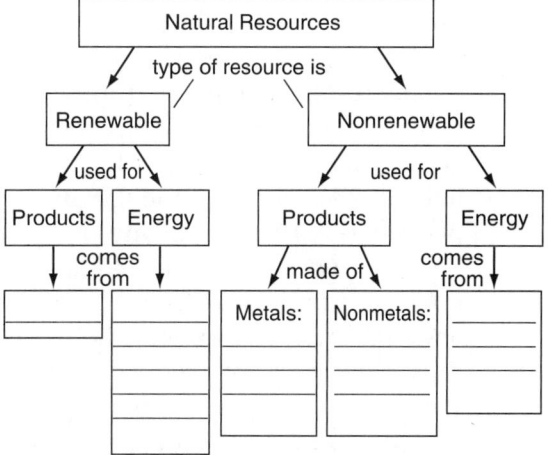

24. Explain what is meant by the term "sustainable development." Give *one* example of sustainable use of a natural resource.

25. Refer to Figure 21-2 on page 266 to answer this question: How might the activity these young people are involved in actually help to conserve forests and wildlife?

26. In what way does the protection of habitats help enforce the Endangered Species Act?

27. Refer to Figure 21-6 on page 269 to answer this question: How does the method of "clear-cutting" shown here damage habitats both on land and in the water?

28. Why are people in developing nations sometimes more directly affected by conservation programs than are people in more developed (industrialized) nations?

29. How do "parks for people" programs help local people and their environment at the same time?

30. Give at least one reason why *all* nations—developed and developing—should be involved in global efforts to save the biosphere.

31. List *five* changes that could be made in your own home that could help protect the environment.

32. List *five* changes that could be made in your community that could help protect the environment.

Reading Comprehension

Base your answers to questions 33 to 37 on the information below and on your knowledge of biology. Use one or more complete sentences to answer each question.

> The location of a former fuel storage depot and packaging operation in the industrial port of Toronto, Canada, was the proposed site of a sports arena and entertainment complex. The problem was that the soil in this area was contaminated with gasoline, diesel fuel, home heating oil, and grease from the operation of the previous facility. Unless these substances were removed, the project could not proceed.
>
> The traditional method of cleaning up such sites is the "dig-and-dump" method, in which the contaminated soil is removed, deposited in landfills, and replaced with clean soil. This dig-and-dump method is messy and costly, and it adds to landfills that are already overloaded. A technique known as *bioremediation*, which was used to help in the cleanup of the *Exxon Valdez* oil spill in Alaska, offered a relatively inexpensive way of dealing with this pollution problem.
>
> The bioremediation cleanup process cost $1.4 million, one-third the cost of the dig-and-dump method, and involved encasing 85,000 tons of soil in a plastic "biocell" the size of a football field. This plastic-encased soil already contained naturally occurring bacteria that would have cleaned up the area after 50 years or more with the amounts of oxygen and nutrients normally found in the soil. But air, water, and fertilizer were piped into the biocell, stimulating the bacteria to reproduce rapidly and speed up the process. The cleanup by this technique was begun in August and completed in November of the same year. The bacteria attack parts of the contaminating molecules by breaking the carbon-to-carbon bonds that hold them together. This helps to change these molecules in the soil into carbon dioxide and water.
>
> Although this method is effective for cleaning up some forms of pollution, bioremediation is not effective for inorganic materials, such as lead or other heavy metals, since these wastes are already in a base state that cannot be degraded any further.

33. The use of bioremediation by humans is an example of
    A. interfering with nature so that natural processes cannot take place
    B. using a completely unnatural method to solve a natural problem
    C. solving a pollution problem by speeding up natural processes
    D. using naturally occurring bacteria to cure a human disease

34. State an ecological problem associated with the use of the dig-and-dump method.

35. The bacteria convert the contaminants into
  A. carbon dioxide and water
  B. other toxic substances
  C. proteins and fats
  D. diesel fuel and grease

36. Explain why the cleanup of the proposed sports and entertainment site took only four months.

37. Bioremediation is *not* an effective method for breaking down
  A. grease and heating oil
  B. gasoline for vehicles
  C. fuel for diesel engines
  D. heavy metals such as lead

PRACTICE TEST 1

PART 1

Directions for Questions 1 through 15 For each of the questions or incomplete statements below, choose the best of the answer choices given. Write your answers on a separate sheet of paper.

1. The complementary mRNA sequence for the DNA sequence GAACCT is
   A. GAACCU
   B. CTTGGA
   C. CUUGGA
   D. GAACCT

2. What is the first step that occurs during protein synthesis?
   A. The mRNA attaches to a ribosome.
   B. Amino acids are linked together.
   C. The tRNA attaches to a codon.
   D. DNA serves as a template for mRNA.

3. Recombinant DNA is used in genetic engineering to
   A. finally eliminate all human viruses
   B. create individuals that show incomplete dominance
   C. increase the frequency of sexual reproduction
   D. synthesize proteins needed by humans

4. Which parts of DNA would be held together by weak hydrogen bonds?
   A. phosphate and sugar
   B. phosphate and adenine
   C. cytosine and sugar
   D. cytosine and guanine

Base your answers to questions 5 and 6 on the figure below.

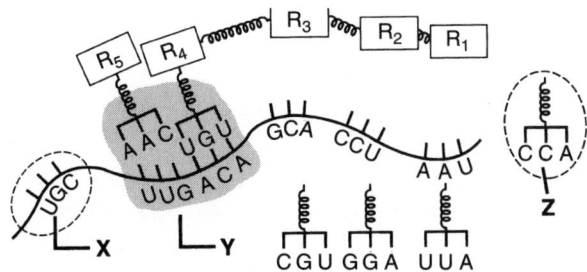

5. What structure is represented by the letter *Y*?

A. mitochondria   C. nucleus
B. chloroplast   D. ribosome

6. The original template for this process would be
   A. ribosomal RNA   C. transfer RNA
   B. a DNA strand   D. messenger RNA

7. There is a loss of energy at each trophic level. The process that brings new energy into the ecosystem and sustains it is
   A. decomposition
   B. photosynthesis
   C. nitrogen fixation
   D. ecological succession

8. Mice, insects, shrubs, owls, and trees are found in a forested area. The population that is smallest in numbers would be the
   A. insects   C. trees
   B. mice   D. owls

9. Which pyramid below is the most energy efficient?

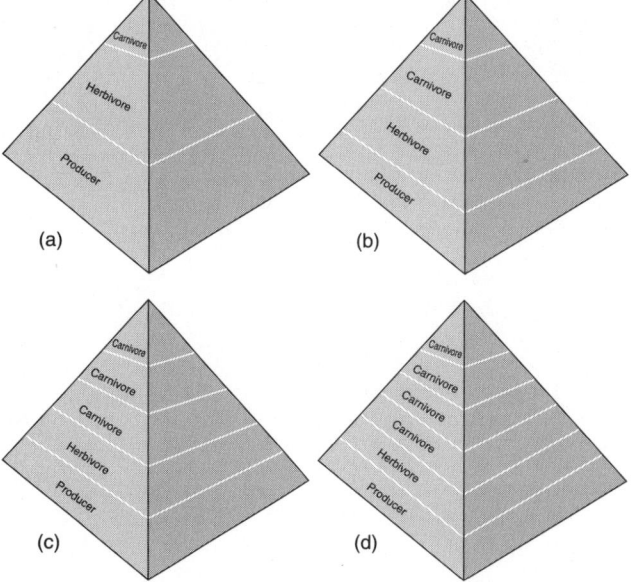

Practice Test 1   **277**

A. pyramid *a*   C. pyramid *c*
B. pyramid *b*   D. pyramid *d*

10. The most important limiting factor for organisms in a desert would be
    A. predators   C. water
    B. temperature   D. wind

11. Which kingdom contains only prokaryotic organisms?
    A. Protista   C. Fungi
    B. Monera   D. Plantae

12. Which statement is true about a chromosomal disorder?
    A. It never affects an organism's genotype.
    B. It always produces a recessive genotype.
    C. It may affect an organism's phenotype.
    D. It never affects an organism's phenotype.

13. A change in a species' genetic code that may affect its survival can result from
    A. a struggle for food
    B. competition
    C. a lower birth rate
    D. a mutation

14. The segment of DNA that determines a specific trait of an organism is called a
    A. homolog   C. gene
    B. chromatid   D. factor

15. The Principle of Independent Assortment states that genes for different traits
    A. exchange alleles during gamete formation
    B. separate during gamete formation
    C. stay together during gamete formation
    D. separate after gamete formation

End of Part 1

You may check your work on this part only.

Do not go to the next page.

PART 2

Directions for Questions 16 through 30 For each of the questions or incomplete statements below, choose the best of the answer choices given. Write your answers on a separate sheet of paper.

16. Two black mice mated and produced 24 black and 8 white offspring. Black hair is dominant to white hair. What can you determine about the genes of both parents?
    A. Both parents are homozygous dominant.
    B. Both parents are homozygous recessive.
    C. Both parents are heterozygous.
    D. One parent is heterozygous; the other is homozygous dominant.

17. Which statement concerning amino acids is correct?
    A. They are the waste products of protein synthesis.
    B. They are the molecular building blocks of starch.
    C. They are stored as fat molecules in the human liver.
    D. They are the molecular building blocks of proteins.

18. The diagram below represents the evolutionary relationships of several species. Which statement about these species is correct?

    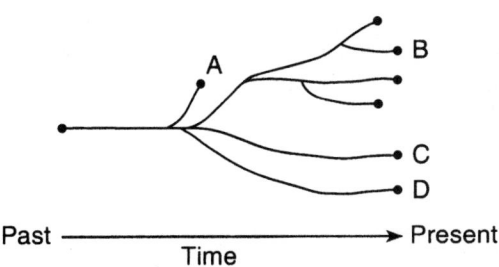

    A. Species *A, B, C,* and *D* came from totally different ancestors.
    B. Species *B* evolved directly from ancient species *A*.
    C. Species *A, B,* and *C* can now interbreed successfully.
    D. Species *C* is more closely related to species *D* than to *B*.

19. A brown mink and a silver mink mated and produced brown offspring. When the brown offspring were crossed with each other, the phenotypic ratio of their offspring was 3 brown to 1 silver. These results are best explained by
    A. independent assortment and crossing-over
    B. dominance, segregation, and recombination
    C. codominance, segregation, and recombination
    D. incomplete dominance inheritance

20. Which process results in two new cells, each with half the number of chromosomes as the original parent cell?
    A. meiosis   C. development
    B. fertilization   D. mitosis

21. A certain organic compound contains only the elements carbon, hydrogen, and oxygen in a ration of 1 : 2 : 1. This compound is probably a
    A. nucleic acid   C. protein
    B. monosaccharide   D. fatty acid

22. ATP is a compound that is synthesized when
    A. chemical bonds are formed between carbon atoms during photosynthesis
    B. energy stored in chemical bonds is released during cellular respiration
    C. energy stored in nitrogen is released, forming proteins
    D. digestive enzymes break down DNA into smaller parts

23. Haploid gametes are produced by the process of
    A. meiosis
    B. mitosis
    C. fertilization
    D. fission

24. Biologists have determined that the cell membrane is composed mostly of
    A. proteins and starch
    B. proteins and cellulose
    C. lipids and starch
    D. lipids and proteins

25. Which end product of respiration is of the greatest benefit to an organism?
    A. glucose molecules
    B. carbon dioxide
    C. ATP molecules
    D. water molecules

Base your answer to question 26 on the information and graph below.

The graph shows the relative amounts of product formed by the action of an enzyme in a solution with a pH of 6 at different temperatures.

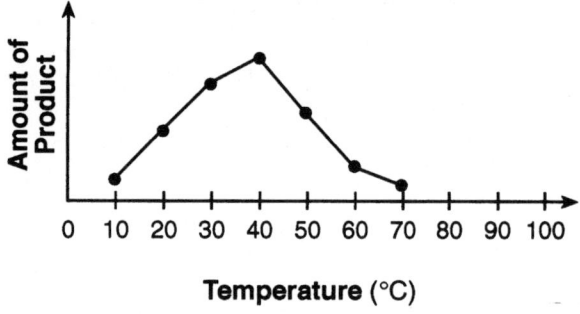

26. Based on the data given, you could predict that if the experiment is repeated at a pH of 4, the amount of product formed at each temperature will be
    A. equal to the amount produced at pH 6
    B. greater than the amount produced at pH 6
    C. less than the amount produced at pH 6
    D. impossible to predict based on this graph

27. What feature do chloroplasts and mitochondria have in common?
    A. they produce carbohydrates
    B. they contain their own DNA
    C. they recycle cellular materials
    D. they give structure to the cell

28. What type of energy conversion occurs during photosynthesis?
    A. light energy into heat energy
    B. chemical energy into light energy
    C. light energy into chemical energy
    D. heat energy into light energy

29. Which change would increase the rate of photosynthesis in a plant?
    A. a decrease in the amount of CO_2 available
    B. an increase in the amount of CO_2 available
    C. a decrease in the amount of H_2O available
    D. an increase in the amount of O_2 available

30. To which group of organic compounds do starch and glycogen belong?
    A. peptides
    B. polypeptides
    C. disaccharides
    D. polysaccharides

End of Part 2

You may check your work on this part only.

Do not go to the next page.

PART 3

Directions for Questions 31 through 45 For each of the questions or incomplete statements below, choose the best of the answer choices given. Write your answers on a separate sheet of paper.

31. A cell that is in the process of aerobic respiration
    A. uses less oxygen than in anaerobic respiration
    B. produces more ATP than in anaerobic respiration
    C. uses less carbon dioxide than in anaerobic respiration
    D. produces more alcohol than in anaerobic respiration

32. Which word equation represents the process of photosynthesis?
    A. carbon dioxide + water → glucose + oxygen
    B. glucose → alcohol + carbon dioxide
    C. carbon dioxide + water → maltose + glucose
    D. glucose + oxygen → carbon dioxide + water

33. What type of organic molecule is the preferred energy source for a cell?
    A. lipids
    B. proteins
    C. carbohydrates
    D. nucleic acids

34. The figure below could be used to illustrate all of the following types of transport *except*

    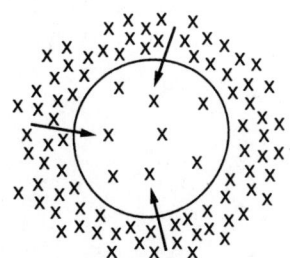
    Figure A Substance moving into

    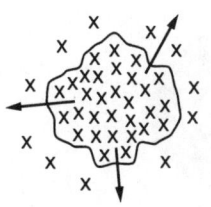
    Figure B Substance moving out

    A. diffusion
    B. osmosis
    C. active transport
    D. passive transport

35. The complementary nitrogenous base for thymine on a strand of DNA would be
    A. adenine   C. guanine
    B. cytosine   D. uracil

36. All the organelles in a cell work together to carry out
    A. passive transport
    B. active transport
    C. information storage
    D. metabolic processes

37. While viewing a slide of rapidly moving sperm cells, a student concludes that these cells probably have many organelles called
    A. vacuoles   C. chloroplasts
    B. ribosomes   D. mitochondria

38. Using a microscope, a botanist observed a plant cell in a drop of water. She added a 20 percent salt solution to the slide and then observed it again. She drew a diagram of the plant cell before (*A*) and after (*B*) the salt water was added. The plant cell looks different in diagrams *A* and *B* because

    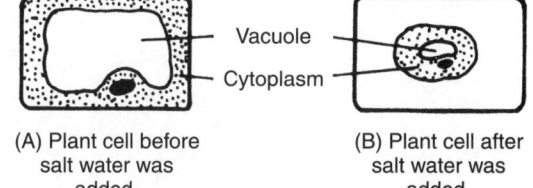
    (A) Plant cell before salt water was added.
    (B) Plant cell after salt water was added.

    A. more salt flowed out of the cell than into the cell
    B. more salt flowed into the cell than out of the cell

Practice Test 1   **281**

C. more water flowed into the cell than out of the cell
D. more water flowed out of the cell than into the cell

39. The type of cell that contains a membrane-bound nucleus is a
   A. prokaryotic cell   C. viral cell
   B. bacterial cell   D. eukaryotic cell

40. Molecules will move across a cell membrane until their concentration
   A. is hypotonic inside
   B. is hypertonic inside
   C. reaches equilibrium
   D. causes plasmolysis

41. The transport of glucose across a cell membrane requires
   A. cellular energy
   B. carbohydrates
   C. active transport
   D. a carrier protein

42. Which structures in the figure below enable an observer to identify it as a plant cell?

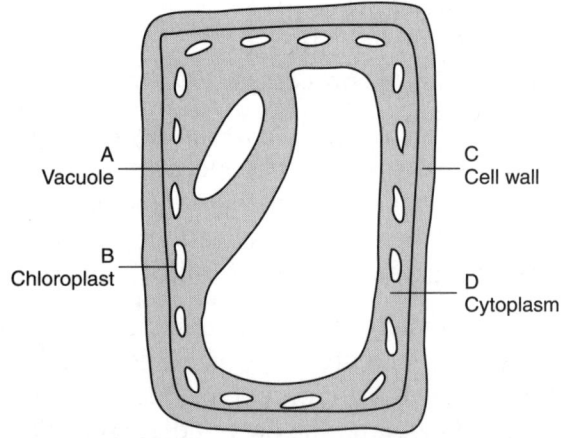

A. *A* and *B*   C. *A* and *C*
B. *B* and *C*   D. *B* and *D*

43. A cell is put into a solution that has a higher concentration of dissolved particles than does the inside of the cell. Relative to the cell, the outside solution is
   A. polytonic   C. hypotonic
   B. isotonic   D. hypertonic

44. During protein synthesis, the amino acid sequence is determined by the
   A. speed at which translation occurs
   B. number of mitochondria in the cell
   C. genetic code in the DNA molecule
   D. number of ribosomes in the cell

45. When you finish your lab experiment, you notice that you have extra chemicals. These left-over chemicals should be
   A. disposed of in the wastebasket
   B. returned to their original bottles
   C. slowly flushed down the sink drain
   D. disposed of as instructed by your teacher

Directions for Question 46 Respond fully to the open-ended question that follows. Use one or more complete sentences to answer the question. Write your answer on a separate sheet of paper.

46. Enzyme molecules are affected by changes in conditions within organisms. Explain how a prolonged, excessively high body temperature during an illness could be fatal to a person. Your answer must include:
   • the role of enzymes in the human body;
   • the effect of this high body temperature on enzyme activity;
   • the reason this high body temperature can result in death.

End of Part 3

You may check your work on this part only.

PRACTICE TEST 2

PART 1

Directions for Questions 1 through 15 For each of the questions or incomplete statements below, choose the best of the answer choices given. Write your answers on a separate sheet of paper.

1. Five identical bean plants were planted in five different types of soil. Each plant received the same amount of sunlight and water. Four of the plants continued to grow. One plant failed to grow. Which statement is a valid conclusion that can be drawn from the results of this experiment?
   A. All five different types of soil support plant growth.
   B. The amount of sunlight was different for each plant.
   C. Soil type is not as important as the amount of water.
   D. One type of soil inhibited the growth of a bean plant.

2. According to the graph below, about how many cells per milliliter (mL) of culture would be present after 90 minutes?

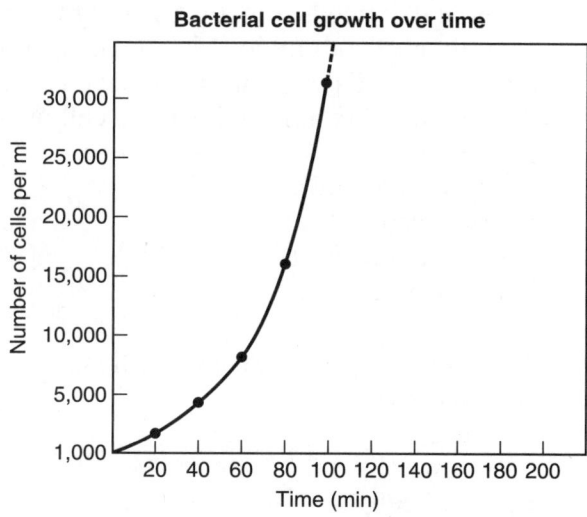

   A. 12,000
   B. 15,000
   C. 22,000
   D. 29,000

3. To test the effect of an antibiotic on bacterial growth, scientists grew six cultures of the same strain to a concentration of 1000 cells per mL. When the cultures reached this concentration, the same concentration of antibiotic was added to each culture. After two days of growth at 37°C, the cell concentrations in each culture were measured. The proper control for this experiment would be to
   A. add a different concentration of antibiotic to each culture
   B. add the same antibiotic concentration to six cultures of the same strain and incubate them at room temperature
   C. grow another six cultures of the same strain under the same conditions and then add sterile water only
   D. measure the cell concentration of just one of the cultures after one day

4. The graph below shows the world's fish catch in millions of metric tons (mmt) from years 1950–1990. According to the graph, over that period of time the world's catch increased approximately

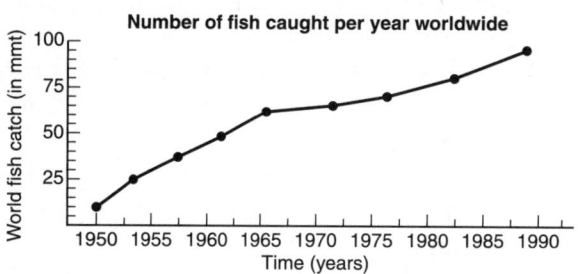

   A. three times
   B. five times
   C. seven times
   D. nine times

5. Andy conducted a science experiment and was upset that the results from the experiment did not support his original hypothesis. Andy should

A. perform a different experiment to support his hypothesis
B. ignore the results and redo the same experiment
C. repeat the experiment and modify his hypothesis if necessary
D. publish the results of the experiment without repeating it

6. Rachel knows that her cat and her friend's cat are both afraid of lightning. She concludes that all cats are afraid of lightning. Rachel's conclusion is based on
   A. a controlled experiment
   B. observation and reasoning
   C. scientific research
   D. emotional intuition

Base you answer to question 7 on the information and graph below.

The graph shows the relative amounts of product formed by the action of an enzyme in a solution with a pH of 6 at different temperatures.

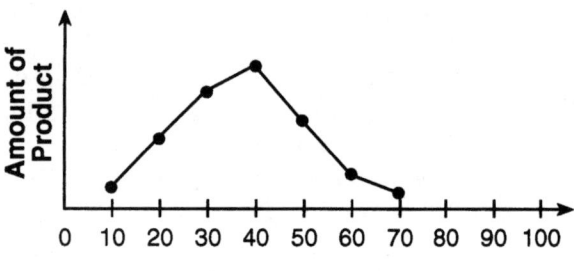

7. Based on the data given, you could predict that if the experiment is repeated at a pH of 8, the amount of product formed at each temperature will be
   A. equal to the amount produced at pH 6
   B. greater than the amount produced at pH 6
   C. less than the amount produced at pH 6
   D. impossible to predict based on this graph

8. A plant cell is placed on a slide with salt water. After several minutes, the cell shrinks due to the diffusion of
   A. salt from the plant cell to the environment
   B. salt from the environment into the plant cell
   C. water from the plant cell to the environment
   D. water from the environment into the plant cell

9. When placed in an isotonic solution, a cell's volume will
   A. stay the same
   B. swell
   C. shrink
   D. shrink, and then swell

10. The movement of molecules into a cell is most dependent on the
    A. selectivity of the cell membrane
    B. selectivity of the cell wall
    C. number of vacuoles in the cell
    D. number of mitochondria

11. Which substance would be *least* likely to pass through a cell membrane by diffusion?
    A. carbon dioxide
    B. glucose
    C. oxygen
    D. water

12. ATP is essential for the process of
    A. passive transport
    B. active transport
    C. facilitated diffusion
    D. osmotic diffusion

13. A nerve cell needs to maintain a higher concentration of sodium ions inside the cell than outside of it. The sodium ions move across the cell membrane into the nerve cell by a process called
    A. facilitated diffusion
    B. passive transport
    C. active transport
    D. diffusion

14. An example of a prokaryotic cell is a
    A. bacterium
    B. virus
    C. yeast cell
    D. plant cell

15. The following pie chart shows the proportion of elements in the human body. Which conclusion is best supported by this chart?

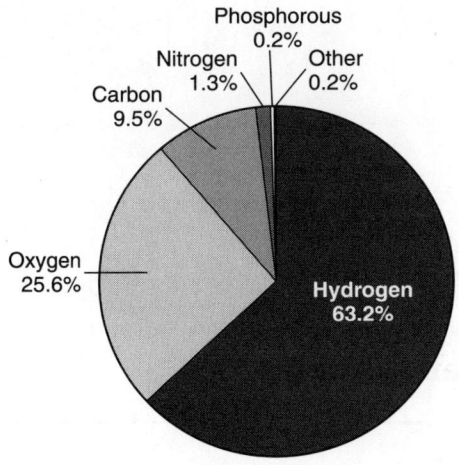

A. The human body is composed of six different elements.
B. Oxygen and carbon are the most abundant elements in the human body.
C. There is more nitrogen than phosphorus in the human body.
D. Hydrogen has a greater mass than carbon in the human body.

End of Part 1

You may check your work on this part only.

Do not go to the next page.

PART 2

Directions for Questions 16 through 30 For each of the questions or incomplete statements below, choose the best of the answer choices given. Write your answers on a separate sheet of paper.

16. Which of these cells do *not* have cell walls?
    A. plant cells
    B. fungal cells
    C. bacterial cells
    D. animal cells

17. The cellular organelle that functions in the intracellular transport of cellular molecules is the
    A. endoplasmic reticulum
    B. cell membrane
    C. ribosome
    D. cell wall

18. The substances that most directly control the rate of reaction during cellular respiration are known as
    A. enzymes
    B. phosphates
    C. monosaccharides
    D. disaccharides

19. Which process forms carbon dioxide and water as waste products?
    A. digestion
    B. protein synthesis
    C. cellular respiration
    D. photosynthesis

20. Where do the final steps of aerobic cellular respiration occur?
    A. along the endoplasmic reticulum
    B. throughout the cytoplasm
    C. on the surface of the ribosomes
    D. inside the mitochondria

21. In animal cells, the energy used to convert ADP to ATP comes directly from
    A. hormones
    B. sunlight
    C. organic molecules
    D. inorganic molecules

22. What determines the specific shape of a protein molecule?
    A. whether it is organic or inorganic
    B. the sequence of its amino acids
    C. the number of lipids inside the protein
    D. the number of chromosomes in the cell

23. Which molecule stores the largest amount of energy?
    A. fat
    B. protein
    C. starch
    D. glycogen

24. Which substance stores glucose in the human liver?
    A. chitin
    B. starch
    C. glycogen
    D. cellulose

25. The process shown in the following diagram results in the formation of
    A. diploid cells
    B. haploid cells
    C. sex cells
    D. gametes

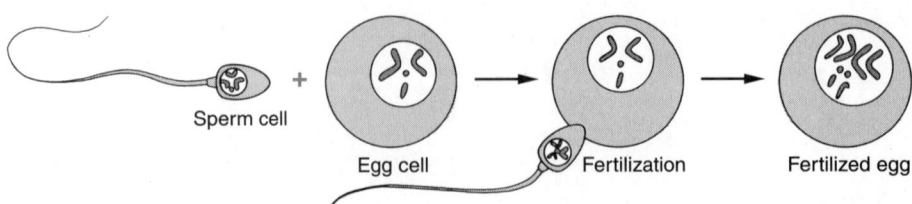

26. Before a cell divides, its DNA and associated proteins condense to form the
    A. chromosomes
    B. genes
    C. centrioles
    D. centromeres

27. Of the events listed below, which mitotic event happens last?
    A. the replication of chromosomes
    B. disintegration of the nuclear membrane
    C. appearance of the spindle fibers
    D. separation of the chromatids by the spindles

28. Bacteria reproduce by means of the process called
    A. mitosis
    B. cytokinesis
    C. binary fission
    D. conjugation

29. Clones are produced during the process of
    A. spermatogenesis
    B. gametogenesis
    C. asexual reproduction
    D. sexual reproduction

30. DNA transcription takes place inside a cell's
    A. nucleus
    B. vacuole
    C. ribosome
    D. lysosome

End of Part 2

You may check your work on this part only.

Do not go to the next page.

PART 3

Directions for Questions 31 through 45 For each of the questions or incomplete statements below, choose the best of the answer choices given. Write your answers on a separate sheet of paper.

31. RNA receives information from DNA by
    A. binding with the DNA and forming a triple helix
    B. matching with the complementary bases of a DNA strand
    C. making an exact copy of the DNA molecule
    D. accepting proteins through the nuclear membrane pores

32. The process that involves the pairing of a codon with an anticodon on a tRNA is called
    A. transcription
    B. synthesis
    C. replication
    D. translation

33. In the following diagram, what substance is represented by the letter *X*?

    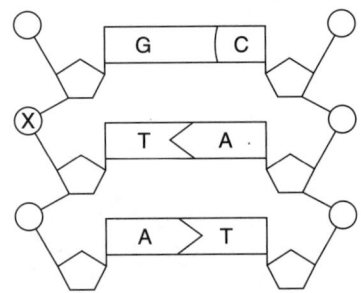

    A. hydrogen
    B. deoxyribose
    C. phosphate
    D. thymine

34. A DNA nucleotide unit consists of a
    A. phosphate group, a six-carbon sugar, and a hydrogen base
    B. phosphate group, a five-carbon sugar, and a nitrogenous base
    C. phosphate group, a six-carbon sugar, and a nitrogenous base
    D. hydrogen group, a five-carbon sugar, and a nitrogenous base

35. One way in which RNA differs from DNA is that
    A. RNA is double stranded, while DNA is single stranded
    B. RNA contains thymine, while DNA contains uracil
    C. RNA is found only in the nucleus of eukaryotes
    D. RNA contains ribose, while DNA contains deoxyribose

Base your answers to questions 36 and 37 on the graph below.

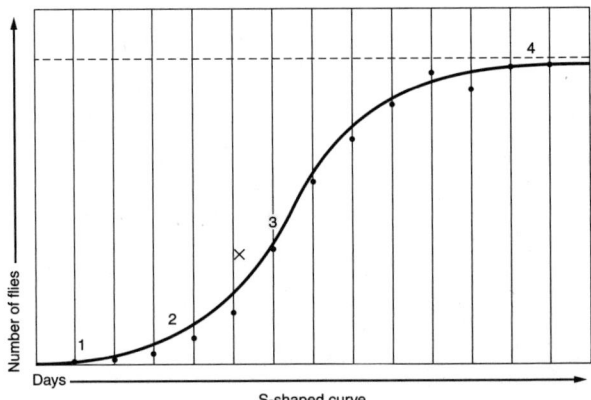

S-shaped curve.

36. Which number represents the population in its most rapid phase of growth?
    A. 1   C. 3
    B. 2   D. 4

37. At which point has the population reached the carrying capacity?
    A. 1   C. 3
    B. 2   D. 4

38. The changes that occur in an ecosystem over a long period of time are called
    A. natural selections
    B. geological succession
    C. ecological succession
    D. biological diversity

39. Which set of body parts represents homologous structures?
    A. wings of an insect and wings of a hummingbird
    B. tentacles of an octopus and flippers of a whale
    C. front legs of a butterfly and leg bones of a human
    D. front leg bones of a dog and wing bones of a bat

40. The type of organism *not* included in the pyramid below is the

    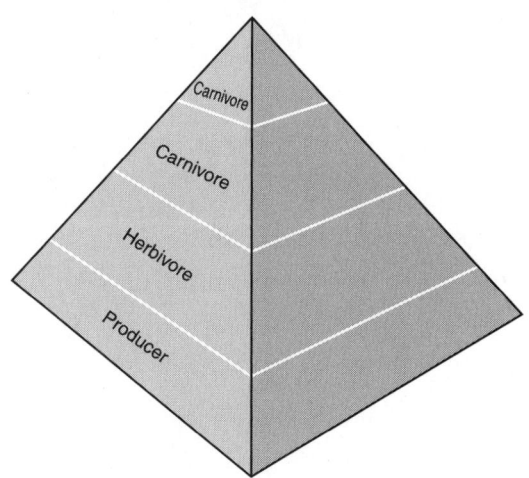

    A. plant or alga
    B. plant eater
    C. meat eater
    D. decomposer

41. Which organisms in the food web below occupy similar trophic levels?

    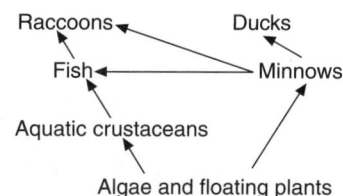

    A. algae and minnows
    B. fish and crustaceans
    C. raccoons and minnows
    D. minnows and crustaceans

42. A population of squirrels is divided in two by a large valley. Over time, they become so different that they can no longer interbreed. This change in the two groups is due to
    A. hybridization
    B. fossilization
    C. geographic isolation
    D. primary succession

43. A current idea in the field of classification divides life into three broad categories, as shown below. Which concept is supported by this diagram?

    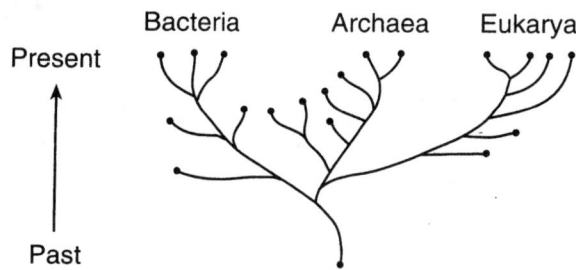

    A. All evolutionary pathways eventually lead to present-day organisms.
    B. Evolutionary pathways proceed in only one set direction over time.
    C. All evolutionary pathways continue for the same total length of time.
    D. Evolutionary pathways diverge and proceed in many different directions.

44. Some concepts included in Darwin's theory of natural selection are represented in the chart below. Which concept would be correctly placed in box *X*?

    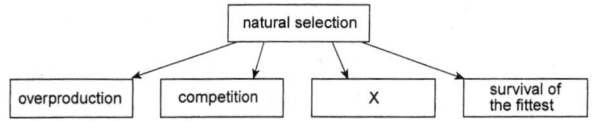

    A. use and disuse
    B. variation in traits
    C. changes in nucleic acids
    D. inheriting acquired traits

45. Two nucleotide sequences found in two different species are almost exactly the same. This fact suggests that these two species probably
    A. are evolving into the same species
    B. are members of one large gene pool
    C. share a common evolutionary history
    D. have the same number of mutations

Directions for Question 46 Respond fully to the open-ended question that follows. Use one or more complete sentences to answer the question. Write your answer on a separate sheet of paper.

46. Just like complex organisms, cells are able to survive by coordinating various activities. Complex organisms have a variety of systems, and cells have a variety of organelles that work together for survival. Describe the roles of two organelles. In your answer be sure to include:
    - the names of two organelles and the function of each;
    - an explanation of how these two organelles work together;
    - the name of an organelle and the name of a system in the human body that have similar functions.

End of Part 3

You may check your work on this part only.

Glossary

A

abiotic describes the nonliving factors in an organism's environment, e.g., soil and water

acid rain a type of far-ranging air pollution that forms when sulfur dioxide and nitrogen oxides combine with water droplets in the atmosphere; harms forests and lakes

active transport movement of substances across a membrane from an area of lower concentration to an area of higher concentration; requires energy

adaptations special characteristics that make an organism well suited for a particular environment

AIDS (acquired immunodeficiency syndrome) an immunodeficiency disease, caused by HIV in humans

algae plantlike organisms, often single-celled, that carry out photosynthesis

alleles the two different versions of a gene for a particular trait

allergic reactions conditions caused by an overreaction of the immune system

alternation of generations the life cycle of plants that includes sexually reproducing and asexually reproducing generations

amino acids organic compounds that are the building blocks (subunits) of proteins

antibiotics chemicals that kill specific microorganisms (usually bacteria); frequently used to combat infectious diseases in people and animals

antibodies molecules that the immune system produces to defend against foreign objects in the body; antibodies bind to specific antigens

antigens protein molecules on the surface of a foreign object that are detected by the immune system, causing it to produce antibodies

antihistamines drugs that are used to treat allergies because they stop the release of histamines

artificial selection see *selective breeding*

asexual reproduction the process that requires only one parent to pass on genetic information, e.g., budding

atmosphere the blanket of gases that covers Earth; usually called "air"

atoms the smallest units of an element that can combine with other elements; make up all matter

ATP (adenosine triphosphate) the substance (derived from energy stored in glucose) used by cells as an immediate source of chemical energy

autoimmune diseases occur when an overactive immune system starts to attack its own normal body tissues

autotrophic describes a self-feeding organism that obtains its energy from inorganic sources, e.g., plants (producers)

B

bacteria single-celled organisms that have no nuclear membrane to contain their DNA molecule

bacteriophages refers to viruses that infect, and destroy, bacterial cells

B cells special kinds of white blood cells that respond to specific antigens by producing antibody proteins that bind to it; include memory B cells

behavior every action that an animal takes, either learned or instinctive; usually to aid survival

binomial nomenclature the scientific system for naming organisms using two-word Latin terms

bioaccumulation occurs when the level of a chemical in each organism increases, and becomes more harmful, as it moves up the food chain

biochemistry the chemistry of living things; can provide evidence of common ancestry (evolution)

biodiversity the variety of different species in a community or ecosystem

biology the study of living things

biomes very large geographic areas that are characterized by a certain climate and ecosystem

biosphere the total area of land, water, and air on Earth's surface where life is found

biotechnology describes new procedures and devices that utilize discoveries in biology; see also *recombinant DNA technology* and *genetic engineering*

biotic describes the other living factors in an organism's environment

budding a form of asexual reproduction in which the offspring grows out of the side of the parent

C

cancer a disease that results from uncontrolled cell division, which damages normal tissues

carbohydrates a group of organic compounds (includes simple sugars and starch), made up of hydrogen, oxygen, and carbon

carbon one of the six most important chemical elements for living things; carbon atoms form the backbone of nearly all organic compounds

carbon dioxide the inorganic molecule from which plants get carbon for photosynthesis; waste product of cellular respiration; a greenhouse gas

carnivores animals that obtain their energy by eating other animals; see also *consumers* and *heterotrophic*

carrying capacity the size of a population that can be supported by an ecosystem

catalysts substances that increase the rate of a chemical reaction, but are not changed during the reaction

cell the smallest living unit of an organism; all organisms are made up of at least one cell

cell membrane a selectively permeable membrane that separates the inside and outside of a cell and regulates the substances that pass through it; see also *plasma membrane*

cell theory the idea that all organisms are made up of one or more cells, which are the basic unit of structure and function, and which come only from other cells

cellular respiration the process that uses oxygen to create ATP for energy use by cells

central nervous system the brain and nerve (spinal) cord

centrifuge equipment used to separate materials of different densities from one another

cerebellum below and behind the cerebrum, it processes sensory information and coordinates body movements

cerebrum the most highly developed, and most recently evolved, part of the human brain

chloroplasts the organelles in plants and algae that contain chlorophyll pigment and carry out photosynthesis

chromosomes structures composed of DNA that contain the genetic (hereditary) material

cilia in protists, short hairlike organelles that cover their outer surface, enabling locomotion

circulation the movement of blood throughout the body of an animal

cladistics an approach to classifying organisms based on the specific characteristics of a group

classification the grouping and naming of organisms according to their evolutionary relationships and shared characteristics

cloning the production of identical genetic copies by asexual reproduction (or from the cell of another individual)

closed circulatory system materials and wastes are dissolved in blood within a system of vessels

codon a sequence of three nucleotide bases that makes up the "word" coding for an amino acid

coevolution the process by which species change in relation to changes in other species

community all the populations of different species that interact within a particular area

competition the struggle between organisms for limited resources such as food and space

compound light microscope tool that uses two lens types to magnify the image of a specimen

conditioned response a type of behavior that occurs when an animal learns to respond to a particular stimulus, such as pet fish swimming to the top of a tank at feeding time

consumers organisms that obtain their energy by feeding on other organisms; heterotrophs

coordination the means by which body systems work together to maintain homeostasis

cyclosis the natural streaming of cytoplasm that occurs, and moves materials, within all cells

cytoplasm the watery fluid (gel) that fills a cell, surrounding its organelles

D

decomposers heterotrophic organisms that obtain their energy by feeding on decaying organisms

deforestation the cutting down and clearing away of forests; clear-cutting

density the number of individuals in a population in a given area; affects population growth

dependent variable in an experiment, the change that occurs because of the independent variable

deplete (depletion) to use up natural resources that cannot be replaced within our lifetimes

development the changes in an organism that occur from fertilization until death

deviations changes in the body's normal functions that are detected by control mechanisms, which maintain a balanced internal environment

dichotomous key a classification tool composed of a list of observable, alternative characteristics that leads, step-by-step, to the correct identification of an organism

differentiation the formation of specialized cells from less specialized parent cells through controlled gene expression

diffusion the movement of molecules from an area of higher concentration to an area of lower concentration

digestion the process of breaking down food particles into molecules small enough to be absorbed by cells

disease a disruption of homeostasis; a condition in a living body that impairs normal functioning

diversity the variety of different traits in a species or different species in an ecosystem

DNA (deoxyribonucleic acid) the nucleic acid molecules that contain the hereditary material and instructions for all cellular activities in all organisms

dominant the allele that is expressed, i.e., that shows its trait

dominant homozygous describes when a genotype has the two dominant alleles for a trait

dynamic equilibrium in the body, a state of homeostasis in which conditions fluctuate yet always stay within certain limits

E

ecology the biological study of the interactions of living things and their environment

ecosystem includes all the living and nonliving factors that interact in a specific area

egg the female gamete that supplies half the genetic information to the zygote

embryo an organism in an early stage of development before it is hatched, born, or germinated

endoskeleton in vertebrates, the internal skeleton (for support and movement)

energy what is required continuously by all living things to stay organized and remain alive

energy pyramid describes the flow of energy through an ecosystem; most energy is at the base (producers) and decreases at each higher level (consumers)

environment the physical surroundings of an organism, with which it interacts

enzymes proteins that act as catalysts to increase the rate of chemical reactions in living things

erosion refers to the loss of soil; can be caused by exposure to wind and rain

estivation behavior in which an animal burrows in the ground to sleep during the hot, dry months

eukaryotes refers to cells that contain a nucleus and membrane-bound organelles

evolution the change in organisms over time due to natural selection acting on genetic variations that enable them to adapt to changing environments

excretion the process by which metabolic wastes are removed from the body

exoskeleton in some invertebrates, the external skeleton (for support and movement)

expression the use of genetic information in a gene to produce a particular characteristic, which can be modified by interactions with the environment

external development occurs when an embryo develops in the outside environment, not in the female's body

external fertilization occurs when the sperm and egg nuclei fuse in the outside environment, not in the female's body

extinct describes when a group of organisms is no longer alive, e.g., all dinosaurs

extinction the complete disappearance of all living members of a species (or group of species)

F

feedback mechanism a system that reverses an original response to a stimulus when the desired condition is reached; important for homeostasis

fertilization in sexual reproduction, process by which the nuclei of an egg cell and a sperm cell unite to form a zygote

fetus a developing embryo after the first three months of development

fieldwork the scientific study of living things in their natural habitat

flagella in some protists and algae, long hairlike organelles that can pull the cell through the water

flukes parasitic flatworms that cause intestinal, urinary, and blood infections

food chain the direct transfer of energy from one organism to the next

food web the complex, interconnecting food chains in a community

fossils the traces or remains of dead organisms that have been preserved by natural processes

fungi (*singular,* fungus) the kingdom of heterotrophic organisms that obtain their energy by feeding on decaying organisms, e.g., yeast and mushrooms

G

gametes the male and female sex cells that combine to form a zygote during fertilization

ganglia in the earthworm, the fused nerve cells that make up its brain; connect to the nerve cord

gel electrophoresis technique by which DNA fragments are separated to compare DNA from one person with that of another

genes the segments of DNA that contain the genetic information for a given trait or protein

gene expression see *expression*

genetic code the triplet code; i.e., the sequences of three nucleotide bases (codons), which code for specific amino acids

genetic engineering recombinant DNA technology, i.e., the insertion of genes from one organism into the genetic material of another; see also *biotechnology*

genetic variation the differences among offspring in their genetic makeup, due to recombination

genome all the genes that are present in an organism

genomics the detailed study of the genetic material and particular traits of different organisms

genotype the genetic combination that determines a trait

genus a group that has one or more closely related, different species classified within it

geographic isolation the physical separation of related populations, by a natural barrier such as a river or mountain, that prevents interbreeding and leads to speciation

geologic time Earth's history divided into vast units of time by which scientists mark important changes in Earth's climate, surface, and life-forms

global warming an increase in the average atmospheric temperature of Earth due to more heat-trapping CO_2 in the air, which causes the "greenhouse effect"

glucose a simple sugar that has six carbon atoms bonded together; a subunit of complex carbohydrates

H

habit a common type of behavior that results from repeating an action over and over again

habitat the place in which an organism lives; a specific environment and its living community

habituation a simple learned behavior in which an organism learns to stop responding to an unimportant stimulus

herbivores animals that obtain their energy by eating plants; see also *consumers* and *heterotrophic*

hereditary describes the genetic information that is passed from parents to offspring

heterotrophic describes an organism that obtains its energy by feeding on other living things, e.g., animals (consumers)

heterozygous describes when the two alleles for a trait are different, i.e., hybrid

hibernation process by which animals slow their metabolic functions and sleep during the winter

homeostasis the ability of living things to detect external changes and to maintain a constant internal environment

homozygous describes when the two alleles for a trait are the same

hormones chemical messengers released into the blood, which bind with receptor proteins to cause long-lasting changes in the body

host the organism that a parasite uses for food and shelter; the cell in which a virus reproduces

Human Genome Project worldwide effort to analyze the three billion base pairs of human DNA
hydrogen one of the six most important chemical elements for living things

I
immune system recognizes and attacks specific invaders, e.g., bacteria, to protect the body against infection and disease
immunity the ability to resist or prevent infection by a particular microbe
immunodeficiency disease occurs when the body's immune system is underactive because it has been weakened, e.g., by HIV
imprinting a type of learned behavior that occurs shortly after birth, e.g., when a baby goose follows the first moving object it sees
incomplete dominance when the heterozygous offspring shows an intermediate phenotype for a trait
independent variable in an experiment, the one factor that might explain the observation
infectious diseases refers to diseases caused by pathogenic microorganisms that may be passed from one person to another; see also *disease*
inflammation the second line of defense against infection, in which the affected area becomes warm, reddened, and aided by white blood cells
inheritance the process by which traits are passed from one generation to the next
inorganic in cells, substances that allow chemical reactions to take place; in ecosystems, substances that are cycled between living things and the environment
innate behavior refers to behavior that is inborn and does not have to be learned
instincts complex, inborn behaviors that aid survival, e.g., reflexes and fixed action patterns
insulin substance secreted by the pancreas that maintains normal blood sugar levels
intercellular fluid (ICF) the fluid that surrounds all cells and contains dissolved substances
internal development occurs when the embryo develops within the female's body
internal fertilization occurs when the sperm fertilizes the egg cell within the female's body

K
karyotpye a photograph showing all 23 chromosome pairs from a human fetal cell
kingdom the largest taxonomic group into which scientists categorize related living things

L
life functions the basic processes carried out by all living things
life span the length of time between the birth and death of an organism
limiting factors the specific environmental requirements that can limit where an organism lives
linked occurs when the genes for one trait are inherited along with genes for another trait, e.g., red hair and freckles, because they are on the same pair of chromosomes
lipids the group of organic compounds that includes fats and oils; store energy for long-term use
lysosomes organelles involved in breaking down food for the cell

M
malfunction occurs when an organ or body system stops functioning properly, which may lead to disease or death
Malpighian tubules the excretory organs of the grasshopper
medulla oblongata the lowest part of the brain stem, it controls involuntary activities
meiosis the division of one parent cell into four daughter cells; reduces the number of chromosomes to one-half the normal number, i.e., produces gametes
membrane see *cell membrane*
metabolism the chemical reactions (building up; breaking down) that take place in an organism
microbes microscopic organisms that may cause disease when they invade another organism's body; microorganisms, e.g., bacteria and viruses
migration a form of social behavior in which groups of animals travel long distances together to feed and to breed
mitochondria the organelles in which the cell's energy is released
mitosis the division of one cell's nucleus (i.e., the chromosomes) into two identical daughter cell nuclei

molecules the smallest unit of a compound; made up of atoms held together by a chemical bond

movement the flow of materials between the cell and its environment, i.e., transport; a property of living things, i.e., locomotion

multicellular describes organisms that are made up of more than one cell

mutation an error in the linear sequence (gene) of a DNA molecule; a change in genetic material

N

natural selection the process by which organisms having the most adaptive traits for an environment are more likely to survive and reproduce

nerve cells in animals, the cells (neurons) that receive and send nerve impulses to other nerve cells and to other types of cells

nerve net in the hydra, a network of modified neurons that transmits the nerve impulses

niche an organism's specific role in, or interaction with, its ecosystem; how an organism survives

nitrogen one of the six most important chemical elements for living things

nucleic acids the organic compounds that include DNA and RNA

nucleotides the building blocks, or subunits, of DNA and RNA; they include four types of nitrogen bases, which occur in two pairs

nucleus the dense region of a (eukaryotic) cell that contains the genetic material

nutrients important molecules in food, such as lipids, proteins, and vitamins

nutrition the life processes by which organisms take in and utilize nutrients

O

observation what is made when you notice a natural event; part of the scientific method

open circulatory system materials and wastes pass into large cavities, or sinuses, and bathe tissues

organ a structure made up of similar tissues that work together to perform the same task, e.g., the liver; describes a level of organization in living things

organelles structures within a cell that perform a particular task, e.g., the vacuole

organic describes compounds that contain carbon and hydrogen

organic compounds describes compounds that contain carbon and hydrogen (in living things)

organisms living things; life-forms

organ system a group of organs that work together to perform a major task, e.g., the digestive system

osmosis the diffusion of water molecules across a membrane

ovaries the female reproductive organs that produce the mature egg cells

oxygen one of the six most important chemical elements for living things; released as a result of photosynthesis; essential to cellular (aerobic) respiration

ozone shield the layer of ozone gas that surrounds Earth high in the atmosphere and blocks out harmful ultraviolet (UV) radiation

P

parasites organisms that live in or on another organism (the host), causing it harm

passive transport movement of substances across a membrane; requires no use of energy

pathogens microscopic organisms that cause diseases, such as certain bacteria and viruses; see also *microbes*

pecking order a social behavior that determines the importance of an individual in a group

pedigree a chart that shows the inheritance patterns of certain genetic traits in a family for several generations

peripheral nervous system network of nerves that carries signals between the central nervous system and all parts of the body

pesticides chemicals used to kill agricultural pests, mainly insects, some of which have evolved resistance to the chemicals

pH a measurement (on a scale of 0 to 14) of how acidic or basic a solution is

phenotype the physical appearance of a trait in an organism

pheromones chemicals produced by animals that function in communication

phloem in plants, the vascular tissue that carries dissolved food and other substances from the leaves to the rest of the plant

photosynthesis the process that, in the presence of light energy, produces chemical energy (glucose) and water

placenta the structure that forms in the uterus of mammals to nourish a developing embryo and remove its waste products

plasma membrane separates the inside and outside of a cell; see also *cell membrane*

pollutants harmful substances that are not normally found in the environment

polygenic inheritance describes a trait is determined by more than one gene, such as height

polymer a large molecule, such as blood protein hemoglobin, made up of smaller molecules

population all the individuals of the same species that live in the same area at one time

predation the interaction in which members of one population are the food source for members of another population

predator an organism that feeds on another living organism (the prey); a consumer

predator-prey a relationship which the prey is usually killed right away by the predator

pregnancy in animals, the condition of having a developing embryo within the body

prey a living organism that is eaten by another organism (the predator)

producers organisms on the first trophic level, which obtain their energy from inorganic sources, e.g., by photosynthesis; autotrophic life-forms

prokaryotes refers to cells that do not contain a nucleus or any membrane-bound organelles

proteins a group of organic compounds that are made up of chains of amino acids

protists the kingdom that includes eukaryotic unicellular and multicellular organisms that live in aquatic or moist environments; algae and protozoa

pseudopods in protists, extensions of the cell that engulf food particles and that aid in movement

Punnett square a box-shaped chart that is used to show all possible combinations of gametes and their probabilities when two organisms are crossed

R

radiation a form of energy that can cause genetic mutations in sex cells and body cells

receptors molecules that play an important role in the interactions between cells, e.g., molecules that bind with hormones

recessive the allele that is hidden, i.e., that does not show its trait, unless two copies are present

recessive homozygous describes when a genotype has the two recessive alleles for a trait

recombinant DNA technology describes the methods used to remove and put together (i.e., recombine) genes from one cell to another

recombination the formation of new combinations of genetic material due to crossing-over during meiosis or due to genetic engineering; produces greater genetic variability

recombining during meiosis, the process that causes an increase in genetic variability due to the exchange of material between chromosomes

reduction division during meiosis, diploid cell divides to produce two new cells with haploid number of chromosomes

reflex a type of innate behavior that happens very quickly; aids survival

replicate the process by which DNA makes a copy of itself during cell division and protein synthesis

replication the duplication of a cell's genetic material during its life cycle

reproduction the production of offspring (i.e., passing on of hereditary information), either by sexual or asexual means

reproductive isolation situation in which a population is physically separated from others of its kind and changes so much that it can no longer interbreed with them; leads to speciation

residue the remains of dead organisms, which are recycled in ecosystems by the activities of bacteria and fungi

resistance describes natural (genetic) ability of some bacteria to survive exposure to antibiotics

resource natural materials, living and nonliving, that are valuable to people; e.g., trees and soil

response an organism's reaction to a stimulus; can be inborn or learned

respiration in the lungs, the process of exchanging gases; in cells, the process that releases the chemical energy stored in food; see also *cellular respiration*

restriction enzyme recognizes a specific sequence of four to six base pairs and cuts the DNA

ribosomes the organelles at which protein synthesis occurs, and which contain RNA

RNA (ribonucleic acid) the nucleic acid molecules that function to translate genetic information and carry out protein synthesis

S

sampling a fieldwork technique in which a representative portion of a population is used to determine one specific characteristic of the entire population, such as total numbers

saprophytes organisms that obtain nutrients from dead or decaying organisms, e.g., the fungi

scavenge to gather and feed on the remains of a kill, rather than to hunt living animals

schooling a form of social behavior in which animals, such as fish, move together in large groups

science a body of knowledge about our world

scientific inquiry the nature of scientific thinking and learning

scientific method an organized approach to problem solving

selective breeding the process by which humans encourage the development of preferred traits by allowing only those plants or animals that have those traits to breed

selective permeability the ability of a cell membrane to determine which molecules can pass through it

sense organs the eyes, ears, nose, mouth, and body surface (e.g., skin) of an organism

sensory perception the receiving of information about the environment through the sense organs

setae in the earthworm, the pairs of bristles on most of its segments, which aid locomotion

sex cells the male and female gametes; they have one-half the normal chromosome number as a result of meiosis

sex-linked genes describes traits controlled by genes from the female parent because the genes on the X chromosome have no corresponding alleles on the Y chromosome

sexual reproduction the process that requires two parents (a male and a female) to pass on genetic information

simple sugars single sugars that have six carbon atoms, e.g., glucose

social behavior occurs when animals live in groups, or societies, thus aiding each others survival

solar energy radiant energy from the sun, i.e., sunlight, which is a renewable resource

species a group of similar organisms that can breed and produce fertile offspring

sperm the male gamete that supplies half the genetic information to the zygote

spiracles small holes in the exoskeleton of an insect through which respiratory gases are exchanged

stability the tendency of an ecosystem to resist change and remain healthy; usually, the greater the species diversity, the more stable the ecosystem

starch a complex carbohydrate (polysaccharide) made up of many glucose molecules; used for energy storage in plants

statistical analysis the mathematical process used to determine if experimental results obtained are valid or if they might have been simply due to chance

stem cells human cells that have not yet developed into mature cells; have the ability to become almost any type of tissue

stimulus (*plural,* stimuli) any event, change, or condition in the environment that causes an organism to make a response (i.e., to react)

subunits the four types of nucleotide bases that make up the DNA molecule

succession the gradual replacement of one ecological community by another until reaching a point of stability; usually a series of slow changes

symbiosis a close relationship between two or more different organisms that live together; each partner may either help, harm, or have no effect on the other partner

synthesis the building of compounds that are essential to life, e.g., protein synthesis

system an organized group of structures that works together to perform a task; see also *organ system*

systematics an approach to classifying organisms based on the evolutionary relationships within and between groups

T

tapeworms parasitic flatworms that absorb nutrients from the hosts directly through their bodies

taxonomy the branch of science that deals with the classification of organisms

T cells special kinds of white blood cells that detect and destroy cells that have been infected by microbes (i.e., killer T cells) or assist other B cells and T cells (i.e., helper T cells)

technology the process of using scientific knowledge and other resources to develop new products and processes

template in DNA replication, the original molecule that is used to make a copy

territory the area in which an animal lives, and which it usually defends from other animals

testcross test conducted (cross with homozygous recessive) to determine the true genotype of a plant or animal that shows the dominant phenotype (physical trait)

testes the pair of male reproductive organs that produces the sperm cells

theory a general statement supported by many scientific observations; represents the most logical explanation of the evidence

theory of evolution explains how all organisms have developed and changed over time; see also *evolution*

tissues a group of similar cells that work together to perform the same function; describes a level of organization in living things

toxins chemicals that can harm a developing fetus if taken during pregnancy; also, chemicals that may get passed, and increase, from one trophic level to the next as they move up the food chain

trait a genetic characteristic, determined by the combination of alleles

transcription the process by which a DNA strand is copied into an RNA strand

translation the process by which RNA functions to assemble amino acid sequences at the ribosome

transport involves the absorption and movement of materials throughout an organism's body

trophic levels the feeding levels through which energy flows in a food chain

U

unicellular describes single-celled organisms; i.e., organisms made up of one cell

uterus in mammals, the reproductive organ that holds the developing embryo

V

vaccination method used (usually injection) to prepare the immune system to fight off infection by a specific pathogen

vacuoles the organelles that stores materials, i.e., food and wastes, for the cell

variability see *genetic variation*

vascular refers to specialized conducting tissues that transport materials throughout the bodies of plants and animals

vectors special molecules that can move pieces of DNA from the cell of one organism to another

virulent virus a virus that has the ability to cause disease in its host

virus a microscopic, nonliving particle of genetic material that can replicate only within a host cell

W

wet mount describes technique used for staining and viewing a microscopic specimen

white blood cells several types of cells that work to protect the body from disease-causing microbes and foreign substances

X

xylem in plants, the vascular tissue that carries water and dissolved minerals upward from the roots

Z

zygote the fertilized egg cell that is formed when the nuclei of two gametes (a male and a female) fuse; contains the diploid number of chromosomes

Index

A
Abiotic factors, 87, 236
Acacia tree, 243
Accuracy, 32
Acid rain, 256
Acquired immunodeficiency syndrome (AIDS), 111, 112, 190
Acquired trait, 135
Active site, 68
Active transport, 54–55
Adaptations, 136
  behavioral, 154
  to environment, 153–154
  physical, 154
  types of, 154
Adaptive radiation, 155
Adenosine diphosphate (ADP), 80
Adenosine triphosphate (ATP), 55, 79
African sleeping sickness, 106
Air pollutants, 255
Algae, 107, 127
Alien species, introducing, 267
Alleles, 206
  dominant, 209
  multiple, 209–211
  recessive, 209
Allergic reactions, 112
Alternation of generations, 128
Alveoli, 81
Ameba, 127
Amebic dysentery, 106
Amino acids, 45, 147, 167, 175
Amoeba, 106
Anaerobic respiration, 79–80
Anatomy, 13
  defined, 14
  evidence from comparative, 145–146
Anemia, sickle-cell, 102, 226
Angiosperms, 128
Animals, 129. *See also* specific
Antibiotics, 106
  resistance of, in bacteria, 137–138
Antibodies, 105, 109
Anticodon, 177
Antigens, 109
Antihistamines, 112
Aquatic ecosystems, 237–238
Archaebacteria, 126
Aristotle, 122
Arm, 27
Artificial selection, 138, 214
Asbestosis, 103

Ascaris roundworms, 108
Ascomycota, 107
Asexual reproduction, 188–190
Atoms, 41, 42
Autoimmune diseases, 112
Autosomes, 226
Autotrophs, 76, 77, 127

B
Bacteria, 105–106, 125, 127
  antibiotic resistance in, 137–138
  nitrogen-fixing, 92
Bacteriophages, 104
Balance
  electronic, 29
  triple-beam, 29
Base, 27
B cells, 110–111
  memory, 111
Behavioral adaptations, 154
Benedict solution tests, 30
Binary fission, 188
Binomial nomenclature, 122
Bioaccumulation, 89
Biochemical processes, 43–44
Biochemistry
  defined, 14
  evidence from comparative, 146–147
Biodiversity, importance of, 92–93
Biology
  defined, 13
  early discoveries in, 24–25
  main branches of, 14
Biomagnification, 89
Biomes, 236
Biosphere, 265
  protecting, 265–267
  saving, 270–271
Biotechnology engineering, 221
Biotic factors, 87, 236
Biuret solution, 30
Body, defenses of, against disease, 108–109
Body tube, of microscope, 27
Bonds
  covalent, 43, 55
  ionic, 43
Bone cells, 189
Botany, 14
Breathing, 80–81
Broad-leaved forests, 238
Bromthymol blue, 30
Brown, Robert, 25
Budding, 188

C
Cancer, 189–190
Cape buffalo, 242
Capillaries, 63
Capsid, 104
Carbohydrates, 44
Carbon, 43, 90
Carbon cycle, 91
Carbon dioxide (CO_2), 55
  in atmosphere, 256–257
Carnivores, 88
Carrier proteins, 52
Carriers, 54, 212
Carrying capacity, 240
Catalysts, 68
  organic, 68
Cell division, 186, 187
  rate of, 189
Cell membrane, 51, 52–53, 125
  transport across, 53–55
Cells, 25, 41, 51
  B, 110–111
  bone, 189
  communication between, 56–57
  cycle of, 185
  daughter, 187
  differentiation of, 178–179
  egg, 196
  energy transformations inside, 55–56
  eukaryotic, 52
  homeostasis and, 63–64
  host, 104
  life cycle of, 185–187
  nerve, 56
  parent, 187
  parts of, 51–53
  plant, 128
  red blood, 189
  sex, 196
  sperm, 196
  T, 110–111
  target, 67
  white blood, 109, 110
Cell theory, 25, 51
Cellular respiration, 65, 78–80
Cellulose, 44, 128
Cell wall, 52
Celsius thermometer, 29
Centimeter, 28
Centrifugation, 32
Centrifuge, 32
Cestoda, 107
Characteristics, derived, 123
Chargaff, Erwin, 166

Index **301**

Chemical bond, 43
Chemical indicators, 30
Chemical reactions, 55
Chemicals, handling safely, 30–31
Chemical stains, 30
Chicken pox, 105
Chlorofluorocarbons (CFCs), 257
Chlorophyll, 77, 128
Chloroplasts, 52, 77
Chromatography, 31–32
Chromosomes, 165, 186, 195–196
   double-stranded, 197
   sex, 226
   single-stranded, 197
Ciliophora, 106–107
Circulatory system, 67
Cladistics, 123
Cladogram, 123, 124
Classification, 122
   defined, 121
   early, 122
   five-kingdom system of, 123–124
   modern, 123–124
   six-kingdom system of, 125–129
Clean Air Act (1970), 256
Clean Air Act (1977), 256
Clear-cutting, 269
Climax community, 243
Clips, 27
Clones, 188
Cloning, 188
Cnidarians, 129
Coarse adjustment, 27
Codominance, 211
Codon, 176
Coevolution, 242
Color blindness, 211–212
Commensalism, 242
Communication between cells, 56–57
Community, 239
   interactions of, 240–243
Comparative anatomy, evidence from, 145–146
Comparative biochemistry, evidence from, 146–147
Comparative embryology, evidence from, 146
Competition, 135, 240
Compound light microscope, 25–26
   parts of, 27
Compounds, 43
   organic, 43–44, 76
Concentration gradient, 53
Conformation, 45
Coniferous evergreen trees, 238
Consumers, 87
Control, 56
Control unit, 66
Covalent bond, 43, 55
Crick, Francis, 25, 166
Crossing-over, 197
Cuticle, 107

Cystic fibrosis (CF), 213
Cytology, 14
Cytoplasm, 51

D

Darwin, Charles, 134, 144, 153
   evolution theory of, 25, 134–137
Daughter cells, 187
Deciduous trees, 238
Decomposers, 91
Deep-sea communities, 43
Deforestation, 257
Density, 240
Deoxyribonucleic acid (DNA), 45–46, 67, 104, 147, 165, 186
   genetic information in, 167
   structure of, 25, 166–167
Deoxyribonucleic acid (DNA) fingerprinting, 224
Deoxyribonucleic acid (DNA) replication, 168–169
   errors in, 169
Dependent variable, 16
Derived characteristics, 123
Deserts, 238
Deuteromycota, 107
Developed country, environmental protection in, 269–270
Developing country, environmental protection in, 270
Development
   external, 199–200
   internal, 200
   sustainable, 268–271
Diaphragm, 27, 80
Dichlorodiphenyl trichloroethane (DDT), 89
Dichotomous key, 129–130
Dicot, 130
Diffusion, 53
   facilitated, 54
Dinoflagellates, 107
Disease-causing organisms, 104, 107–108
Diseases, 102–104
   autoimmune, 112
   body's defenses against, 108–109
   factors that cause, 102–104
   of immune system, 112
   immunodeficiency, 112
   infectious, 103
Dominance, 207–208
   incomplete, 211
Dominant allele, 209
Dominant homozygous, 208
Double-stranded chromosomes, 197
Dynamic equilibrium, 63

E

Earth Day, 265
Earth Summit, 270
Ecological succession, 243–244

Ecology, 13, 235
   defined, 14
   ecosystems and, 235–236
Ecosystems, 236
   aquatic, 237–238
   basic characteristics of, 87
   ecology and, 235–236
   energy flow through, 87–89
   environmental factors and, 236–238
   freshwater, 237–238
   land, 238, 253
   recycling of materials in, 90–92
Effector, 66
Egg cell, 196
Electrical voltage, 56
Electronic balance, 29
Electron microscopes, 28
Electrons, 42
Electron shells, 42
Electrophoresis, 32
Elements, 43
Embryo, 200
Embryology
   defined, 14
   evidence from comparative, 146
Endangered Species Act, 270
End brush, 57
Endocrine glands, 67
Endocrine system, 56, 64, 67–68
   human, 67
Endoplasmic reticulum, 51
Energy, 42
   flow of, through ecosystems, 87–89
   geothermal, 268
   potential, 55
   solar, 267
   transformations inside cells, 55–56
Energy pyramid, 89
Engineering
   biotechnology, 221
   genetic, 221
Entamoeba histolytica, 106
Entropy, 42
Environment
   adaptations to, 153–154
   effect of people on, 252–256
   gene expression and, 214
Environmental factors, ecosystems and, 236–238
Environmental Protection Agency (EPA), 269–270
Environmental protection in developing country, 270
Enzyme action, lock-and-key model of, 68
Enzymes, 68–69
   restriction, 222
Equilibrium, dynamic, 63
Erosion, 254
Eubacteria, 127
Euglena, 127
Eukaryotes, 125

Eukaryotic cells, 52
 comparison of organelles in, 126
Eukaryotic organisms, 124
Evolution
 by natural selection, 136–137
 within populations, 147–148
 theory of, 25, 133–137
 types of evidence for, 144–147
Excretory system, 64
Exercise, maintaining homeostasis during, 64–65
Exhalation, 80
External development, 199–200
Extinct, 134
Extinction of species, 156
Eyepiece, of microscope, 25–26

F

Facilitated diffusion, 54
Factors, 206
 abiotic, 87, 236
 biotic, 87, 236
 limiting, 236
Family, 122
 genetic disorders and history of, 225–227
Feedback mechanism, 65, 67–68
 homeostasis and, 65–66
Fertilization, 129, 196
 internal, 199–200
 types of, 199
Field of view, 26
Fieldwork, 244
Finches, 136
Fine adjustment, 27
Five-kingdom system of classification, 123–124
Flatworms, 107
Flowering plants, sex life of, 198–199
Fluid mosaic model, 52
Flukes, 107
Food, 76–77
Food chains, 88–89
Food webs, 88–89
Forests, sustainable development of, 269
Fossil fuels, 255
Fossil record, 121
Fossils, evidence from, 144–145
Franklin, Rosalind, 166
Freshwater ecosystems, 237–238
Fungi, 107, 127–128

G

G-8 Summit, 271
Galápagos Islands, 136, 155
Gametes, 135, 196, 198
Gel electrophoresis, 32, 224
Gene expression, 178–179
 environment and, 214
Genes, 175
 sex-linked, 211

Gene therapy, 221
Genetic code, 176
Genetic disorders, family history and, 225–227
Genetic engineering, 221
Genetic mutation, 169
Genetics, 13
 defined, 14
Genetic screening, 226
Genetic variations, 135, 198
Genome, 207
Genomics, 221
Genotype, 208
Genus, 122
Geographic isolation, 155
Geothermal energy, 268
Global air pollution problems, 256–257
Global warming, 256
Glucose, 44, 77
Glycogen, 44
Glycolysis, 79–80
Golgi complex, 52
Gonorrhea, 106
Graduated cylinder, 28
Grasslands, 238
Groundwater, 255
Group size, 17
Guard cells, 66
Gymnosperms, 128

H

Habitat, 236
 destruction of, 93–94
Habitat loss, 93
Helper T cells, 111
Hemophilia, 211–212
Hepatitis A, 105
Herbivore, 88
Hereditary information, 206
Hereditary traits, 135
Heredity, 165
 fundamental principles of, 25
Heterotrophic, 76
Heterotrophs, 87, 91, 128
Hibernation, 154
High blood pressure, 104
Histamines, 112
Homeostasis, 63–64
 defined, 63
 disruption of, 102
 feedback mechanisms and, 65–66
 plants and, 66
 systems for maintaining, 64–65
Homologous structures, 145
Homo sapiens, 122
Homozygous, 208
Hooke, Robert, 24–25
Hormones, 67–68
Host cell, 104
Human genome project, 225
Human immune system, 109–111

Human immunodeficiency virus (HIV), 111, 112
Human population growth, 257–258
Hybrid, 208
Hydrogen, 43, 90
Hydrolysis, 80
Hydrothermal vents, 43
Hypertension, 104
Hypertonic, 53
Hypothesis, forming testable, 16
Hypotonic, 53

I

Immune system, 64, 109, 190
 diseases of, 112
 human, 109–111
Immunity, 110
Immunodeficiency disease, 112
Incomplete dominance, 211
Independent variable, 16
Industrial Revolution, 252, 253
Infectious diseases, 103
Inference, 18
Inflammation, 108–109, 112
Inhalation, 80
Inheritance, 206
 as factor in causing disease, 102
 Mendel's ideas about, 206–207
 Mendel's laws of, 209
 polygenic, 209–211, 210
Inorganic substances, 76
Integration, 56
Interactions, 235
 community, 240–243
Intercellular fluid, 63–64
Interdependence, 235
Internal development, 200
Internal fertilization, 199–200
Iodine solution tests, 30
Ionic bond, 43
Isolation
 geographic, 155
 reproductive, 154
Isotonic, 53

J

Jellyfish, 129

K

Karyotype, 226
Killer T cells, 111
Kingdom, 122
Krebs cycle, 80
Kyoto Treaty, 271

L

Laboratory techniques and procedures, 31–32
Lactase, 69
Lactose, 69
Land ecosystems, 238, 253
Law of Dominance, 209

Law of Independent Assortment, 209
Law of Segregation and
  Recombination, 209
Leaves, photosynthesis and, 78
Leeches, 108
Leeuwenhoek, Anton van, 24
Length, measuring, 28
Life cycle of cell, 185–187
Life span, 185
Lifestyles as cause of disease, 103–104
Limiting factors, 236
Linear sequence, 166
Linked traits, 211
Linnaeus, Carolus, 122, 123–124
Lipids, 45
Litmus paper, 30
Living things
  classification of, 121
  levels of organization in, 41–42
Lock-and-key model of enzyme
  action, 68
Lugol solution tests, 30
Lupus erythematosus, 112
Lysogenic virus, 104
Lysosomes, 51
Lytic virus, 104

M

Marine biology, 14
Marsupial mammals, 200
Mass, 29
Mass extinctions, 156
Materials, recycling of, in ecosystems, 90–92
Matter, 42
Measurements, taking, 28–29
Meiosis, 196, 197–198
Memory B cells, 111
Mendel, Gregor, 25, 206–211, 214
Meniscus, 28
Messenger RNA (mRNA), 175–176
Metabolic activities, 44
Metabolism, 44
Meter, 28
Metric ruler, 28
Microbes, 108
Microbiology, 14
Microorganisms
  disease-causing, 103, 104
Microscopes
  compound light, 25–26
  electron, 28
  scanning electron, 28
  transmission electron, 28
  using, 25–28
Millimeters, 28
Mirror, 27
Mitochondria, 52
Mitosis, 187
  stages of, 187

Molecules, 43
  receptor, 53
Monera, 124
Monocot, 130
Monomers, 44
Monoploid, 197
Monosaccharides, 44
Morphology, 121
Movement, 52
Multicellular organisms, 41, 63, 121
Multiple alleles, 209–211
Mutation, 135, 178, 226
  genetic, 169
Mutualism, 242–243
Mycobacterium tuberculosis, 106

N

Natural phenomena and events, 13–14
Natural selection, evolution by, 136–137
Needle-leaved forests, 238
Neisseria, 106
Nematoda, 108
Nerve cells, 56
  crossing gap between, 56–57
Nerve impulse, 56
Nervous system, 56, 64
Neurons, 41, 56
Niche, 155, 236
Nitrogen, 43, 90
  in atmosphere, 92
Nitrogen-fixing bacteria, 92
Nomenclature, 122
  binomial, 122
Nonrenewable resource, 267
Nosepiece, of microscope, 27
Nucleic acids, 45–46
Nucleotides, 45–46, 166
Nucleus, 52

O

Objective lens, 25–26, 27
Observation, 14–15
Ocular lens, 25–26, 27
Order, 122
Organelles, 41, 51–52
  comparison of, in eukaryotic and
    prokaryotic cells, 126
Organic catalysts, 68
Organic compounds, 41, 43–44, 76
  families of, 44–46
Organisms, 41
  autotrophic, 76, 77
  disease-causing, 107–108
  eukaryotic, 124
  heterotrophics, 76
  multicellular, 41, 121
  pioneer, 243
  prokaryotic, 124
  unicellular, 41, 121

Organs, 41
  malfunction as cause of disease, 103
Organ systems, 41
Osmosis, 54
Oxygen, 43, 90
Oxygen cycle, 91
Ozone shield, 257

P

Paramecium, 127
Parasites, 106, 242
Parasitic worms, 107–108
Parent cell, 187
Parks for people, 270
Passive transport, 53–54
Pathogens, 103
Pedigree charts, 213
Pedigrees, 211–213
Peer review, 33
Penicillin, 137
Peppered moth, 147–148
Peptidoglycan, 127
Permeability, selective, 53
Pesticides, 138
Phenotype, 207
Phenylalanine, 226–227
Phenylketonuria, 226–227
Phosphorus, 43, 92
Photosynthesis, 77–80, 91, 128
  leaves and, 78
  process of, 77
  purpose of, 78–79
pH paper, 30
pH scale, 69
Phylogenetic taxonomy, 123
Phylogenetic tree, 123, 124
Phylum, 122
Physical adaptations, 154
Physiology, 14
Pioneer organisms, 243
Placenta, 200
Plant cells, 128
Plants, 129
  homeostasis and, 66
Plasma membrane, 52
Plasmodium, 106
Plasmolysis, 54
Pneumococcus, 105–106
Poisons as cause of disease, 103
Pollutants as cause of disease, 103
Pollution
  air, 256–257
  water, 254–255
Polygenic inheritance, 209–211, 210
Polymers, 44, 167, 175
Polyp generation, 129
Polysaccharides, 44
Population growth
  factors that affect, 239–240
  human, 257–258

Populations
  defined, 239
  evolution within, 147–148
  study of, 239
Potential energy, 55
Predation, 241–242
Predator, 241
Predator-prey relationship, 242
Prey, 241
Primary succession, 243
Probability, 213
Producers, 87
Prokaryotes, 125
Prokaryotic cells, comparison of organelles in, 126
Prokaryotic organisms, 124
Proteins, 44–45, 147, 167, 175
  carrier, 52
Protists, 127
Protozoa, 106–107, 127
Pseudopodia, 106
Punnett squares, 213

R
Radiation, 135
  adaptive, 155
Reactants, 68
Receptor molecules, 53, 67
Receptors, 53
Recessive allele, 209
Recessive homozygous, 208
Recessiveness, 207–208
Recombinant DNA technology, 221–222
  tools of, 222–225
  uses of, 224–225
Recombination, 135, 197
Recycling, 266
Red blood cells, 189
Red tide, 107
Reduction division, 197
Reliability, 32
Renewable resource, 267
Replication, 186
  deoxyribonucleic acid (DNA), 168–169
Reproduction, 135, 185
  asexual, 188–190
  sexual, 135, 195–196
Reproductive isolation, 154
Residue, 91
Resistance, 137
Resources
  nonrenewable, 267
  renewable, 267
Respiration, 80–81
  anaerobic, 79–80
  cellular, 78–79
Respiratory surface, 81
Respiratory system, 80
Response, 65

Restriction enzyme, 222
Restriction site, 222
Rheumatoid arthritis, 112
Ribonucleic acid (RNA), 45–46, 175
Ribosomes, 175
Roundworms, 108
Ruler, metric, 28
Runners, 188

S
Saccharides, 44
Safety procedures, 29–31
Saltwater ecosystem, 237
Sample size, 17
Sampling populations, 244–245
Saprophytes, 128
Sarcodina, 106
SARS, 105
Scanning electron microscope (SEM), 28
Scavenge, 91
Schistosoma, 107
Schleiden, Matthias, 25
Schwann, Theodor, 25
Science
  answering questions in, 14–15
  defined, 13
Scientific experiment
  conducting, 17
  designing, 16–17
Scientific explanations, 13
Scientific inquiry, 13, 18
Scientific investigations, 32–33
Scientific methods, 13, 15–17
  conducting experiment, 17
  designing experiment, 16–17
  forming testable hypothesis, 16
  making observation and stating question, 16
Scientific results, communicating, 17–18
Scientific thinking, 13
Scolex, 107
Scrubbers, 256
Secondary succession, 244
Selection
  artificial, 138, 214
  natural, 136–138
Selective breeding, 138, 214
Selective permeability, 53
Sensor, 65
Sex cells, 196
Sex chromosomes, 226
Sex life of flowering plants, 198–199
Sex-linked genes, 211
Sex-linked traits, 211–213
Sexual reproduction, 135, 195–196
Shared derived characteristics, 123
Sickle-cell anemia, 102, 226
Single-celled organisms, 41, 63, 121

Single-stranded chromosomes, 197
Six-kingdom system of classification, 125–129
Slime molds, 127
Soil, losing, 253
Solar energy, 267
Solid wastes, 253
Solvent, 31
Speciation, 154
Species, 121, 122, 134
  diversity of, 92
  extinction of, 156
Sperm cell, 196
Spleen, 110
Sporozoa, 106
Stability, 92
Stage, 27
Starch, 44
Statistical analysis, 32
Stem cells, 225
Sterilization, 106
Stimulus, 65
Stinging ants, 243
Stomata, 66
Stomates, 78
Streptococcus, 106
Substances, inorganic, 76
Substrate, 68
Subunits, 44, 166
Succession, primary, 243
Sulfur, 43
Survival of fittest, 137
Sustainable development, 268–271
Symbiosis, 242
Symbiotic relationships, types of, 242–243
Synapse, 57
Syphilis, 106
System, 43
Systematics, 123

T
Tapeworms, 107–108
Target cells, 67
Taxonomic groupings, 122–123
Taxonomists, 121
Taxonomy, 121
  defined, 14
  phylogenetic, 123
Tay-Sachs disease, 213
T cells, 110–111
  helper, 111
  killer, 111
Technology, 25
Temperate forest, 238
Temperature
  importance of, to enzymes, 69
  measuring, 29
Template, 168
Testcross, 208

Theory, 13
  cell, 25
  of evolution, 25, 133–137
Thymus gland, 110
Tissues, 41
Topsoil, 253
Traits, 206
  acquired, 135
  adaptive value of, 153
  hereditary, 135
  linked, 211
  sex-linked, 211–213
Transcription, 175–176
Transfer RNA (tRNA), 176
Transformations of energy, 42
Translation, 176–178
Transmission electron microscope (TEM), 28
Transport
  across cell membrane, 53–55
  active, 54–55
  passive, 53–54
Transport proteins, 52
Trematoda, 107
Treponema pallidum, 106
Trichinosis, 108

Triple-beam balance, 29
Triplet code, 176
Trophic levels, 87
Tropical rain forests, 238
Trypanosoma, 106
Trypanosomiasis, 106
Tuberculosis, 103
Tundra, 238
Turgid, 54

U
Ultracentrifuge, 32
Unicellular organisms, 41, 63, 124
Uracil, 176

V
Vaccinations, 105, 109–110
Vacuoles, 52
Validity, 32
Variables
  dependent, 16
  independent, 16
Variations, genetic, 135
Vectors, 222
Vestigial structure, 145
Viral infections, treatment for, 105

Virulent virus, 104
Viruses, 104–105
  lysogenic, 104
  lytic, 104
  virulent, 104
Volume, measuring, 28

W
Wastes, solid, 253
Water pollution, 254–255
Watson, James, 25, 166
Watson-Crick model, 166
West Nile virus, 105
Wet mount, steps in preparing, 26–27
White blood cells, 109, 110
Wilkins, Maurice, 166

X
X-ray diffraction patterns, 166

Z
Zero growth, 240
Zooflagellates, 106
Zoology, 14
Zygote, 129, 196, 198

Photo Credits

Photographs are provided courtesy of the following:

National Aeronautics and Space Administration (NASA): 265
N.Y. Public Library/Science, Industry, and Business Library: 210
N.Y.S. Department of Environmental Conservation: 256
Photo Researchers, Inc.: 76, Van Bucher; 92, Carl Frank; 93, Karl Weidmann; 103 (top), Meckes/Ottawa; 103 (bottom), Biophoto Associates; 111, Francis Leroy, Biocosmos/ Science Photo Library (artwork based on SEM); 134, Biophoto Associates; 138, Mary Eleanor Browning; 156, Tom McHugh; 166, A. Barrington Brown; 186, Omikron; 195, Biophoto Associates; 214, Mary Eleanor Browning; 222, Laguna Design (computer artwork); 235, Tom McHugh; 239, Richard Parker; 242 (left), Eric Husking; 242 (right), Tom McHugh; 253, Michael Hayman; 254, Michael P. Gadomski; 258, Rafael Macia; 266, Jeff Isaac Greenberg; 268, Georg Gerster; 269, Simon Fraser/ Science Photo Library; 270 (top), Doug Plummer; 270 (bottom), William & Marcia Levy.
Visuals Unlimited, Inc., © SIU: 224